Methods of surface analysis

Edited by J.M. WALLS

Methods of surface analysis

TECHNIQUES AND APPLICATIONS

Published by the Press Syndicate of the University of Cambridge
The Pitt Building, Trumpington Street, Cambridge CB2 1RP
40 West 20th Street, New York, NY 10011–4211, USA
10 Stamford Road, Oakleigh, Victoria 3166, Australia

© Cambridge University Press 1989

First published 1989
First paperback edition 1990
Reprinted 1992

Printed in Great Britain at the University Press, Cambridge

British Library cataloguing in publication data

Methods of surface analysis: techniques
and applications
1. Surfaces (Physics)   2. Spectrum
analysis
I. Walls, J.M.
541.3'453   QC176.8.S8

Library of Congress cataloguing in publication data

Methods of surface analysis/edited by J.M. Walls.
   p. cm.
Includes index.
ISBN 0 521 30564 0
1. Surfaces (Technology)—Analysis.   I. Walls, J.M.
TP156.S95M48   1988
620.4—dc19   87-20319

ISBN 0 521 30564 0 hardback
ISBN 0 521 38690 X paperback

TM

Contents

List of contributors

Preface

1. Methods of surface analysis — 1
   J.M. Walls

2. Ion erosion in surface analysis — 20
   R. Smith and J.M. Walls

3. Electron and ion energy analysis — 57
   M.P. Seah

4. Auger electron spectroscopy — 87
   H.E. Bishop

5. X-ray photoelectron spectroscopy — 127
   A.B. Christie

6. Static secondary ion mass spectroscopy — 169
   J.C. Vickerman

7. Dynamic secondary ion mass spectrometry — 216
   D.E. Sykes

8. Ion scattering spectroscopy — 263
   D.G. Armour

9. Rutherford back-scattering spectrometry — 299
   W.A. Grant

Index — 338

Contributors

D.G. Armour
Department of Electronic and Electrical Engineering, University of Salford, Salford M5 4WT, UK.

H.E. Bishop
Materials Development Division, AERE Harwell, Didcot, Oxon OXII ORA, UK.

A.B. Christie
VG Scientific, Imberhorne Lane, East Grinstead, West Sussex RH19 IUB, UK.

W.A. Grant
Department of Electronic and Electrical Engineering, University of Salford, Salford M5 4WT, UK.

M.P. Seah
Division of Materials Applications, National Physical Laboratory, Teddington, Middlesex TW11 OLW, UK.

R. Smith
Loughborough University of Technology, Loughborough, Leicestershire LE11 3TU, UK.

D.E. Sykes
Loughborough University of Technology, Loughborough, Leicestershire LE11 3TU, UK.

J.C. Vickerman
Surface Analysis Research Centre, University of Manchester, Institute of Science and Technology, Manchester M60 1QD, UK.

J.M. Walls
VG Ionex, The Maltings, Burgess Hill, West Sussex RH15 9TQ, UK.

Preface

This book was inspired by a series of short courses organised during my period at Loughborough University. My research group in surface physics was intertwined with a commercial analysis service operated by Loughborough Consultants Ltd. With the broad perspective these activities gave us it was clear that there was a substantial need for a continuing educational programme to inform potential industrial and academic users of the benefits of surface analysis. Hence this book is aimed at filling the needs of the non-specialist who wishes to acquire an overall appreciation of modern surface analytical techniques and the ways they may be used in research, development and practical problem solving. This book will help the potential user to determine what each technique can and cannot do. Hopefully it may also help him decide which is the most appropriate technique for a particular problem.

A bewildering number of techniques have been developed for the analysis of surfaces. This is not surprising, given the enormous and increasing importance of the surface to modern technological processes. However, many of these techniques are aimed at obtaining fundamental data in surface science research. It is important to realise that this book concentrates only on those techniques with broad analytical applications. Readers interested in the more specialised techniques are referred to the excellent text of Woodruff and Delchar entitled *Modern techniques of surface science* and published by Cambridge University Press in 1986.

Although the book is aimed predominantly at introducing the techniques and their applications, some background material is also required. Chapter 1 gives an overview to provide some context for the more specialised chapters that follow. Chapters on 'Ion erosion' and 'Electron and ion energy analysis' are included to provide some of the essential background required. I have been fortunate to enlist an authority in *practical surface analysis* to write each of the chapters. The insights of their combined experience could not have been obtained in a single author book. Sadly, one of the authors, Bill Grant, passed away in October 1987. He will be greatly missed by his friends and colleagues in the field.

J.M. Walls

1

Methods of surface analysis

J.M. WALLS

1 Introduction

The importance of the surface composition of solid materials has been recognised for some time. One of the earliest documented references comes from the manuscript *De Proprietatibus Rerum* (The properties of things) written in AD 1250:

> When a plate of gold shall be bonded with a plate of silver or joined thereto, it is necessary to beware of three things, of dust, of wind and of moisture: for if any come between the gold and silver they may not be joined together....

Bondability problems are an important field of application of surface analytical techniques, and materials technology has progressed a long way since the thirteenth century, but it remains a fact that the manner in which any solid surface interacts with its environment, or with any other surface, is determined by the composition of the outermost atomic layers. Although bulk and surface compositions may be related to a greater or lesser degree, and the bulk composition of solid materials is generally easily controlled, it is well known that the nature of the surface is influenced more by its environmental history than by its bulk composition. Information on the composition of the topmost atomic layers of materials is crucial in the understanding of many technologically important processes. Examples of such processes include all chemical reactions (including oxidation and corrosion), catalysis, adhesion, thermionic emission, crystal growth, wear, etc... Hence surface analysis can be used as a diagnostic technique to determine why a surface does or does not possess the desired optical, electrical, mechanical or decorative properties. The ability to analyse the surface composition of materials is playing an increasingly important role in a wide range of industries. Semiconductor manufacture, with its drive towards ever smaller geometries and its multi-layer approach, is an obvious pace setter, but surface analysis techniques are being used with increasing frequency in a variety of other industries.

A number of techniques are now available for surface analysis. However, many of these techniques are highly specialised and are used mostly for fundamental research in surface science. The purpose of this book is to introduce the principles and capabilities of the five techniques with the widest range of applicability, viz Auger electron spectroscopy (AES), X-ray photoelectron spectroscopy (XPS or ESCA), secondary ion mass spectrometry (SIMS), ion scattering spectroscopy (ISS) and Rutherford Backscattering (RBS) (see Tables 1.1–1.6 at the end of this chapter for further information on these techniques). Each of these techniques has its advantages and disadvantages, and some judgement is necessary before choosing the best one to use for a particular problem. This book provides the background information together with examples of applications, which should allow a better informed judgement to be made.

2 Ultra-high vacuum

The commercial availability of reliable ultra-high vaccum (UHV) technology in the late sixties was a critical step for the development of surface analysis techniques. Even under conventional high vacuum, surfaces are exposed to constant interaction with atoms and molecules in the residual vacuum, so that monolayers of contamination can form in the course of a measurement. From the kinetic theory of gases, the arrival rate of N molecules cm^{-2} of molecular weight M at a temperature T K at a pressure of p Torr is given by

$$N = 2.89 \times 10^{22} p(MT)^{1/2} \text{ molecules cm}^{-2}\text{s}^{-1}.$$

Hence at a pressure of 10^{-6} Torr, the arrival rate of nitrogen molecules at room temperature would be 3×10^{14} cm^{-2}s^{-1}. A typical metal has about 10^{15} atoms cm^{-2} in a monolayer, so assuming unit sticking probability, it would take only 3 seconds for a monolayer to form. However, at pressures below 10^{-9} Torr it would take several hours.

Normally, in surface analysis, the sample is introduced from air into an ultra-high vacuum analysis chamber via a fast entry air lock. A third intermediary chamber is sometimes used to allow sample outgassing. The sample is then cleaned by ion bombardment to enable the analysis to proceed under UHV conditions. Most as-received samples in surface analysis carry a surface film of contamination deposited in the atmosphere, by handling or by numerous other mechanisms. Carbon is the main contaminant species and the presence of O, Cl, S, Ca and N are also

Methods of surface analysis 3

commonly detected. An Auger spectrum typical of an as-received Ni surface is shown in Fig. 1.1(a). The contamination layer is removed by argon ion bombardment and Fig. 1.1(b) is a spectrum obtained from the same surface following ion cleaning for 10 seconds. Most of the contaminant species have been removed or reduced and the nickel surface is now exposed.

Ultra-high vacuum systems are now used routinely not only in surface analysis but also in an increasing number of process technologies. UHV

Fig. 1.1. Real samples introduced from air are covered with a film of surface contamination. The Auger spectrum shown in (a) is from an ostensibly pure nickel sheet. In fact, nickel produces only a small signal and the surface consists mainly of carbon probably from a hydrocarbon deposit. Following a short period of ion etching the Auger spectrum in (b) reveals the nickel although the oxygen signal has also increased indicating the presence of an intermediate oxide layer. Argon is also present following its implantation during ion etching.

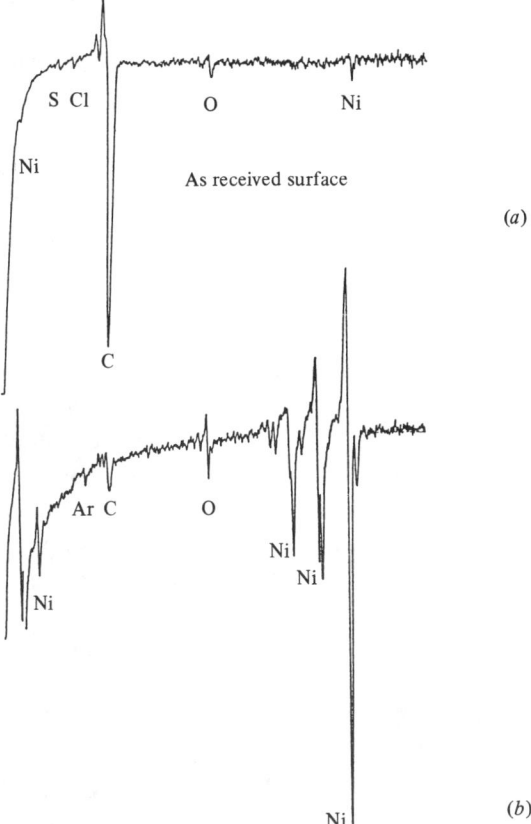

provides the ultimate 'clean room' and the technology is essential, for example, to the growth of low-dimensional structures by molecular beam epitaxy (MBE). UHV systems are usually fabricated from stainless steel. The vacuum seals are usually achieved using Conflat* flanges, in which stainless steel knife edges are sealed into flat copper gaskets. The entire vacuum assembly is normally baked to about 230 °C to remove water vapour and promote outgassing of adsorbed gas species. Many types of vacuum pumps may be used including cold-trapped diffusion pumps, ion pumps, turbomolecular pumps and cryopumps. The choice is dictated by the specific application and personal preference.

In practice, all surface analysis instruments are constructed to UHV standards. One exception is Rutherford backscattering (RBS) where the analysis vessel is very often built only to high-vacuum specifications (10^{-6} Torr). This is because while the other techniques are sensitive to the outermost atomic layers, RBS takes its information from about 100 Å or more. Hence although it is less surface sensitive it is also less sensitive to monolayer coverages of contamination. The other possible exception is secondary ion mass spectrometry (SIMS), but only when used in the dynamics SIMS mode. In this mode the residual gas arrival rate is lower than the sputter-removal rate. However, in this case many argue that the high sensitivity of SIMS may only be properly exploited in UHV conditions since, otherwise, residual levels of carbon, nitrogen, hydrogen and oxygen set an artificially high background for these important elements. The emergence of UHV dynamic SIMS instruments will assist us to test for an improvement in detection limits to these elements.

3 Composition-depth profiling

The term 'surface analysis' to describe the use of the techniques covered by this book can be misleading. Although the techniques derive their usefulness from their true surface sensitivity, they can also be used to determine the composition of much deeper layers. This is normally achieved by sequential (or simultaneous) removal of surface layers by ion beam sputtering and surface analysis. In this way, the composition of a material may be dissected layer by layer to build up what is termed a 'composition-depth profile'. This is probably one of the most important practical modes of surface analysis and the ability of a technique to provide compositional information with depth is a very important guide to its effectiveness. The provision of composition-depth profiles by AES has had an enormous impact on the evaluation of thin films and surface coatings produced by a whole variety of methods for use in optical coatings, materials protection and for semiconductor metallisation. Concentration-depth profiling by SIMS and RBS of dopants in semiconductors has become essential in the evaluation of the various stages in semiconductor processing. The techniques are now used widely for quality control.

The techniques used to obtain composition information with depth are covered in individual chapters. Ion erosion is treated separately in Chapter 2 in view of its importance in sputter-depth profiling. Techniques using taper-sectioning techniques are discussed in Chapter 4.

4 Overview

It will be of value to obtain some overall perspective of the strengths and weaknesses of each method of surface analysis. This will provide a framework on which we can build the detail in the subsequent specialised chapters on each technique. Table 1.1 provides a list of the basic measurement parameters, but we also need to appreciate the real practical constraints. Hence the following pages gives a synopsis of each technique together with their special advantages and disadvantages.

Spatial resolution is often a very important requirement for practical problem solving. For example, some features on microelectronic devices can only be identified with submicron resolution. Small-area stains and particulate contamination must first be located before analysis can proceed. To date these kinds of problems have been addressed using scanning Auger electron spectroscopy. The technique allows a secondary electron image to be formed at TV rates and the same probe can then be focussed into a particular area for analysis. A spatial resolution of about 2000 Å is achieved routinely (see Joshi, Davis & Palmberg, 1975). Recently, however, the emergence of highly focussed gallium ion beams has brought forward a rival technique in Imaging SIMS. Indeed, SIMS has some real advantages. The probe size more accurately determines the area of analysis since ion scattering effects are negligible. Also, element mapping in SIMS is considerably faster than AES due to the much higher signal to background ratio. Both AES and SIMS are ideal for sputter-depth profiling since the probing beam can focus into the ion eroded crater to avoid edge effects. None of the other techniques are serious rivals to AES and SIMS where spatial resolution is required.

Chemical bonding information is often a key requirement in practical surface analysis. For example, oxidation state information is useful in corrosion studies, and chemical bonding is important in the characterisation of catalysts. Only XPS and SIMS are considered to be capable of providing detailed chemical information. The use of XPS to derive functional group information from chemical shifts in photoelectron energies is well known and there is now a wealth of experience and data in almost all areas of materials science (see Briggs & Seah, 1983; Walls & Christie, 1982). Although several years behind in development, it is now understood that SIMS is capable of supplying perhaps even more information about the chemical state of the surfaces from the molecular fragmentation patterns in the SIMS spectra (see Benninghoven, Rudenauer & Werner, 1987). Many leading laboratories use XPS and SIMS to provide an alternative reference technique and to provide information which is often complementary. XPS and SIMS are also techniques amenable to the widest range of materials including polymers, biological materials and organic systems.

Absolute surface sensitivity can sometimes be necessary especially with fundamental studies of adsorption on surfaces. Ion scattering spectroscopy is often used in these circumstances. However, the surface sensitivity of the electron spectroscopies (AES and XPS) can be improved using angular-resolved methods and SIMS is also usually sufficiently surface sensitive. Rutherford backscattering (RBS) would be used only rarely for these types of analysis.

SIMS is by far the most sensitive of the techniques for surface analysis. One of its most important applications is the characterisation of ion implanted dopant profiles in semiconductors with detection limits at the part per million (10^{-6}) and part per billion (10^{-9}) levels. Only RBS among the remaining techniques is capable of comparable sensitivities and only then for heavy elements in light element matrices. None of the other techniques is capable, in practice, of sensitivity below 0.1 atomic percent.

RBS is easily the most quantitative technique. It stands alone as the only technique capable of absolute quantitative measurement. Accuracy is routinely in the range $\pm 5\%$. XPS and AES can both be used to give reasonable quantitative accuracy (better than $\pm 10\%$) provided good calibration standards are used. Compilations of standard spectra are now readily available to assist in this (Davis *et al.*, 1976; Wagner *et al.*, 1979). In general, SIMS is still at best a semiquantitative technique. This aspect will improve as calibration standards become available and the secondary ion yield mechanisms become understood. In dilute systems such as those that occur in ion-implanted semiconductors the technique may already be used to give quantitative analysis ($\pm 5\%$) over several decades of dynamic range.

5   Other techniques

The burgeoning interest in surface science has led to the development of a plethora of new techniques. Many of these are very specific in the types of information made available and the aim of this book has been to introduce those techniques with general analytical applicability. Surface structural analysis is considered only when it is a useful by-product of an analysis technique (such as ion scattering spectroscopy (Chapter 8) and Rutherford backscattering (Chapter 9)). If the reader's interest is centred on surface structure he is referred to techniques such as low energy electron diffraction (LEED), reflection high energy electron diffraction (RHEED) and scanning tunnelling microscopy (STM).

Some of the less well-known methods can be added to the major techniques at modest cost. For example, the addition to an XPS instrument, of a differentially pumped gas discharge lamp can produce low-energy lines at He^1 (21.2 eV) and He^{11} (40.8 eV) which are used as the excitation source for ultra-violet photoelectron spectroscopy (UPS). This technique is useful for studying the electronic band structure of metals, alloys and semiconductors. Likewise, appearance potential spectroscopy (APS) may be used on AES instruments.

In addition, many techniques which are not generally described as methods of surface analysis, may be used in certain circumstances for surface analysis. For example, the scanning transmission electron microscope (STEM) may be used to obtain analytical information from extremely thin sections using ionisation loss spectroscopy (ILS) or energy-dispersive X-ray detection. In some circumstances the electron probe microanalyser (EPMA) may be used to provide surface sensitive information (see Chapter 4).

Three other techniques deserve relatively more attention either because they have general applicability but are only recently developed, or because they have tremendous analytical capability but with specialised application. Sputtered neutral mass spectrometry (SNMS) and the laser microprobe come into the first category while the atom probe field-ion microscope is in the second.

5.1 Sputtered neutral mass spectrometry (SNMS)

In essence, sputtered neutral mass spectrometry (SNMS) is very similar to SIMS. A low-energy ion beam is used to sputter the surface and a mass spectrometer is used to analyse secondary particles (Oechsner, 1984). However in SNMS, the true secondary ions are deliberately discarded and the secondary neutrals are post-ionised for subsequent detection. Post-ionisation can be achieved using a low-pressure plasma, an electron beam or a laser. In each case the ionisation cross-section is predictable and is independent of any matrix effects. As a result SNMS is a quantitative technique with a usefully narrow range of sensitivities.

The use of an rf discharge plasma can result in a post-ionisation efficiency

Fig. 1.2. SIMS and e-beam SNMS taken from the same low-alloy steel sample using a 10 keV argon ion beam with 10 μA current. The maximum count rate on the Fe$^+$ signal is 2×10^7 cps in SIMS and 6×10^6 cps in SNMS. Although both spectra show molecular ion peaks, the metal oxide and hydroxide ion intensities are lower for SNMS. The SNMS data allows easy quantification with monolayer sensitivity in the range 0.01% (data courtesy A. Benninghoven).

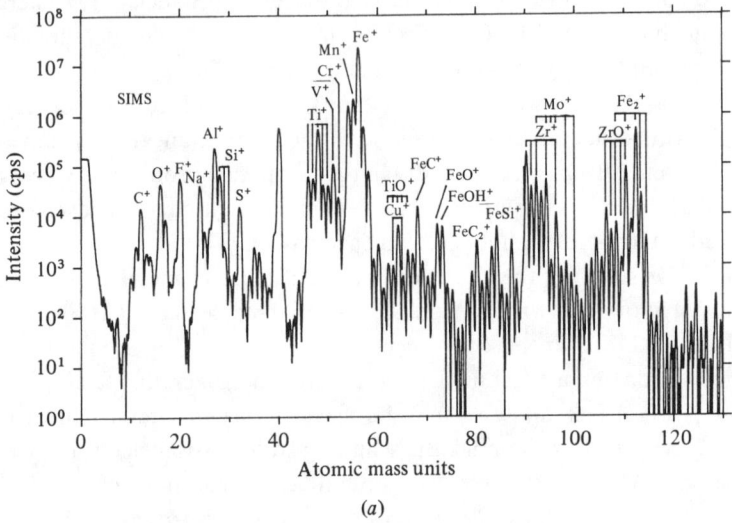

(a)

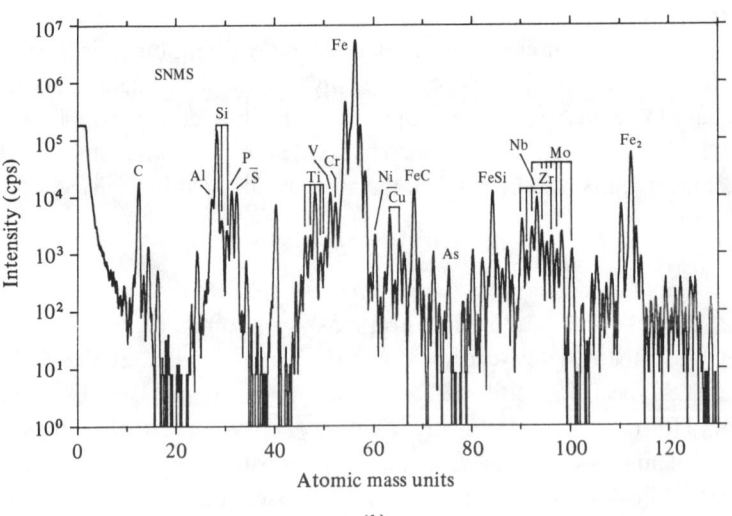

(b)

of 10^{-2}. A significant advantage of this approach is that the plasma not only post-ionises, but it can also be used to sputter the target. The low energy of the impinging ions (100 eV to 1 keV) results in a minimisation of atomic collision mixing (see Chapter 2) and hence the technique may be used to provide depth profiles with extremely good depth resolution. Such depth profiles may be obtained rapidly since the ion current density can exceed 1 mA cm^{-2} over a 0.5 cm^2 area. One disadvantage of plasma-post-ionisation is the occurrence of a background of ions generated by residual gas particles.

A method for electron beam post-ionisation with some useful advantages has been described by Tumpner et al. (1987). This method uses deflection plates to suppress the original secondary ions and a series of apertures with a high-energy bandpass to discriminate against the transmission of residual gas ions. With an electron beam post-ionising current of 3 mA and sputtering ion gun with 20 μA current, then bulk analysis of sub-ppm is possible. Surface monolayer analysis with a sensitivity in the 0.01 atomic percent is also possible with quantification. This method is entirely compatible with conventional SIMS analysis and rapid switching between the two modes (Fig. 1.2).

Pulsed lasers are available which can provide power densities up to 10^{12} W cm^{-2} and with repetition rates exceeding 10 Hz. These lasers may be used to post-ionise sputtered neutrals by a non-resonant multiphoton process with subsequent detection in a time-of-flight mass spectrometer (Becker & Gillen, 1984). Developments are in the early stages but sensitivities for certain elements in the 10^{-7} monolayer have been reported.

5.2 Laser microprobe

The laser microprobe (or laser ionisation mass analyser (LIMA)) utilises a pulsed laser usually operating at 0.26 μm wavelength which is focussed onto the area of interest on the sample in an ultra-high vacuum chamber. The intense pulse causes the evaporation of a small volume of material from the sample surface and analysis of this material is then carried out by measuring the mass-to-charge ratio of ions extracted from the resulting microplasma in a time-of-flight mass spectrometer. The technique provides a means of rapid survey analysis since the entire process from laser pulse to full mass analysis is performed in milliseconds. Elemental sensitivities are in the ppm range with a reasonably narrow range across the periodic table. The laser microprobe provides spatial resolution in the range 1 to 5 μm with a depth of analysis of about 0.5 μm (Fig. 1.3). Defocussing of the laser pulse can result in better surface sensitivity with layers as thin as 300Å being removed. By directing repeated

Fig. 1.3. An example of the use of the laser microprobe for true surface analysis. (*a*) shows a laser microprobe spectra taken from a GaAs surface with a native oxide overlayer while (*b*) shows a spectra from the same GaAs surface freshly etched in 50% HCl solution and rinsed in propanol. XPS measurements from the same sample indicated an oxide thickness of about 15 Å. XPS did not detect chlorine. These spectra show that by using the correct conditions, the laser microprobe has a surface sensitivity comparable to that of XPS (courtesy D.D. Hall and I. Sutherland).

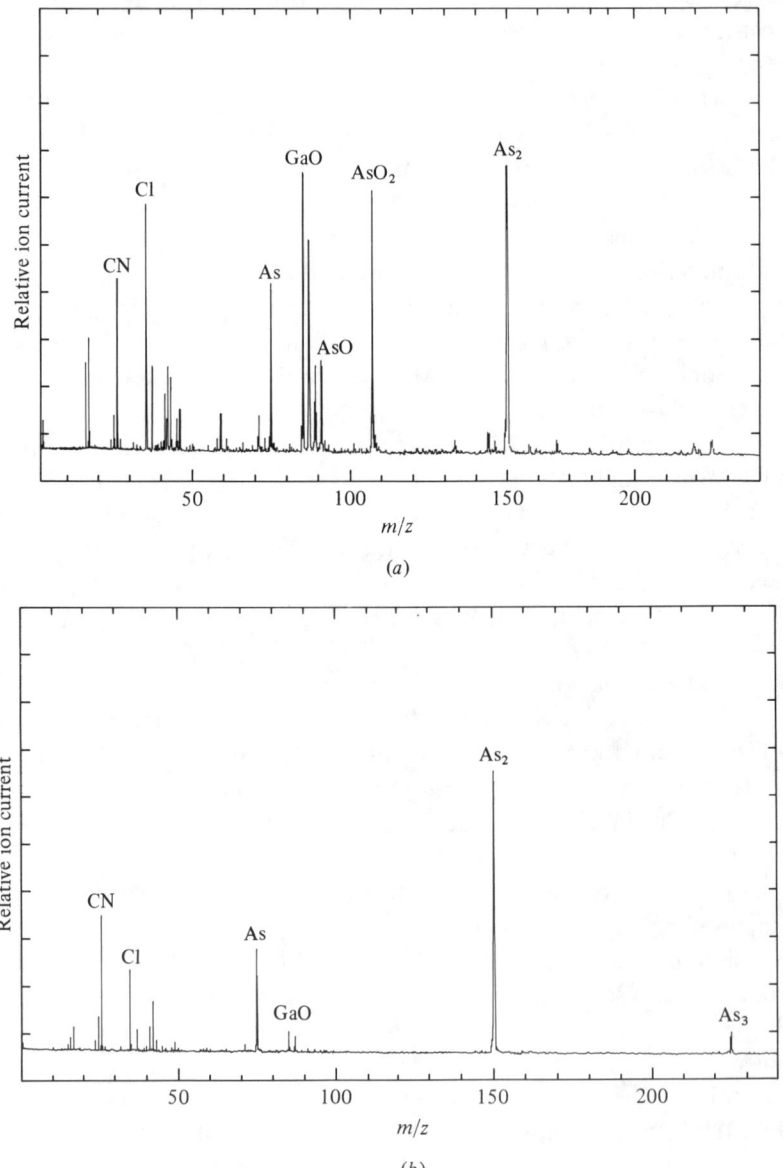

Methods of surface analysis

laser pulse onto the same area of the sample it is possible to produce rapid composition – depth profiles with depths ranging from less than one micron to tens of microns. The laser microprobe is a useful method for identifying organic contamination in small areas and recent instruments combined with time-of-flight SIMS should make some impact in the area of spatially-resolved organic analysis. The main drawback of the technique, at present, is a lack of reproducibility; the intensity of spectra tend to vary from pulse to pulse. Quantification of spectra is also in the early stages of development.

5.3   Atom probe microanalysis

The atom probe has often been described as the ultimate method for surface microanalysis. It combines *atomic* resolution imaging with the ability to analyse individual surface atoms. Its development goes back to 1951 and the invention by Erwin Muller of the field-ion microscope (see Muller & Tsong, 1969).

In the field-ion microscope, the specimen takes the form of a sharp needle usually about 100 Å in diameter. The needle is held in vacuum a few millimetres from a channel plate intensifier and phosphor screen. A positive voltage (5–30 kV) is applied to the needle creating an enormous electric field (typically $> MV\,cm^{-1}$) over the apex (Fig. 1.4). Helium gas is leaked

Fig. 1.4. These three images illustrate the power of atom probe microanalysis. (a) is a field-ion image of a molybdenum surface. Each white spot corresponds to a single atom on the specimen surface. The arrows point to the position of grain boundaries which intersect at a node near the centre of the image. (b) is an imaging atom probe image time-gated for molybdenum. Note the dark lines along the grain boundaries. (c) shows an imaging atom probe micrograph time-gated for oxygen revealing substantial segregation of this element to the grain boundaries (micrographs courtesy A.R. Waugh).

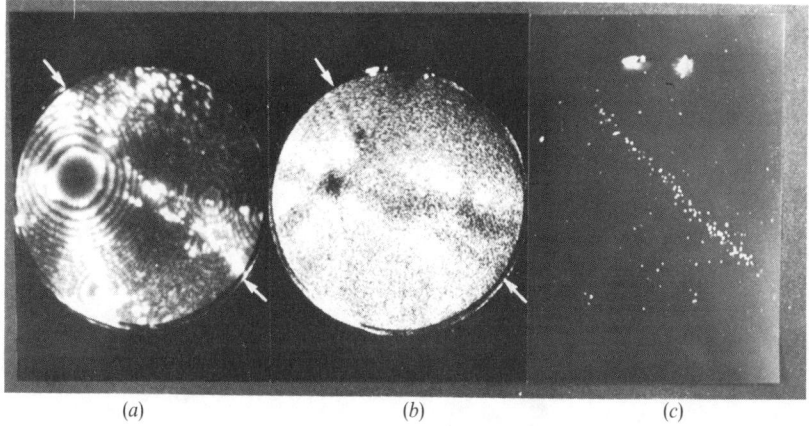

(a)   (b)   (c)

into the vacuum and individual helium atoms are ionised over atomic steps and kinks on the emitter surface where the field is highest. These ions then form the image on the phosphor screen. Image resolution is improved if the needle is cooled to temperatures of 78 K and below since the thermal vibration of the temperature-accommodated helium ions is reduced. By raising the field, individual surface atoms can be field evaporated to reveal the sub-structure of the specimen in atomic detail. By applying an electrical pulse the mass to charge ratio of the individual atom can be resolved in a time-of-flight mass spectrometer. A commercially available instrument capable of this type of performance is shown in Fig. 1.5.

Unfortunately the atom probe is not a technique suitable for general applicability. The sample must take the form of a needle and the material must be capable of withstanding high electrical stress during imaging and field evaporation. Recent work using pulsed lasers to provide the evaporation pulse has extended the range of materials amenable to the technique to include not only the traditional metals (tungsten, iridium,

Fig. 1.5. A commercially available atom probe microanalyser. This instrument is capable of fast specimen insertion into UHV and provides facilities for both field-ion microscopy and imaging atom probe techniques. An energy compensating 'Poschenreider' time-of-flight analyser is also available to enable single atoms to be identified and then analysed (courtesy VG Ionex).

Table 1.1. *Methods of surface analysis*

	Auger electron spectroscopy (AES)	X-ray photoelectron spectroscopy (XPS)	Static SIMS	Dynamic SIMS	Ion scattering spectroscopy (ISS)	Rutherford backscattering (RBS)
Incident particle	Electrons (1–20) keV	X-rays (1254 eV and 1487 eV)	Ions (Ar, Xe$^+$, Ga$^+$)(100 eV –30 keV)	Ions (Ar$^+$, Cs$^+$ O$_2^+$ O$^-$ Ga$^+$)	Ions (He$^+$, Ne$^+$ Ar$^+$, Li$^+$, Na$^+$, K$^+$) (100 eV–5 keV)	Ions (H$^+$ He$^+$) (1–3 MeV)
Emitted particle	Auger electrons (20–2000) eV	Photoelectrons (20–2000) eV	Sputtered ions	Sputtered ions	Scattered primary ions	Scattered primary ions
Element range	> Li ($Z = 3$)	> Li ($Z = 3$)	> H ($Z = 1$)	> H ($Z = 1$)	> H ($Z = 1$)	> H ($Z = 1$)
Detection limit	10^{-3}	10^{-3}	10^{-6}–10^{-9}	10^{-6}–10^{-9}	10^{-1}–10^{-4}	10^{-1}–10^{-4}
Depth of analysis	2 nm	2 nm	1 nm	1 nm	<1 nm	20 nm
Lateral resolution	>200 Å	>150 μm	(500 Å–10 mm)	(500 Å–50 μm)	>10 μm	>2 μm

steel, aluminium) but also semiconductors such as gallium arsenide and silicon in whisker form.

6 Future trends

Methods of surface analysis are still in a phase of intense development, although each technique is at a different stage. Auger electron spectroscopy is in a highly developed phase and it is difficult to predict further increases in spatial resolution or sensitivity. Likewise, Rutherford backscattering and its allied techniques is well understood and developed. Any reduction in the size and cost of the equipment involved could see a more rapid increase in its adoption. Although X-ray photoelectron

Table 1.2. *Auger electron spectroscopy*

Acronyms
AES (Auger electron spectroscopy)
SAM (scanning Auger microprobe)

Summary
In AES a beam of electrons is used to excite Auger electrons from the surface. Since electron beams can be focussed into fine spots, scanning electron microscope techniques can be used to obtain an image of the region under analysis. This makes the technique suitable for the analysis of small areas. The technique is also well suited to composition-depth profiling and hence for the analysis of thin films and surface coatings.

Auger electron spectroscopy

Advantages
High spatial resolution and scanning electron image display of the sample.
Element mapping capability.
Good depth resolution in composition-depth profiling.
Rapid collection of spectra (typically less than 5 minutes).
Quantification to better than $\pm 10\%$ using calibration standards.
High reproducibility of results.
Narrow range of sensitivities; variation in sensitivity across the element range is about a factor of 10.
Large user base and good support data available.

Disadvantages
Electron beam damage can cause artefacts especially on insulating samples.
Severe charging problems occur on some insulator surfaces.
Element mapping is slow due to the high background signal (typically 20 minutes).
Area of analysis is about twice the spot size due to electron scattering effects.
Although some chemical information is available in principle, the technique is largely used for elemental analysis only.

spectroscopy appears to be equally as advanced in its development, more ideas are being generated that could lead to a significant improvement in its ability to analyse small areas, and even to image with spatial resolution in the 10 μm range. Developments in this area could help overcome the most serious limitation of the technique.

Comparatively little effort is being devoted to the development of ion scattering spectroscopy. Although the technique found favour in the late 1970s in many industrial laboratories particularly in the USA, the technique is less well suited to general analysis than AES, XPS and SIMS. The technique is likely to continue to have an important role in laboratories devoted to more fundamental studies of adsorption and surface structure.

Secondary ion mass spectrometry is the technique in which the

Table 1.3. *X-ray photoelectron spectroscopy*

Acronyms
XPS (X-ray photoelectron spectroscopy)
ESCA (electron spectroscopy for chemical analysis)

Summary
In XPS, the sample is illuminated with X-rays which excite photoelectrons from the surface. A major advantage of the technique is that the photoelectron energy is dependent on the precise chemical configuration of the surface atoms and pronounced chemical shifts are produced in the position of the peaks in the XPS spectrum. XPS is amenable to virtually all vacuum-compatible samples since the incident X-rays do not normally cause surface damage. Hence XPS may be used to analyse surfaces such as those that occur on delicate powder materials, polymers and organic coatings.

X-ray photoelectron spectroscopy

Advantages
Chemical information obtained from the chemical shifts in the photoelectron energies.
Minimal beam damage. The technique can be used with very delicate materials.
Minimal sample charging.
Rapid collection of spectra (typically less than 5 minutes).
Quantitative to better than $\pm 10\%$ using calibration standards.
High reproducibility of results.
Narrow range of sensitivities; variation in sensitivity across the element range is about a factor of 10.
Large user base and good support data available.

Disadvantages
Essentially a broad area technique with only limited capability for small area analysis.
Poor depth resolution in composition-depth profiling.

Table 1.4. *Secondary ion mass spectrometry*

Acronym
SIMS (secondary ion mass spectrometry)

Summary
In SIMS a beam of low energy ions is used to sputter surface atoms into the vacuum where the ionised fragments are detected directly using a mass spectrometer. There are two distinct modes of analysis. In static SIMS a fast atom beam or an ion beam of low current density is used so that the analysis is confined to the outermost layers. The analysis can be used to detect sputter-molecular fragments which provide details of the chemical state of the surface. If focussed ion beams are used, SIMS provides a physical image of the sample using secondary electrons, or an elemental (or molecular) map using selected secondary ions. Systems based on microfocussed liquid metal ion sources can provide sub-500 Å resolution. Each secondary ion image corresponds to about a monolayer consumption. In dynamic SIMS, high ion current densities are used to erode successive atomic layers at a relatively fast rate. The technique is capable of detecting as little as 10^{13} atoms cm^{-3}. The high sensitivity of the technique combined with its ability to measure concentration with depth, means it is ideal for the characterisation of dopants and other impurities in semiconductor materials.

Secondary ion mass spectrometry

Advantages
Chemical information obtained from the detection of sputtered molecular fragments.
Ability to detect hydrogen.
Ability to distinguish isotopes.
High sensitivity (typically in the ppm range).
High spatial resolution and scanning electron image display of the sample.
Rapid element and molecular mapping capability (typically < 1 minute for a matrix element).
Good depth resolution and high dynamic range in composition depth profiling.
Rapid collection of mass spectra.
Charging problems overcome using fast atom primary beams or electron beam charge neutralisation.
The technique can be used with very delicate materials.
A developing user base and reasonable support data available.
Quantification to better then 10% with dilute systems (e.g. impurities in semiconductors).

Disadvantages
Essentially a destructive technique.
Wide range of sensitivities (overcome to some extent by using separate ion sources for electropositive and electronegative elements).
Not quantitative for non-dilute systems.

Methods of surface analysis 17

Table 1.5. *Ion scattering spectroscopy*

Acronyms
ISS (ion scattering spectroscopy)
LEISS (low-energy ion scattering spectroscopy)

Summary
In ISS, a mono-energetic primary ion beam is scattered into an analysing spectrometer with an energy which is defined by the mass of the surface target atom. The main advantage of the technique is its high surface sensitivity. The technique is also capable of elucidating surface structure by taking intensity measurements whilst varying the polar and azimuthal angles of the primary ion beam. The technique is a more specialised one and does not have the broad applicability of AES, XPS, SIMS and RBS.

Ion scattering spectroscopy

Advantages
Excellent surface sensitivity; analysis confined to the outermost atomic layer.
Virtually non-destructive.
Surface structure determination possible on single crystal samples.

Disadvantages
A specialist technique with a declining user base.
Essentially a broad base technique with only limited capability for small area analysis.
No depth profiling capability.
Limited commercial development.

Table 1.6. *Rutherford backscattering*

Acronym
RBS (Rutherford backscattering)

Summary
In Rutherford backscattering an energetic (05–3 MeV) beam of light ions (H^+, D^+, $^3He^+$, $^4He^+$) impinges on a target and a small proportion of the ions undergo unscreened collisions with the nuclei of the target atoms. Measurement of the energy of the scattered particles gives the mass of the target atom. Since the cross-section for the scattering event is accurately known the technique provides quantitative analysis without standards. The technique has relatively poor depth resolution (typically 200 Å), but can provide concentration-depth profiling without the need for ion erosion. Although RBS is more sensitive to the heavy elements, its related techniques recoil analysis and nuclear reaction spectrometry provide good sensitivity to the light elements including hydrogen. Ion channelling may be used for site identification of impurity atoms in crystals and for measuring the degree of crystalline disorder.

Table 1.6 (*Cont.*)

Rutherford backscattering

Advantages
Ability to detect hydrogen using recoil analysis.
Concentration-depth profiles without the need to strip the sample.
Absolute quantification to better then 5%. Increased accuracy with calibration standards.
High reproducibility of results.
Virtually non-destructive (except for polymers and biological materials).
Excellent support data available.
Associated technique of ion channelling provides useful structural information on single crystals.

Disadvantages
Essentially a broad area technique with only limited capability for small-area analysis.
Comparatively poor surface sensitivity (50 Å at best, but typically 200 Å).
Elemental analysis only (no chemical bonding information).
Wide range of sensitivities (overcome to some extent by using associated techniques such as nuclear reaction analysis).
Comparatively high capital investment and limited commercial availability.

---

developments are currently most rapid. SIMS has always been acknowledged to have certain key advantages. The technique combines high spatial resolution (500Å), with high sensitivity and an ability to provide detailed surface *chemical* analysis. Efforts are now being made to compile libraries of standard spectra to enable quantitative analysis to be performed. Also the emergence of electron beam SNMS and an additional feature to a quadrupole SIMS instrument could lead to routine and parallel quantitative analysis.

Clearly, each of the techniques covered in this book has its own strengths and weaknesses. Ideally, results from one method should be correlated against those from another. AES, XPS and SIMS are all compatible within the same instrument and multitechnique instruments will continue to have an important part to play in surface analysis.

References

Becker, C.H. & Gillen, K.T. (1984). *Anal. Chem.*, **56**, 1671.
Benninghoven, A., Rudenauer, F.G. & Werner, H.W. (1987). *Secondary Ion Mass Spectrometry* (Wiley-Interscience, New York).
Briggs, D. & Seah, M.P. (1983). *Practical Surface Analysis* (Wiley, Chichester)
Davis, L.E., MacDonald, N.C., Palmberg, P.W., Riach, G.E. & Weber, R.E. (1976). *Handbook of Auger Electron Spectroscopy* (Physical Electronic Industries, Eden Prarie).

Joshi, A., Davis, L.E. & Palmberg, P.W. (1975). In *Methods of Surface Analysis* Ed. A.W. Czanderna (Elsevier, Amsterdam), p. 159
Muller, E.W. & Tsong, T.T. (1969). *Field-Ion Microscopy* (Elsevier, New York).
Oechsner, H. (1984). *Adv. Solid State Phys.*, **24**, 269.
Tumpner, J., Wilsch, R. & Benninghoven, A. (1987). *J. Vac. Sci. Tech.* **5**, 1186.
Walls, J.M. & Christie, A.B. (1982). In *Surface Analysis and Pretreatment of Plastics and Metals*, ed. D.M. Brewis (Applied Science, London), p. 13.
Wagner, C.D., Riggs, W.M., Davis, L.E., Moulder, J.F. & Muilenberg, G.E. (1979). *Handbook of X-ray Photoelectron Spectroscopy* (Perkin Elmer Corp, Minnesota.

*Conflat is a trademark of Varian Associates Inc.

2
Ion erosion in surface analysis

R. SMITH & J.M. WALLS

1 Introduction

The composition of almost all surfaces is different from that of the bulk. In many cases the surface consists of an oxide, nitride, carbide, or some other highly adherent layer. Practical surfaces also carry with them a history of the various processing stages they have been subjected to, and residues of cleaning acids, solvents, polishing compounds and other environmental impurities are often found. Removal of these outermost layers *in situ* by ion beam sputtering has become an essential first step in the application of surface analysis techniques. Removal of further successive atomic layers to provide compositional information with depth has become an equally indispensible tool to the surface analyst. Sputter-depth profiling can be useful at two levels. On the finest scale, a depth profile through < 10 nm can enable the surface and subsurface to be compared. This has been used, for example, to identify segregant species on fracture surfaces, catalyst surfaces, electrical contact failures, etc.... Over thicknesses of 1 μm and beyond the technique is now established as a method used to characterise thin films, multilayer structures and dopant profiles in semiconductors. In short, sputtering provides the finest scale microsectioning technique currently available, and effectively adds a third dimension to many surface analytical methods.

Ion beams are also central to those surface analytical techniques which use ion bombardment as the primary excitation mechanism. These techniques include secondary ion mass spectroscopy (SIMS), ion scattering spectroscopy (ISS), and Rutherford backscattering spectroscopy (RBS). In SIMS alone, up to five different types of primary ion source are used to optimise the various modes of operation. A range of UHV compatible ion sources has been developed over recent years to meet the requirements of surface analysis and some of these are discussed in Section 2.2.

Although ion bombardment is essential to the effectiveness of surface analytical techniques, the process is not perfect. Ion bombardment can alter the composition of the surface by a variety of mechanisms including preferential sputtering, ion implantation and ion-induced association and disassociation. The process can also cause ion-induced topography, atomic mixing and radiation-enhanced diffusion, all of which can affect the depth scale of a sputter-depth profile.

Clearly, the surface analyst requires a knowledge of the way in which the sputtering process affects the compositional integrity of the sample. This chapter provides an introduction to the mechanisms of sputtering, the various parameters that control erosion rates, and the important factors which influence ion-induced changes in structure and composition.

2 Ion sources

A number of different types of ion source are employed in the various methods of surface analysis. Essentially ions are created by any process that imparts to an atom an energy greater than its first ionisation potential. Processes used include plasma discharge, electron impact, field ionisation and field evaporation. Ion energies are formed by a separate acceleration system and can vary from a few hundred eV used in a gentle static SIMS analysis through to an MeV beam used in Rutherford backscattering and its associated techniques. A list of ion guns used in surface analysis and their applications is given in Table 2.1.

UHV compatibility is a general requirement for all sources except those used in the accelerator-based methods. Materials are limited to those with low vapour pressures and to those that may be baked to 250 °C or otherwise easily demounted. For this reason, electrostatic ion optics are preferred to magnetic. Where wound magnets are used they must be capable of being decoupled from the vacuum system during bakeout. Source gases used are normally high-purity, research grade to avoid unnecessary surface contamination. Gas feed lines via bakeable UHV leak valves direct to the source and differential pumping of the ion optical column are often used to maintain high vacuum conditions in the vicinity of the sample.

All ion guns require regular maintenance since the ion source and its associated accelerating column (especially beam-limiting apertures) will necessarily suffer from erosion by sputtering. Even in the best-designed ion columns, insulators become coated with sputter-deposited conducting layers. Solid source materials also require periodic replenishment. Ion sources used on a routine surface analysis system require regular

Table 2.1. *Ion guns used in surface analysis*

Type of source	Gas species	Minimum spot size	Applications
Cold cathode discharge ion source	Argon, other inert gases, O_2^+, N_2^+	10 mm	Surface cleaning and depth profiling in AES and XPS (1–10 keV)
Twin-anode electrostatic ion source	Argon, other inert gases, O_2^+, N_2^+	10 mm	Surface cleaning and depth profiling in AES and XPS (1–10 keV)
Kaufmann ion source	Inert gases	1 mm	Broad-area surface cleaning 500 eV–3 keV
Hot filament ion source	Inert gases	10 μm	Surface cleaning and depth profiling in AES and XPS (1–10 keV)
			Primary source for static SIMS (1–5 keV)
Duoplasmatron ion source	O_2^+, N_2^+ inert gases	1 μm	Surface cleaning and depth profiling in AES and XPS (1–10 keV)
			Primary ion source for dynamic SIMS (O_2^+, Ar^+, 1–20 keV)
Surface ionisation ion source	Cs^+, other alkali metals	1 μm	Source for ion accelerators used in RBS primary ion source for dynamic SIMS (Cs^+, 1–10 keV)
Liquid metal ion source	Ga^+, In^+, Cs^+ other metals and alloys	500 Å	Primary ion source for imaging SIMS (Ga^+, Cs^+, 1–30 keV)

preventative maintenance to ensure continuous use to original performance specification.

2.1 Cold cathode ion sources

These sources are relatively inexpensive and produce high-current (50 μA) ion beams in a typical spot diameter of about 10 mm with energies variable in the range 1–10 keV. They are used extensively for surface cleaning and composition-depth profiling and are particularly suitable for XPS-based instruments where broad-area etching is important. Since cold cathode sources do not employ a heated filament, the sources may in principle, be used with a range of both inert and chemically active gases.

There are two popular versions of the cold cathode source, each obtains high-ion currents by promoting available electron path lengths in the source; one uses an electrostatic field and the other uses a magnetic field. A schematic diagram of the twin anode electrostatic ion gun is shown in Fig. 2.1. The gas species is fed directly into the source and a differential pressure then exists between the source and the vacuum system. The two anodes are held at a positive potential V with respect to the earthed cathode. A cold cathode discharge is maintained down to pressures below 10^{-5} mbar and the electrons tend to oscillate transversely between the anodes. Ions from the discharge are allowed to emerge through an aperture in the cathode and these are used to bombard the earthed target. The energy of each individual ion depends on its precise position on formation within the source, and hence a source of this type produces a wide energy distribution. It also produces neutrals approximately equivalent in number to the ions together with a significant proportion of multiple-charged species and electrons. The source can be used for fast atom bombardment by using a simple deflector for the charged particle fraction of the beam.

A commercially available electromagnetically confined cold cathode discharge source is shown in Fig. 2.2. In this case a magnetic field is applied

Fig. 2.1. A schematic diagram of the twin anode electrostatic ion gun.

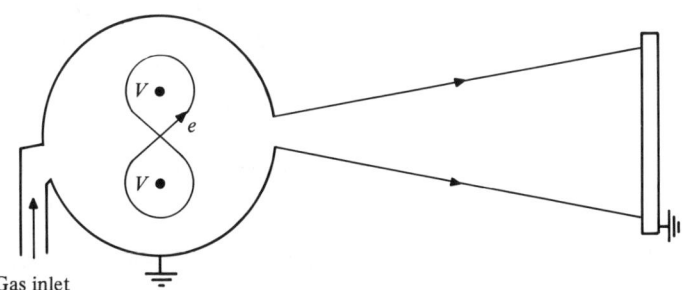

parallel to the hollow anode cylinder axis constraining electrons to move in oscillating helical trajectories between the two cathodes at either end. This gun produces a performance similar to the electrostatic twin anode source. Each source may be fitted with a single-lens focussing system and aperture to improve the ion current spatial distribution.

2.2 Kaufman or broad area ion source

Although the cold cathode ion sources described above are currently the most popular type of source in surface analysis for broad-area ion-beam etching, there is a growing requirement for a source with a uniform ion flux density across a large area (~ 3 cm) and a well-characterised set of beam parameters. This requirement can be met using the Kaufman source (Kaufman *et al.* (1982)). Originally intended for space propulsion applications, the Kaufman source produces a low-energy (500 eV–3 keV), high-current density (> 1 mA cm^{-2}), collimated ion beam with low-energy spread, with a uniform ion flux over several centimetres.

Fig. 2.2. An electromagnetically confined cold cathode discharge source (courtesy VG Ionex).

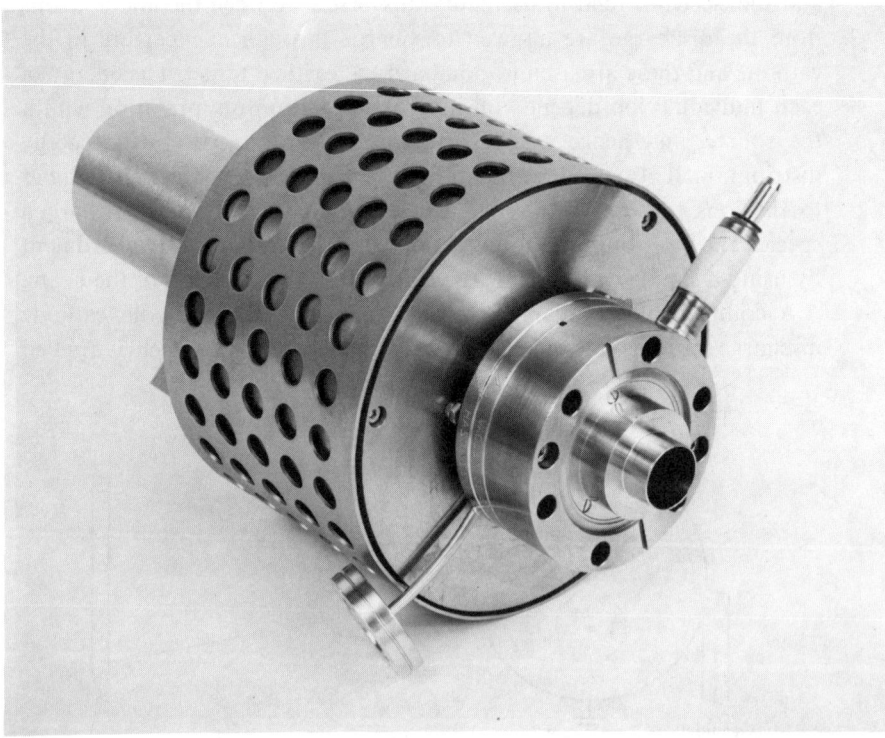

Ion erosion in surface analysis

In operation, the gas is introduced into the discharge chamber to be ionised by electron-impact from a thermionic cathode to an array of anodes which form rings around the chamber walls. A multipole magnetic field is arranged to shield the anodes to provide a uniform ion flux across the extraction grid region. Since the source has a large open structure, the gas flow into the analysis region is relatively high even with differential pumping. The Kaufman source is already used extensively in thin film deposition and although it is relatively expensive, its characteristics provide it with many potential advantages for uniform depth profiling in XPS.

2.3 Electron impact ion sources

A commercially available electron impact source is shown in Fig. 2.3. Electrons from a heated filament are accelerated into a cylindrical grid where they have sufficient energy ($\sim 160\,\text{eV}$) to ionise gas atoms on collision. The ions are extracted from the grid by fringing fields from an adjacent anode which accelerates the ions into a focussing lens system, via deflection plates to the sample held at earth potential. The ion energy is controlled by the positive voltage applied to the grid (typically $0.5\,\text{keV}{-}5\,\text{keV}$).

Since most of the ions are formed within an equipotential, this source has the advantage that it produces a narrow energy spread (although there is usually a small fraction of multiple-charged species and neutrals). The beam can be focussed into a spot with a Gaussian ion current

Fig. 2.3. The AG61 small spot electron impact ion source with differential pumping and gas feed line (courtesy VG Ionex).

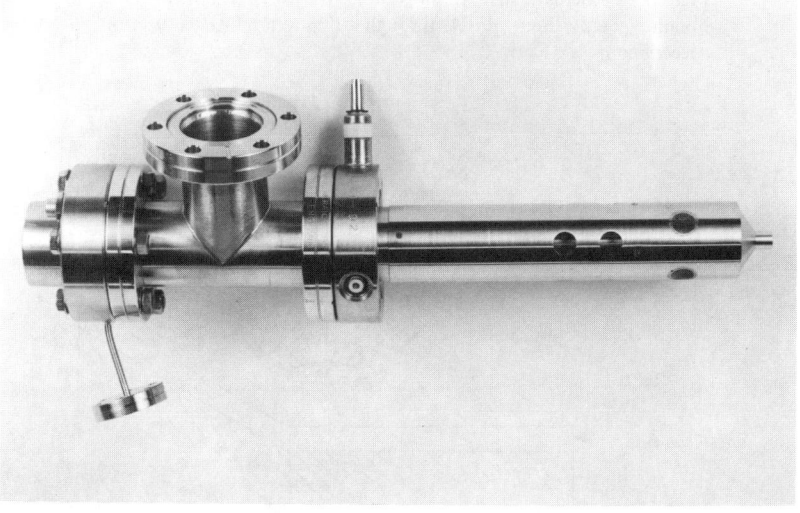

distribution. Ion guns with spot diameters from 50 μm to several mm are available commercially.

The number of ions produced per second in the grid region is given by:

$$i_t = \frac{i^- \lambda p \sigma}{e},$$

where p is the operating gas pressure (typically 5×10^{-5} mbar), i^- is the electron current at the grid (typically 30 mA), σ is the differential ionisation coefficient for the gas species and λ is the mean free path of the electrons. The source produces a well-defined beam with good ion current stability. A current of 20 nA in a spot diameter of 50 μm can be achieved (current density 250 μA cm^{-2}). Its use is restricted to inert gases since its operation with other species (O_2, N_2) leads rapidly to filament failure.

The electron impact source is the most popular choice for dedicated AES instruments. The ion beam can be raster-scanned thereby producing a square etchpit in the sample surface or alternatively the ion beam can be used in static mode to produce a Gaussian-shaped crater. This latter mode of operation results in higher ion current density and shorter sputtering times; etching rates of 25 nm min^{-1} (on Ta_2O_5) are typical. In both cases the electron beam used for Auger analysis is very much smaller than the crater and the analysis can be performed at the bottom of the crater without significant edge effect contributions from the crater walls as illustrated in Fig. 2.4. In X-ray photoelectron spectroscopy (XPS) applications it is necessary to scan the beam, reducing the effective ion current density to about 10 μA cm^{-2} with sputtering rates of about

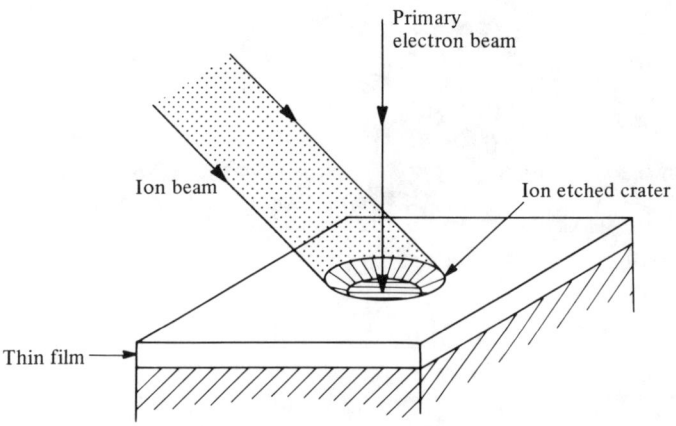

Fig. 2.4. A schematic diagram illustrating the use of a static ion beam to obtain a composition – profile through a thin film by Auger electron spectroscopy.

1 nm min^{-1}. For this reason in combined AES/XPS instruments, a cold cathode ion source is often preferred.

Since the electron impact source produces a focussed, well-characterised beam of low-energy inert gas ions at relatively low cost, it is often used in other techniques such as static SIMS and ion scattering spectroscopy. In both these applications it is used as the primary excitation agent and a Wien filter is often added to remove impurities from the beam.

2.4 Duoplasmatron ion source

The duoplasmatron was first devised by von Ardenne (1956). It uses a magnetically constricted arc to produce a dense plasma with an ion density of about 10^{14} cm^{-3} from which the ion beam is extracted. The arc is confined through a small aperture in an electrode at an intermediate potential situated between the anode and cathode. Power consumption is high and it is normally necessary to water cool the anode. Electrons are obtained using a filament or a cold cathode and hence the duoplasmatron is a reliable, well-developed source which provides intense, well-characterised ion beams with relatively low energy spreads (~ 10 eV). The combination of a high-brightness source with low energy spread makes the duoplasmatron suitable for small-spot focussing.

Versions of the duoplasmatron are now available commercially both for large-area ion etching and for small-spot imaging and microanalysis. A duoplasmatron used as a primary ion source for dynamic SIMS is shown in Fig. 2.5. This ion gun incorporates a Wien filter to remove beam impurities and a small angle deflection (typically 1°–4°) to eliminate neutrals. The gun is also equipped with two stages of differential pumping to reduce the working pressure in the vacuum chamber and to reduce the likelihood of neutral formation near the target. This type of gun may be used to provide a spot diameter < 5 μm with ion current densities greater than 10 mA cm^{-2}.

Fig. 2.5. A mass filtered, small spot duoplasmatron for high-performance dynamic SIMS (courtesy VG Ionex).

2.5 Surface ionisation source

Caesium ion sources are used in dynamic SIMS to optimise detection sensitivity to the electronegative elements. This is usually achieved using a surface ionisation source which produces a comparable performance with the duoplasmatron for alkali metal ions. The source usually consists of two elements; one provides the source material by evaporation and the other is heated to ionise the material on contact. The ioniser surface is heated to sufficiently high temperature to ensure that the desorption rate exceeds the arrival rate. The ratio of the numbers of ionised atoms n^+ to neutrals n^0 is given by:

$$\frac{n^+}{n^0} A \exp(\varphi - Ei)/kT,$$

where A is a constant, Ei is the ionisation potential of the source material, φ is the work function of the ioniser, k is Boltzmann's constant and T is the absolute temperature of the ioniser. Clearly, the ionisation probability is high when $\varphi - Ei > kT$. Thus, for example, ($\varphi = 5.3\,\text{eV}$) at 2443 K results in virtually complete ionisation of caesium.

Since the surface ionisation process does not rely on energetic collisions between atoms and electrons, no ionisation of residual gas takes place, and this results in high beam purity. The sources provide a supply of ions with a low-energy spread with thermal dimensions only. The sources are also UHV compatible. Since the source materials are usually highly chemically reactive, a beam line isolation valve is normally employed in the ion optical column. This also facilitates periodic replacement of the source material.

2.6 Liquid-metal ion sources

The development of high-brightness liquid ion sources over the past decade has resulted in focussed ion-beam systems with submicron spot sizes and current densities in excess of $1\,\text{A}\,\text{cm}^{-2}$. The use of these

Fig. 2.6. A schematic diagram illustrating the construction of a liquid-metal ion source.

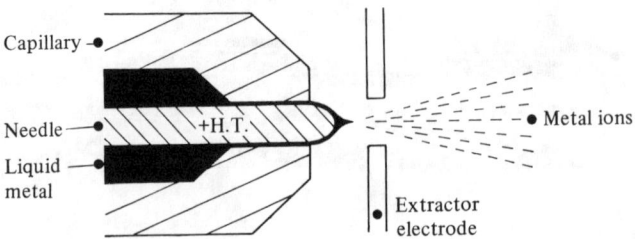

sources has led to a dramatic improvement in the lateral spatial resolution of SIMS (Waugh et al., 1984). Systems with spatial resolution below 500Å are now available commercially. Other applications of microfocussed ion guns include ion-beam machining, ion-beam repair of optical or X-ray lithographic masks and ion beam resist exposure.

The construction of a liquid metal ion source is illustrated schematically in Fig. 2.6. In the most popular (and reliable) configuration a solid needle is used as the source substrate. The needle is normally fabricated from tungsten and the radius at the apex is usually about 5 μm. The needle is set within a cylindrical capillary and the intermediate region holds a reservoir of the source material. A high potential is placed on the needle to provide an intense electric field at the apex. The source material is indirectly heated and the liquid film distorts at the apex to form a stable jet-like protrusion on the end of a Taylor cone shape (Kingham & Swanson, 1984). Ions are then emitted by field evaporation from an extremely small emitting area. The ion energy spread is approximately 5 eV and the axial current intensity is typically $20\,\mu A\,Sr^{-1}$.

Gallium is the most popular source material. It has a low melting point (28 °C), a vapour pressure below 10^{-9} mbar, and it is available in high-purity form. Since it possesses only two intense isotopes it also causes minimal interference in the SIMS spectrum. Gallium ion sources produce stable and reliable beams and are ideal for routine surface analysis. Other materials include indium, tin, gold, bismuth and caesium. Alloy sources employing heated hairpin substrates are also under development for semiconductor implantation applications.

The characteristics of the liquid metal ion source, viz, high brightness, source equipotential and relatively low energy spread, make it ideal for use with electrostatic focussing systems. Spot sizes below one micron are readily achieved at 10 keV and 500 Å spot-size specifications at 30 keV are available commercially.

Liquid metal ion sources are fully compatible with UHV vacuum systems. They present no gas load to the vacuum and hence are ideal for bolt-on SIMS additions to an existing system. In dedicated SIMS instruments, the guns are often differentially pumped to avoid needle oxidation (and consequential dewetting) when used in oxygen flood conditions which are often employed to maximise secondary ion yields.

Liquid-metal ion guns may be scanned at TV rates to provide ion-induced secondary electron imaging. Sample analysis may be obtained by conventional mass spectrometry while SIMS maps are obtained using the elemental (or molecular) ion signal from the mass spectrometer to modulate a slow-scan oscilloscope display. SIMS maps obtained in this way can have similar spatial resolution to the best Auger maps, but they are obtained with higher analytical sensitivity and in a fraction of the time (Bayly et al., 1985).

3 Sputtering

3.1 Sputtering mechanisms

In the sputtering process, either physical or chemical effects can dominate the interactions of the ion beam with the surface and this has led to the terms physical and chemical sputtering, depending on the dominant effect. With physical sputtering, the ion beam incident on the material sets up a collision cascade in the surface layers of the solid with

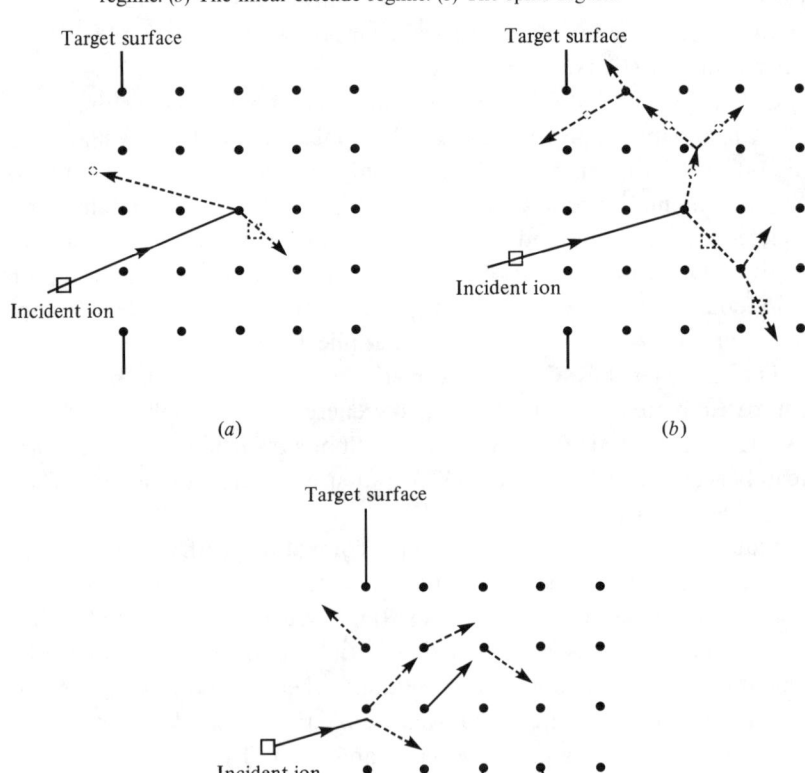

Fig. 2.7. Sputtering regimes by elastic collisions. (a) The single knock-on regime. (b) The linear cascade regime. (c) The spike regime.

energy transferred from the ions to the atoms of the target. A recoil is created when more energy is transferred in the collision than the binding energy of the target atom. A surface atom or molecule becomes sputtered if the energy transferred to it is greater than its surface binding energy and the imparted momentum has a component normal to the surface. Subsurface recoils can also contribute to the sputtering process provided they sustain sufficient momentum in the collision cascade to reach and escape from the surface of the material.

For physical sputtering, three qualitatively different situations for elastic collisions of random materials (Sigmund (1983)) are generally identified, which are categorised as the single knock-on regime, the linear cascade regime and the spike regime, see Fig. 2.7.

In the single knock-on regime, the energy transfer to the target is sufficient to generate primary recoils but these generally have insufficient energy to generate further recoils. These single knock-on recoils may be ejected from the surface if they are energetic enough to escape and sputtering takes place. The linear cascade regime is attained when recoils from ion-target collisions receive sufficiently high energy to generate further recoils and thus a cascade, but of sufficiently small density so that collisions between moving atoms are infrequent. In the spike regime the collisions are such as to set in motion a high density of moving particles within a certain volume of the target material.

For low-energy ion beams $\lesssim 1\,\text{keV}$ the collisions fall within those of the single knock-on regime category, but this can extend to the lower keV range for light ions. For higher energies the linear cascade regime best describes the process except for heavy ions which stop quickly and generate spikes.

As the energy of the beam increases from the threshold region for sputtering to occur, a linear region develops where the sputtering yield is proportional to the beam energy. This occurs because the amount of energy T transferred between two colliding spheres in an elastic collision is

$$T = (4m_1 m_2/(m_1 + m_2)^2)E,$$

where E is the energy of the primary beam species of mass m_1 and the target has mass m_2. At higher energies the beam is no longer interacting with surface atoms and atoms energised from deeper within the material undergo energy-losing collisions, thus reducing their probability of ejection.

The sputtering yield exhibits also a periodicity as a function of atomic number both of the target species and the ion beam species. This is shown in Fig. 2.8. This periodicity is interpreted as dependent on the number of electrons in the outer shell of the atom, with the inert gases appearing to give the higher sputtering yields.

In many surface analytical techniques bombardment of the target material by inert ions is common but reactive ions can also be used (e.g. SIMS) and chemical sputtering may occur. In this case the surface layer can become a compound with different properties to that of the target material. The sputtering yield of the target can thus be either increased as a result of chemical combination if a volatile compound is formed, or decreased if a very stable solid compound such as an oxide is formed. Chemical effects such as these are an important area of current research and can occur even if a beam of inert ions is used with a reactive residual gas present in the vacuum chamber – something which must always happen in practice to a greater or lesser degree.

Sputtering may also occur by electronic excitation where, for example, in insulators, the excitation energy of excited electronic states may be transferred to atomic motion. Sigmund (1983) suggests that chemical sputtering might more aptly be categorised as a special case of sputtering by electronic excitation. However, in surface analysis, physical (ballistic) sputtering occurs irrespective of whether or not chemical and electronic processes also need consideration. Chemical sputtering, usually characterised by strong temperature variations in the sputtering yield, has been reviewed by Roth (1983), but in this chapter it will be the important features of physical sputtering that will receive most consideration.

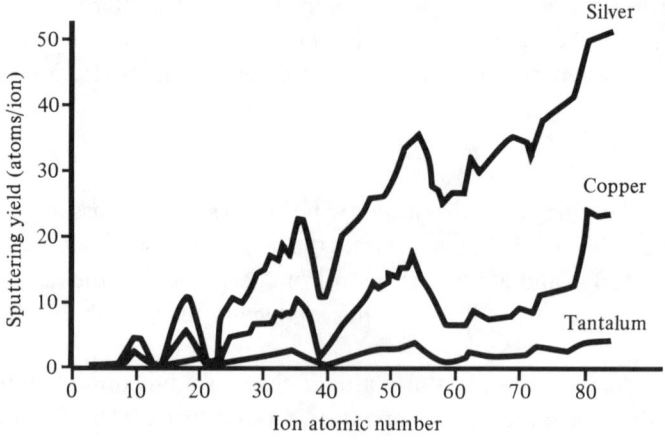

Fig. 2.8. Sputtering yield as a function of atomic number (Almen & Bruce (1961)).

3.2 Structural effects in sputtering

Although the physical process of sputtering is intrinsically the same for amorphous, polycrystalline and crystalline materials, the regular array of atoms in a crystalline structure produces effects which are not present in randomly ordered materials. Perhaps the most important consideration when using an ion beam to depth profile, is the effect of this structure on the sputtering yields. The sputtering yield Y is defined as the average number of atoms removed from a surface per incident ion. This is a meaningful definition in the linear cascade or single knock-on regimes where the number of sputtered atoms is also proportional to the number of incident particles. The sputtering yield is a function of the angle of incidence for random materials, but for crystals it is also a function of the particular crystal plane undergoing ion bombardment. For single crystals where bombardment is close to the close-packed crystal axes, the yields are much lower than for other directions of incidence.

These reduced sputtering yields can be due to channelling where an incident ion is channelled down the spaces in the crystal structure and penetrates much farther into the material than would otherwise occur. This means that the energy transmitted to the crystal is deposited over a greater depth and in the subsequent collision cascade fewer target atoms have sufficient energy to reach the surface of the crystal. However, for irradiation perpendicular to the close-packed crystal planes the sputtering yield can be much higher than for polycrystalline materials. The channelling process occurs when an ion moves at a small angle to a row of target atoms and such that after each collision, this angle decreases. In the contrary case, where successive angles increase, the process is known as dechannelling. Lindhard (1965) has examined the behaviour of ions channelled along a crystal axis and has shown that channelling will occur if the velocity vector of an ion makes an angle with the direction of an atomic row less than a critical angle ψ_c which is energy dependent.

In addition to channelling effects, a focussed collision sequence can occur (Thompson, 1969) where momentum is transferred from atom to atom along an atomic row. A simple model shows that this can occur if $d < 4R_0$, where d is the lattice spacing, and R_0 is the elastic hard sphere radius which can be ascertained from the interaction potential V, i.e. that value which gives $V(R_0) = E_0$ the pre-collision ion energy, see Fig. 2.9.

The regular lattice structure also effects an angular distribution of sputtered atoms from monocrystalline materials and this spatial distribution depends on the energy and the direction of incidence of the ion beam, see Fig. 2.10. For crystalline materials, the angular distribution of sputtered atoms can manifest itself as a well-defined ejection pattern which has been

Fig. 2.9. Focussing in a crystalline material. (a) A focussed collision sequence. (b) The simple focussing process. The relation between angles θ_{n-1} and θ_n from collisions between the $(n-1)$th and the nth atoms in a row is given by $\theta_n = (1 - d/(2R))\theta_{n-1}$, if the angles are small. The angle θ_n converges to zero as $n \to \infty$ if $d < 4R$. This is a simple focussed collision sequence.

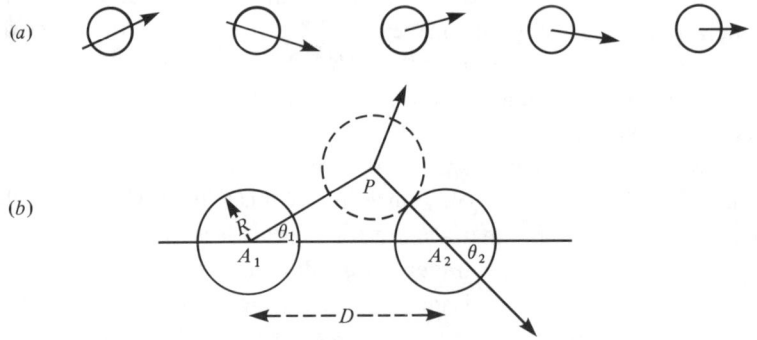

Fig. 2.10. Ejection pattern from the (111) face of a gold crystal under 40 keV Ar^+ ion bombardment.

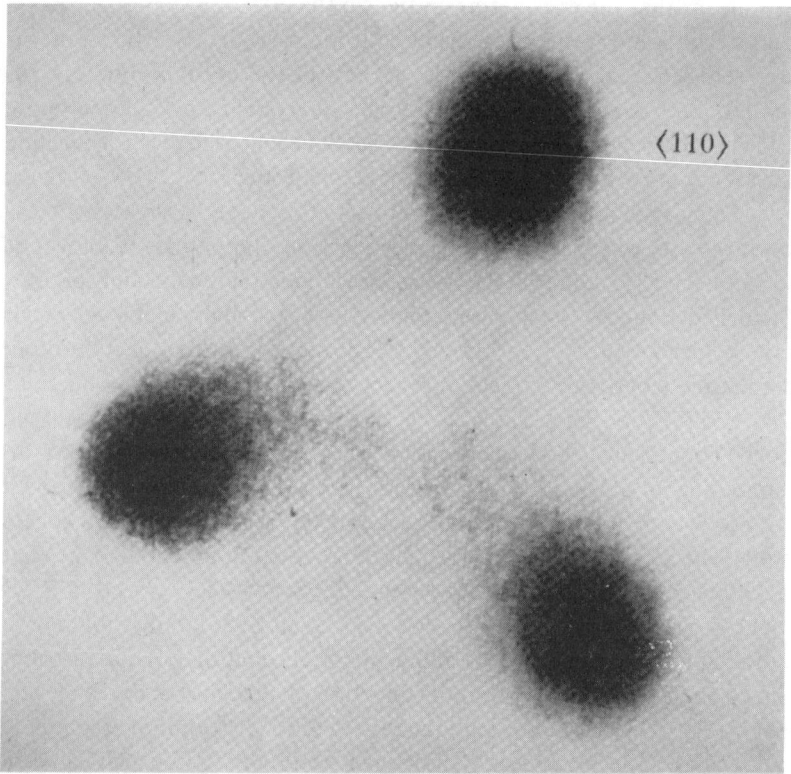

interpreted in a number of different ways. Focussing has been suggested as one possible explanation for these spots although Lehmann & Sigmund (1966) have derived a theory which shows how the spots could arise from the regularity of the surface structure combined with the energy spectrum of the sputtered particles. The precise mechanism for the spot formation is still open to doubt.

The structure of the crystal surface can be damaged by the ion beam. Etch pits and dislocations can form and if the ion doses are sufficiently high, the crystalline structure can be severely disrupted. It is known, for example, that silicon becomes amorphous under ion bombardment and the low-energy electron diffraction (LEED) patterns can be degraded when low-energy ion bombardment is used to prepare clean monocrystalline material. The effect seems to be less for metals than for semiconductor materials. Thus radiation damage that occurs can affect surface analytical measurements in a number of different ways. It can induce unwanted topography and cause not only structural damage but also ion implantation and the displacement of tracer materials for which the analysis is required. These effects will be the subject of further discussion later in this chapter.

3.3 Ion energy and projectile species

The energy of the incoming ion beam has an important effect upon the internal damage to the target material and the sputtering yield from it. In surface analysis the incoming ion beam generally has an energy in the lower keV range $0.5 \leqslant E_0 \leqslant 15\,\text{keV}$. For this range of energies the energy loss of the bombarding ions is mainly a result of two processes: nuclear interactions resulting from collisions between the screened nuclear charges of the ion and the target atoms, and electronic loss resulting from the interactions of the fast ion with the lattice electrons. In certain circumstances a charge exchange between the moving ion and the target atom can also contribute to the energy loss.

For elastic hard sphere collisions the energy of the colliding ion before a collision, E_0, is related to the energy after collision, E_1, by the laws of classical mechanics

$$E_1 = (1 + 2\rho \cos \theta + \rho^2)/(1 + \rho)^2 E_0,$$

where $\rho = m_1/m_2$ is the mass ratio of the ion and the target atom. The angle θ is the centre of mass scattering angle given by

$$\theta = \pi - 2 \int_0^{u_m} (1 - V(r)(1 + \rho)/E_0 - s^2/r^2)^{-1/2} s\,d(1/r),$$

where s is the impact parameter, u_m the first zero of $(1 - V(r)/E_0(1 + \rho) - s^2/r^2)$ and $V(r)$ is the interaction potential (Goldstein, 1980).

The electronic loss in a collision can be modelled for low-energy ions by (Biersak & Haggmark, 1980)

$$E_1 - E_0 = (0.045k)/(\pi a_F^2) \exp(-0.3s/a_F) E_0^{1/2}$$

where k is the Lindhard velocity stopping parameter and a_F is the Firsov screening length given by $0.8853a_0/(Z_1^{1/2} + Z_2^{1/2})^{2/3}$. In this formula a_0 is the Bohr radius $a_0 = 0.529$Å and Z_1 and Z_2 are the atomic numbers of the ion and target atoms.

Computer simulations enable the depth of damage, lateral spread and sputtering yields to be predicted fairly accurately using these models. Broadly, these programs (Harrison, 1983) fall into two categories: the binary collision and the molecular dynamics approach. The molecular dynamics or multiple interaction simulations follow the history of every particle generated in a collision cascade arising from an individual ion trajectory by a solution of the full equations of motion, whereas in the binary collision approximation the interactions between any two colliding particles are treated as individual events.

Many more simulations are possible using the binary collision approach than using molecular dynamics. In the linear cascade regime the binary collision approximation should suffice for most purposes, although recent work, using molecular dynamics, has shown that for certain crystal structures a small number of impacting ions can be responsible for a considerable proportion of the total sputtering yield arising from collisions

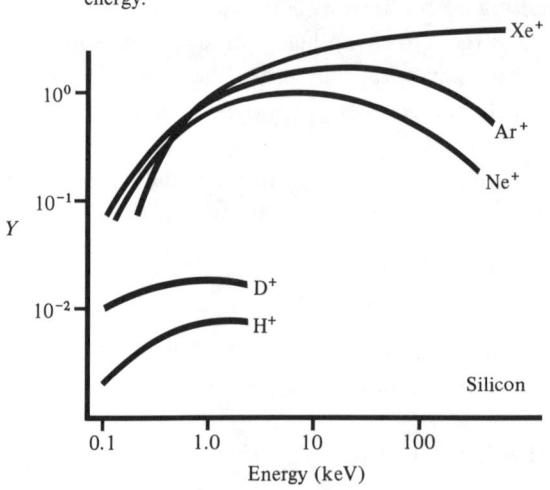

Fig. 2.11. Experimental sputtering yield (Y) data for Si as a function of ion energy.

which are not binary. Small etch pits can form as a result of a single ion bombardment, provided that the ion is incident at a particular point on the crystal plane.

Statistical theories based on transport theory have also been proposed (Sigmund, 1969). These have had some success in predicting the sputtering yield variation with energy for low-energy ions $\geqslant 1$ keV but generally they do not accurately predict the variation of sputtering yield with angle of incidence near grazing incidence where a different formulation of the problem must be used. Nor do they deal adequately with monocrystals, whereas the computer simulations of collisions do succeed in this. Fig. 2.11 illustrates the dependence of the sputtering yield on ion energy for silicon bombarded with a number of different ion species.

At low energies there is an energy threshold for sputtering to occur. Below a certain critical energy the ions transfer insufficient energy to the target to overcome the surface binding energy. The yield then rises as the incident beam energy is increased and eventually at high energies decreases as the recoil cascades are formed deeper within the material and cannot reach the surface.

A collection and comparison of sputtering yield data as a function of ion energy is available covering the literature up to 1977 (Andersen & Day, 1983) for normally incident ions. This data has been compared with the results from transport theory (nuclear stopping only) for energies $\geqslant 1$ keV with good agreement in most cases. The data does contain a spread of values, probably due to the different doses in many of the experiments. It is known that changes in surface topography and ion implantation occur with increasing dose and these effects could alter the measured sputtering yields. Implantation can cause changes in the surface binding energy, and hence sputtering yields, particularly where heavy ions are implanted, e.g. Xe in Si.

The general result from these figures seems to be that the theory predicts high-yield materials well, but overestimates for low-yield. The same is also true for dependence on projectile species with good agreement for light targets but not for heavier targets, particularly if bombarded with heavier species. An advantage of the transport theory over the computer models is that it does allow closed form expressions for the sputtering yield of the target and the ranges of the bombarding ions as a function of ion energy.

3.4 *Angle of incidence*

The angle of incidence of the incoming beam to the surface of the material crucially affects both the sputtering yield and the angular distribution of the sputtered particles. The angular variation has already been discussed with regard to how it is affected by the structure of the material, but even for amorphous and polycrystalline materials the distribution is not generally uniform. Experimental results, however, do indicate that for random materials at normal incidence the sputtered particle distribution is approximately uniform in the range 1–10 keV. This is the so-called 'cosine' distribution since, if we represent the sputtered flux by the radial vector from the point of impact, this varies as the cosine of the polar angle. However, the particular spatial distribution is dependent on both energy and the beam and target composition. For non-normal incidence these distributions are distorted with less material being sputtered back in the direction of beam incidence. Fig. 2.12 illustrates these effects.

The effect of angle of incidence on the sputtering yield is also of crucial importance, and in depth profiling, the beam is often orientated to the sample surface at an angle near to that for maximum sputtering yield. For random materials, see Fig. 2.13(a), (c), the yield generally rises to a maximum at an angle of about 60° to the surface and then drops to zero at grazing incidence where ion reflection takes place. Computer simulation programs of atomic collision processes have had some success at predicting the yields but the most frequently used transport theory models are often only accurate near to normal incidence.

For monocrystalline materials the sputtering yields are even more complicated functions of angle of incidence and also depend on the particular crystal plane under ion bombardment, see Fig. 2.13(b). There is no full compendium of data for sputtering yields as a function of angle of incidence for crystalline materials. Channelling theory has been used

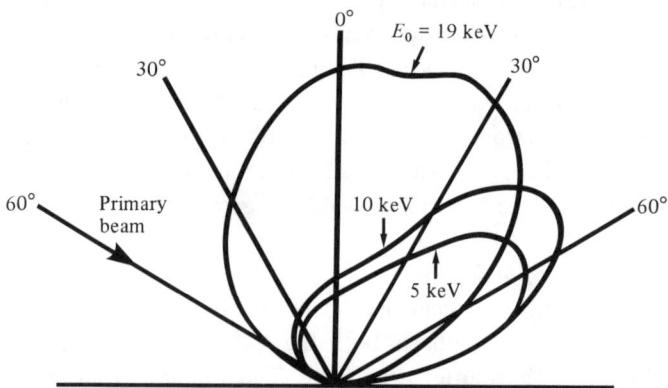

Fig. 2.12. Angular distributions of sputtered material for polycrystalline tungsten bombarded with oblique incidence Kr^+ ions of different energies.

Fig. 2.13. Sputtering yield as a function of angle of incidence. (a) Experimental data illustrating the influence of bombardment angle θ on the normalised sputtering yield of different polycrystalline metals bombarded with Ar$^+$ ions of 1.05 keV. (b) The sputtering yield dependence $Y(\theta, \psi)$ for 20 keV Ar$^+$ ions on the 100 face of various crystals. The rotation axis is the 011 axis. (c) Normalised sputtering yield curves for amorphous Si, under bombardment by low energy Ar$^+$ ions. These theoretical calculations have been carried out by the authors using a binary collision model.

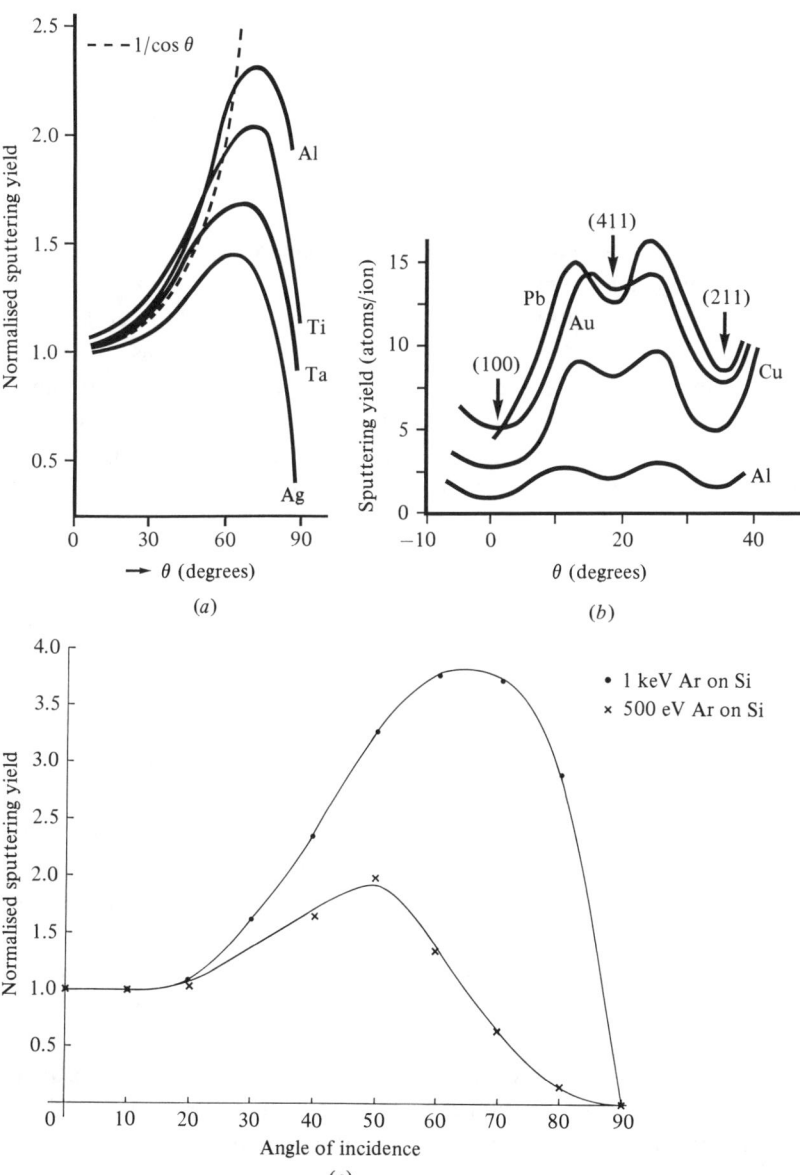

to give a rough approximation to the sputtering yields for high-energy ions (Onderdelinden, 1968) which have been compared to experimental results with reasonable agreement. However, at lower energies, below 10 keV, the relative sputtering yields cannot be accurately predicted by channelling theory. Sputtering yield measurements indicate that the yield from crystal planes is energy dependent with for example the (111) face of Cu giving a greater sputtering yield than the (001) face at ion energies above 1 keV but a lower yield at energies below 500 keV. Angle of incidence is also an important factor in the development of surface topography on a material and also affects the depth of atomic mixing within the solid. In certain circumstances these effects can affect the integrity of composition depth profiles and this will be the subject of further discussion in Section 4.

4   Ion-induced structural and compositional changes
4.1   Ion-induced topography

Since the sputtering process is destructive to the target material, with surface target atoms being removed by the incoming ions, the surface layer of the target undergoes constant change during ion bombardment. The type of change in surface shape that occurs will be dependent upon the detailed parameters of the solid and the incident ion beam.

A number of different mechanisms can be responsible for the growth of topographical features during ion bombardment of a surface. If impurities lie on the surface then these will generally have a different sputtering yield than that of the bulk material. Similarly, if the beam contains impurities, then the surface will sputter at different rates depending upon the nature of the incoming ion. When surface impurities or inclusions are present, and if their sputtering yield is lower than that of the bulk target, then the surface will be shielded by the impurity and features such as cones will appear to grow from the sputtered surface. If the yield is less then etch pits can develop.

Topography is also known to develop in almost contaminant-free conditions and forests of conical structures have been observed on the (11 3 1) face of Cu even with 99.999% pure Cu and with a mass analysed beam (Carter, 1983). Fig. 2.14 shows that these structures are, in fact, pyramids with distinct facets. For the development of topography in the relative absence of contaminants and non-uniformities in the beam, surface and subsurface defects and dislocations are initiated by the ion–atom collisions. This produces variations in the local sputtering yield, usually resulting first in the formation of etch pits which continually evolve and develop as the erosion continues. For relatively heavy ions and doses the surface topography that develops is dominated by the erosion process. In other cases, such as helium ion sputtering where the sputtering yield is lower and which is not erosion dominated, the He gas becomes implanted in the material and builds up below the surface in pockets and eventually

Fig. 2.14. Scanning electron micrograph of the (11 3 1) face of single crystal Cu after bombardment with 40 keV Ar$^+$ ions. (*a*) Showing pit and pyramid coverage of the surface. (*b*) Showing details of one pyramid (courtesy G. Carter).

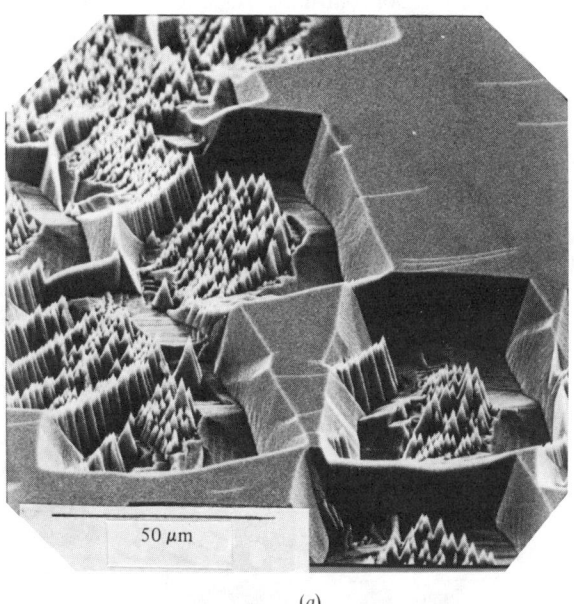

(*a*)

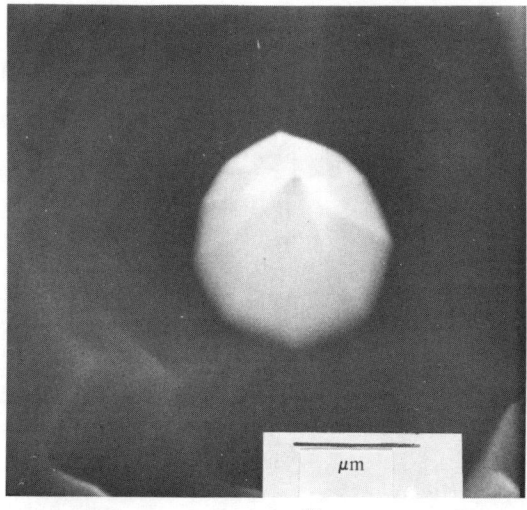

(*b*)

produces blistering of the surface. However, this does not generally happen under the types of conditions encountered using ion bombardment in surface analysis.

There have been a number of recent investigations which have examined both experimentally and theoretically the type of topography generated by ion bombardment. Experimentally, both impurity seeded and pure initially smooth surfaces have been bombarded by different ions under different conditions and the bombarded surfaces examined in the scanning electron microscope. These investigations have illustrated the large variety of different structures which can be present, see Figs. 2.14–2.16. The surface

Fig. 2.15. The development of surface topography on InP (100) following bombardment with 5.5 keV Cs$^+$ ions to a depth of (a) 2.1 μm, (b) 20 μm, (c) 47 μm, (d) 103 μm.

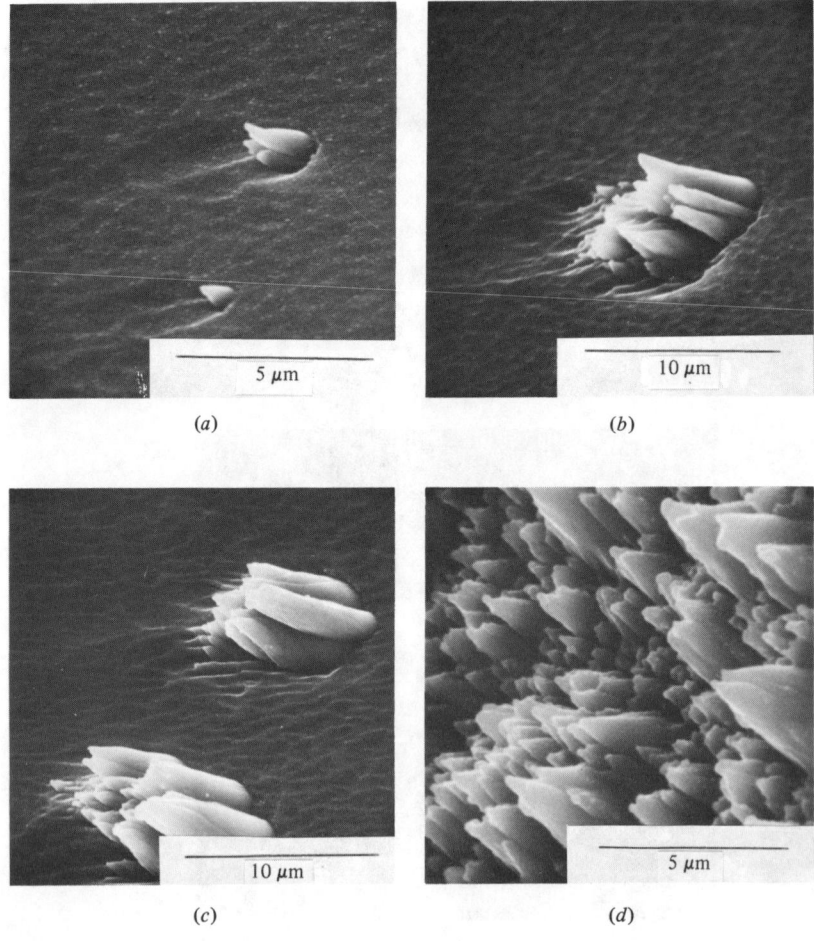

structures which occur under ion bombardment are visible both on metals and semiconductor materials. For crystals, the features exhibit characteristics related to the basic crystallography of the material with pyramids and etch pits having distinct facets in a direction where the sputtering yield has a stationary value. The axes of the conical or pyramid protrusions are observed to be aligned in a direction parallel to the incident ion beam. For compound semiconductor materials, such as InP, the surface can become enriched with one atomic component. Cone formation can occur if one of the atomic components has a lower sputtering yield to the other, see Fig. 2.15. For pure Si under both inert and reactive gas

Fig. 2.16. A sequence of micrographs showing the development of surface topography on Si (100) following 10.5 keV O_2^+ bombardment to a depth of (a), (b) 2.8 μm (c), (d) 13.0 μm.

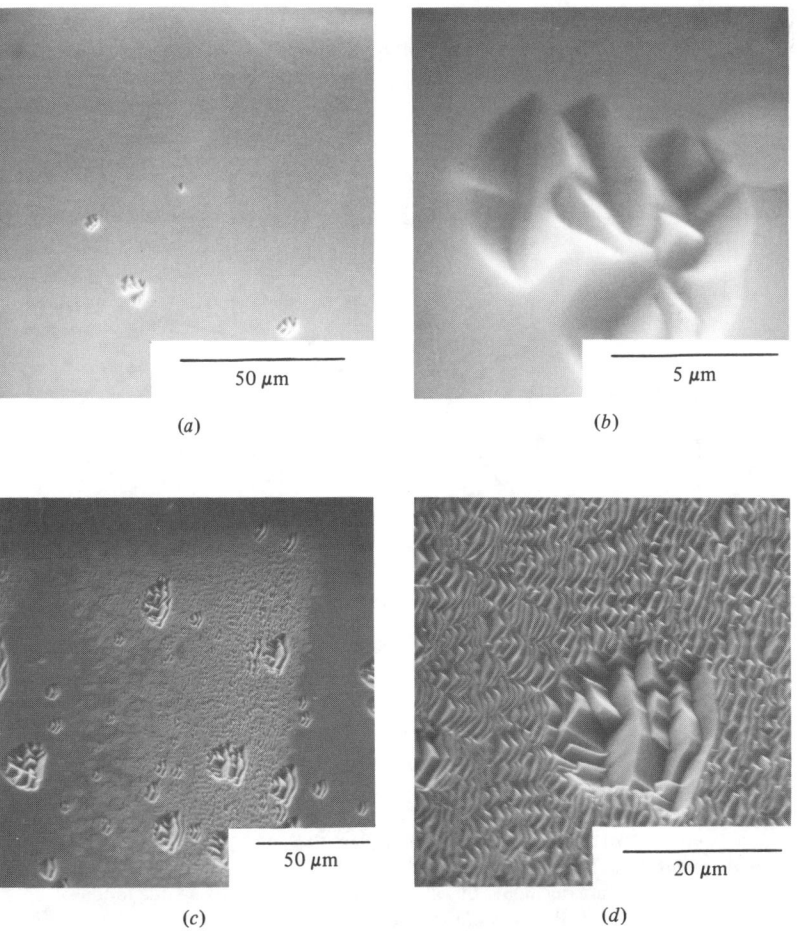

bombardment a facetted surface can develop. Under O_2^+ ion bombardment the entire surface becomes facetted at around a dose of 5×10^{19} ions cm^{-2}, as shown in Fig. 2.16. Under other bombardment conditions ripple structures can develop on Si. These ripple structures have also been observed on other materials, such as glass, and develop either parallel or transverse to the ion beam depending on the angle of incidence. However, experiments bombarding Si with Cs$^+$ ions indicate that an initially flat surface remained flat and featureless up to doses of 10^{20} ions cm^{-2} corresponding to an eroded depth of about 40 μm. This, however, was not the case when using Cs$^+$ ions on other semiconductor materials. Thus the ripples, facets, ridges, furrows, pits and pyramids which form do so under a variety of different conditions with different mechanisms being responsible for their formation. Many of the features which develop are not fully explained. Their importance in surface analysis is in the possible

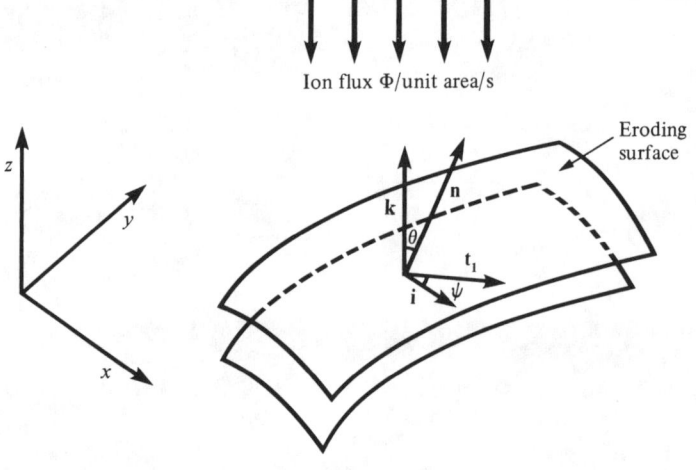

Fig. 2.17. A schematic diagram illustrating the evolution of surface topography during erosion by ion bombardment. The vectors **n**, **k**, are in the surface normal and ion beam directions. The vector **t**$_1$ lies in the tangent plane and the x–y plane and i is in the x direction.

Fig. 2.18. (i) Perspective views of the surface $z = \exp(-(x^2 + y^2/2))$: (a) before bombardment, (b), (c), (d) after successive equal doses. Note the development of the ridge at the top of the structure. (ii) The erosion of a semicircular section: (a) By a normally incident ion beam. (b) The same section but eroded by two beams symmetrically placed at an angle of 30°. (c) 60°. The sputtering yields are for Si under 1 keV Ar$^+$ ion bombardment. (d) Normally incident beam, including the effects of secondary sputtering due to ion reflection from the sides of the hummock.

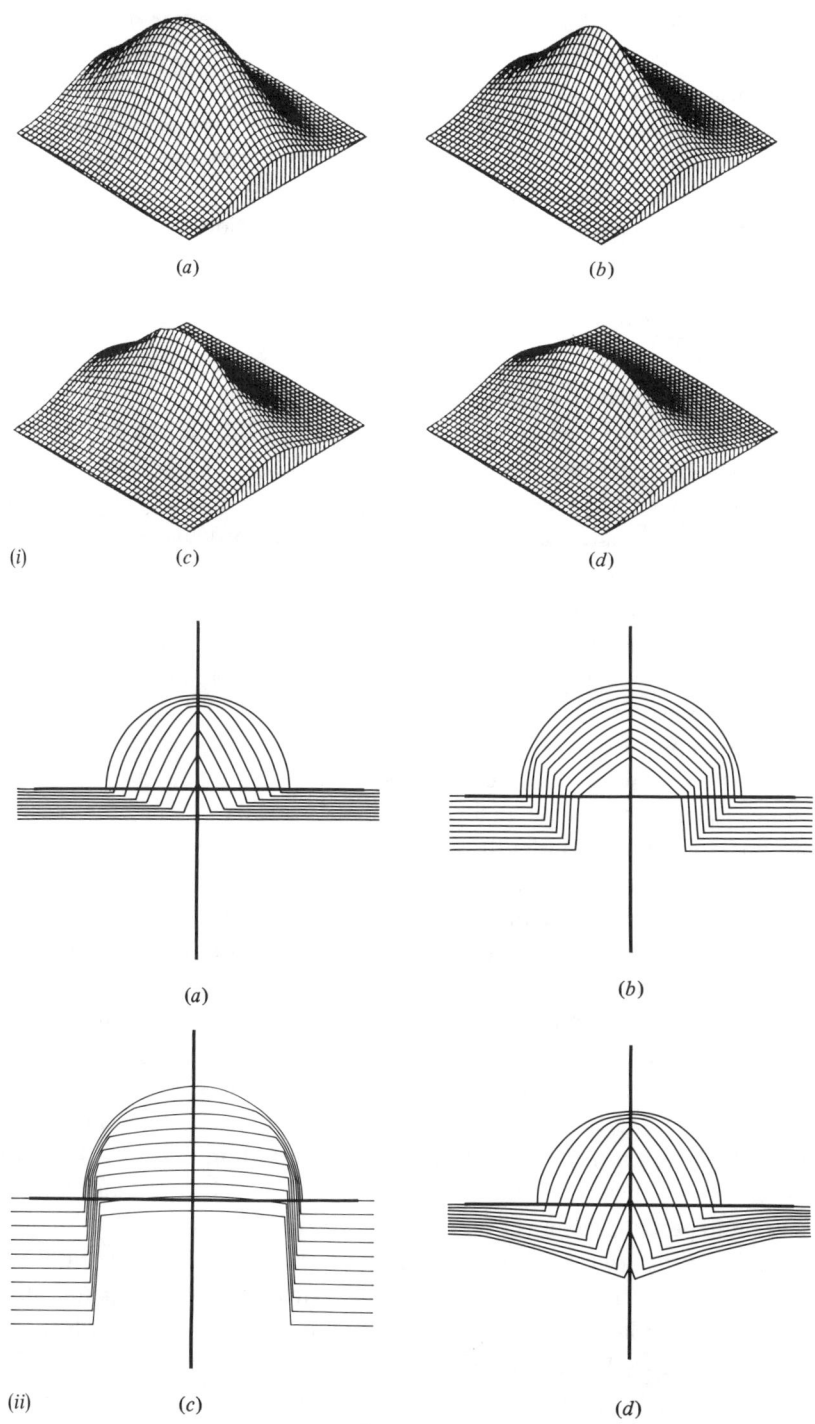

(i)

(ii)

effect that they can have on the depth resolution in sputter profiling if a very rough surface develops.

In order to understand some of these effects more fully, erosion theory has been used to model the development of topography and to examine ways in which unwanted topography can be minimised. The basis of erosion theory is the dependence of the sputtering yield on the angle of incidence, so that the rate of erosion in the direction of the ion beam $\partial \eta / \partial t$ is given by

$$\frac{\partial \eta}{\partial t} = \frac{\varphi}{N} Y(\theta, \psi).$$

Here, the incident flux is ϕ (ions m^{-2}s^{-1}) which may have spatial variation, the atomic density of the target is N, the angle of incidence is θ and ψ is an angle between the unit tangent parallel to the $x - y$ plane and the z direction, see Fig. 2.17. For amorphous materials the sputtering yield Y is a function of the angle θ alone, but for crystals the particular crystal plane under bombardment is also required to be specified.

Computations have been carried out using the theory in both two and three dimensions, solving the equation by the method of characteristics. The theory shows how edges and facets can develop from an initially smooth surface (Smith et al., 1981) but cannot in itself predict the formation of topography from initially flat surfaces. Secondary effects, such as redeposition and ion reflection, have also been considered in the context of this theory. Some simulated surface shapes calculated using the theory are shown in Fig. 2.18 (Smith et al., 1986).

In order to minimise topographical development due to ion-induced effects, a number of mechanisms can be used. Sample rocking and rotation minimise the formation of edges and facets but smooth hummocks can still be present on the surface, particularly if these are impurity generated. Alternatively, the samples can be bombarded by two static ion beams, although directional effects in this case are still not fully eliminated. Both these situations have been modelled theoretically and have confirmed the use of these techniques as possible methods to minimise unwanted topographical growth. Some illustrations are shown in Fig. 2.19. The use of reactive ion beams or the sputtering of a material by inert ions but in a vacuum partial pressure of a reactive gas has also been a useful technique for minimising unwanted topography. For example, if a polycrystalline material is sputtered at an enhanced oxygen partial pressure a dynamically renewed oxide layer can be formed on the polycrystal surface which, if amorphous, avoids the direction structural effects which can appear with crystals. The usually observed relationship of sputtering yield as a function

Ion erosion in surface analysis

Fig. 2.19. (i) Three-dimensional simulations of an axially symmetric exponential hummock $z = \exp(-(x^2 + y^2))$ after successive equal doses. (ii) Illustrating the effect of beam rotation (expanded z scale) for a non-normally incident beam at an angle of $\pi/9$ to the z axis. The rotation frequency is $1/(2\pi)$ secs.

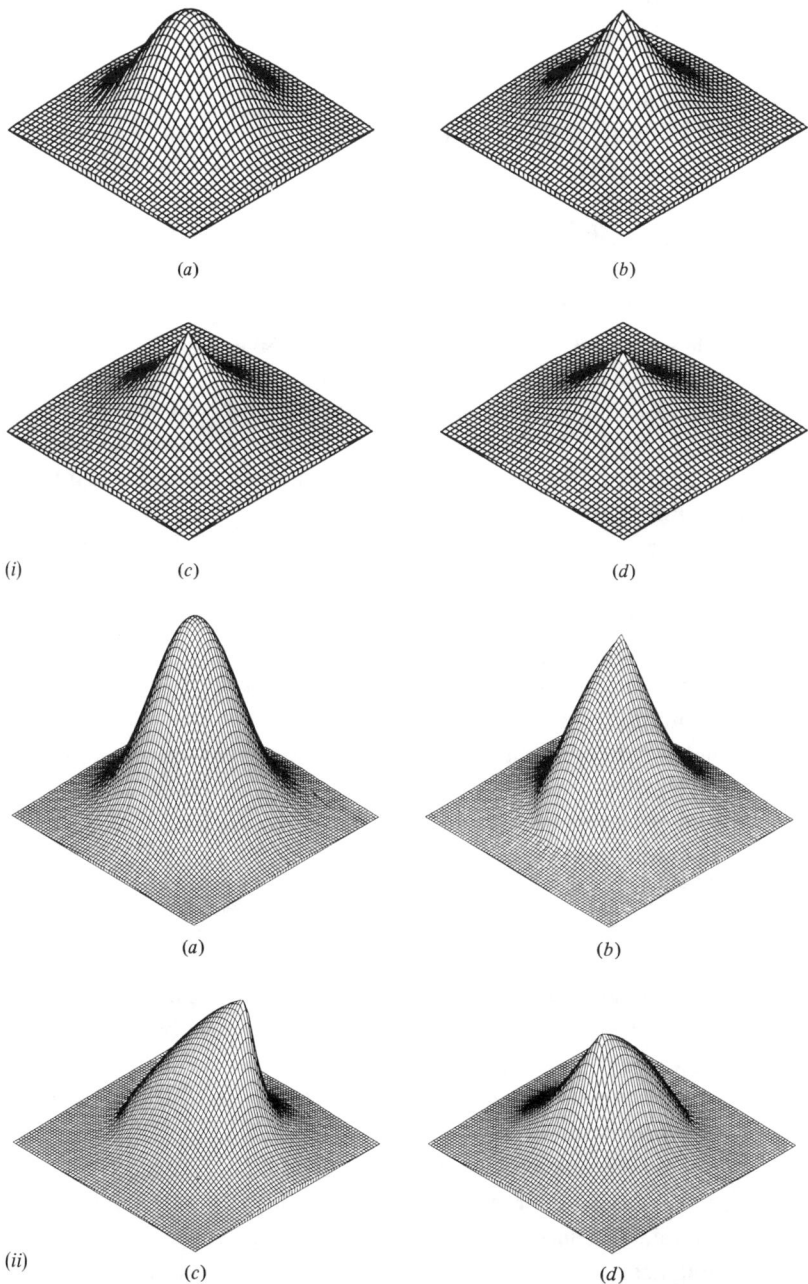

of ion incidence angle, showing a maximum at non-normal incidence for inert gas ions, is not always observed for reactive species and in certain circumstances a maximum at normal incidence is obtained (Maddox & Splinter, 1980), thus confirming the use of such species for reducing unwanted surface topography.

4.2   Distortion of depth profiles by sputtering

The technique of sputter-depth profiling is essentially a destructive one; also the material is never removed layer by layer as would be required for an ideal analysis.

A number of factors can be responsible for the degradation of depth profiles, including ion-induced surface topography, collisional mixing, preferential sputtering, ion implantation and diffusion. The previous section has described how cones and other surface features can appear after ion bombardment of smooth surfaces which contributes to an uncertainty as to the precise location of the reference surface, and hence also to the depth profiles. The other secondary effects can also contribute in certain circumstances. This is not to say that sputter profiling is a poor technique. For many analyses it is the only way and there are many examples where experiments have proved that the measured profiles are extremely accurate and beam-induced effects are negligible. For example, independent experiments have shown that the profiles of B in Si can be measured precisely by means of SIMS if low ion bombardment energies $\leqslant 5\,\text{keV}$ are used. However, in experiments on other materials, even when unwanted surface topography is not generated, a shift and a spreading of the original material distribution has been observed, particularly at higher beam energies. Thus, care is needed in interpreting the results from sputter profiling which can be both material and beam dependent. In this section some of the more important mechanisms that affect the depth profiles are considered.

4.2.1   Collisional mixing

Collisional mixing occurs whenever an incoming ion sets in motion the target atoms. The relocation of a target atom by an incoming ion is known as recoil implantation. In the linear cascade and spike regimes, these recoils can set in motion target atoms which are then relocated and this is generally referred to as cascade mixing. If sputtering were an energy efficient process, i.e., if an incoming ion used up most of its energy in sputtering target atoms then the effect of collisional mixing would be minimal. Unfortunately, for the ion energy range used in many surface analytical techniques, the average incoming ion travels a distance which is an order of magnitude greater than the depth of origin of the sputtered particles, and in losing energy over this distance it sets the recoils in motion. The average fraction of energy of the incoming ion that is transferred to the sputtered particle (neglecting the energy of reflected ions) is a measure of the efficiency of the sputtering process and this has been shown to be dependent on the angle of incidence, with large angles to the normal producing the most efficient process in terms of energy utilisation. However, large angles may not always be practical since, for the most efficient throughput of samples, sputtering around the angle corresponding to the maximum sputtering yield is the best.

It is necessary for good depth resolution that the depth of origin of the sputtered particles is small and this is usually the case for heavy ion bombardment at energies between 0.5 and 25 keV – the range of energies usually used in surface analysis. The mean ion range and depth of damage increases with ion energy. Clearly, lower ion energies are preferable.

The collisional relocation of atoms due to ion-beam erosion has been analysed theoretically in basically three different ways: by a diffusion theory, by transport theory and by Monte Carlo methods. The continuum approach of the diffusion and transport theories have concentrated on modelling the distortion of thin and thick tracer profiles located at some distance below the surface of a target material. The Monte Carlo methods have generally calculated the trajectories of the primary knock-on atoms (recoil implantation). The simplest diffusion model (Andersen, 1979) treats the problem of the spreading of a δ function tracer impurity profile at

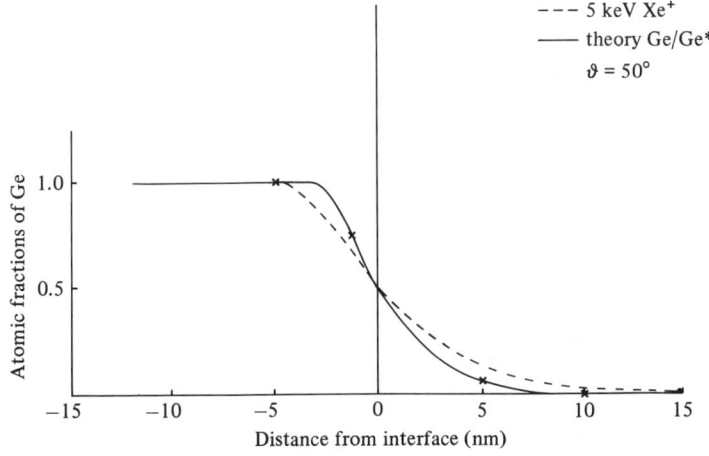

Fig. 2.20. Corrected AES sputter profiles of Ge in a thin sandwich layer of Ge in Si, vapour deposited on a carbon substrate. The theoretical curves illustrate the interface broadening expected as a result of collisional mixing.

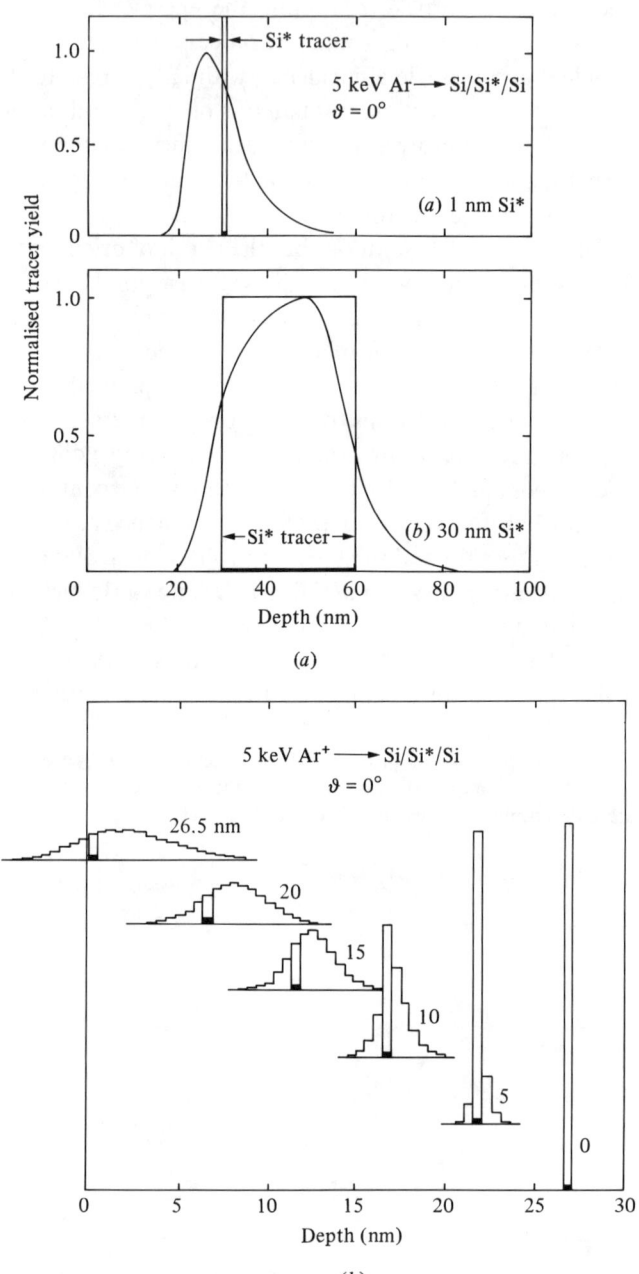

Fig. 2.21. Theoretically calculated beam broadening effects. (a) Calculated sputter profiles using transport theory of Si* markers of Si in Si under 5 keV Ar$^+$ bombardment at normal incidence. (b) Evolution of the broadening of a thin marker of Si* in Si, calculated using Monte Carlo methods.

a given depth. The theory predicts symmetric spreading of the tracer both towards and away from the incoming beam. Nevertheless, good agreement with experimentally observed broadening has been obtained for the case of samples of thin Ge and Si layers deposited on glassy carbon substrates under ion bombardment with 5 keV Ar^+ and Xe^+ ions, see Fig. 2.20. Transport theory is more accurate than the simple diffusional approach (Littmark & Hofer, 1980) and is successful at predicting the shift and the spreading of both thick and thin tracer impurity profiles (see Fig 2.21). The Monte Carlo simulations (Rousch et al., 1981) are particularly useful when groups have access to standard computer packages since different model systems can be analysed merely by changing the input data to the programs. However, the most promising theoretical approach to data is the work of Collins et al. (1985) which combines both diffusion and transport theory and also includes a conservation of mass equation which overcomes the problem of an unphysical build up of material in the target. Initial results look promising and show that an implanted layer can spread either towards or away from the surface depending on the bombardment and material conditions.

4.2.2 Compositional changes

Although a large amount of data exists for measured sputtering yields of elemental targets, the same is not true for alloys or other multicomponent materials. With alloys one component may be preferentially sputtered and the sputtering yield of a multicomponent material may change with time, with a steady state only being achieved after sufficiently high ion doses. Even trace amounts of one element in another can alter the sputtering yields and in these cases can often cause topographical effects where the material present in small amounts has a low sputtering yield. Both increases and decreases in the sputtering yield of alloys compared to the elemental yields have been reported. The steady state surface concentrations depend on the energy and mass of the incident ion in addition to the alloy composition. The experimental results show that for many alloys and compounds a surface enrichment of one of the elemental species is often observed. The observed enrichment is usually in qualitative agreement with the sputtering yields of the pure elements, although for those materials with a large mass ratio of the constituents, surface enrichment of the heavier mass constituent is observed (Liau et al., 1977). Both recoil implantation and preferential sputtering act as processes for the surface enrichment of a species. If preferential sputtering were not important, then recoil implantation would initially predict an enrichment of heavier atoms on the surface and of the lighter component within the material. At larger doses when the surface layer is sputtered away, the region enriched with the light component reaches the surface. In a practical situation where preferential sputtering also needs to be considered, a steady state situation will eventually occur at sufficiently high doses which will

Fig. 2.22. Experimentally observed beam broadening effects. (i) SIMS sputter depth profiles of B implantation distributions in Si. The broadening is illustrated by the separation of the curves at larger depths for increasing beam energy. (a) Little change in the measured profiles between 2 keV and 15 keV O_2^+ bombardment indicates only small mixing effects. (b) Under Ar^+ ion bombardment broadening occurs only at high energies. (ii) Apparent implantation distributions of Bi in Si. Note the bigger shift and broadening compared to the B implants of 82 μm.

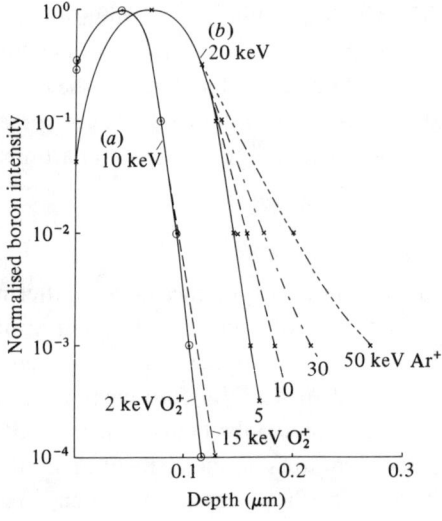

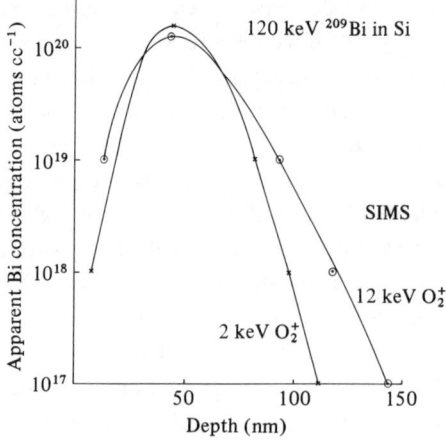

involve both processes. To distinguish between the two effects, it would appear that the effect of preferential sputtering could be measured at low doses before mixing and erosion has substantially changed the material, whereas recoil implantation will only begin to affect the surface after sufficiently high doses for the atomic mixing to be important. Thus the effect of ion bombardment on multicomponent materials is to produce a surface layer which can be different in composition to the bulk material, the thickness and composition of this layer depending on many different parameters associated with the material and the beam.

However, many surface analytical procedures are concerned with the analysis of dilute impurities in a bulk material. SIMS is commonly used

Fig. 2.23. XPS spectra of calcium carbonate. (a) before argon ion bombardment and, (b) after argon ion bombardment ($\sim 7.5 \times 10^{16}$ ions cm^{-2}).

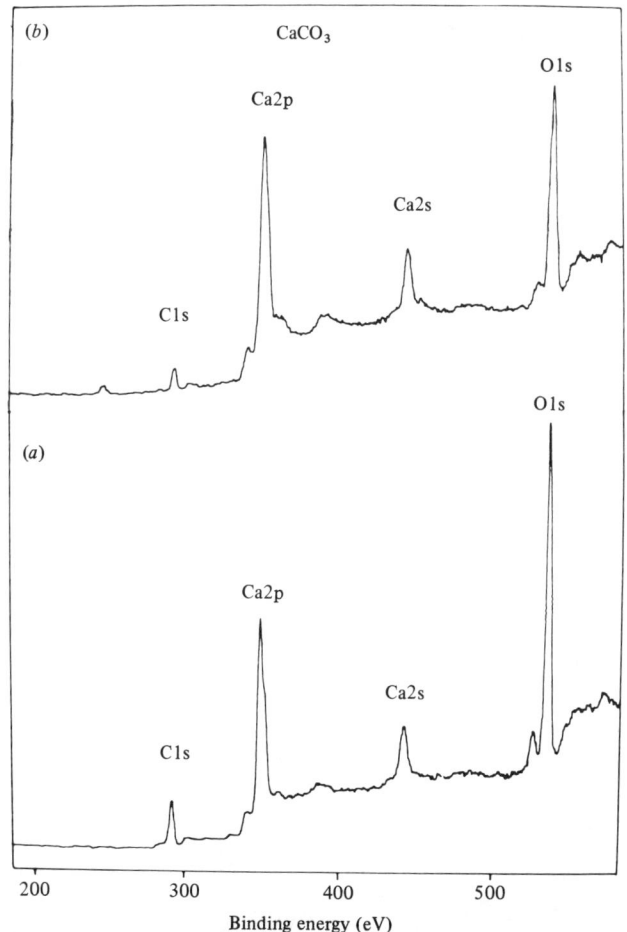

because of its great sensitivity to determine dopant profiles in semiconductors, for example, B and P in Si. Experimental results show that the beam-induced broadening of B concentrations in Si is almost negligible under O_2^+ ion bombardment at energies of 2 keV (Wittmaak, 1984). A comparison between the measured and exact profiles is shown in Fig. 2.22. Other dilute implantation profiles measured using SIMS do not exhibit profile broadening. Under the same bombardment conditions Bi implants in Si become much more rapidly broadened than the B profiles. This has been interpreted as due to radiation-enhanced diffusion. Indeed in many

Fig. 2.24. High energy resolution spectra of the oxygen 1s photoelectron peaks from calcium carbonate: (a) before argon ion bombardment and, (b) after argon ion bombardment.

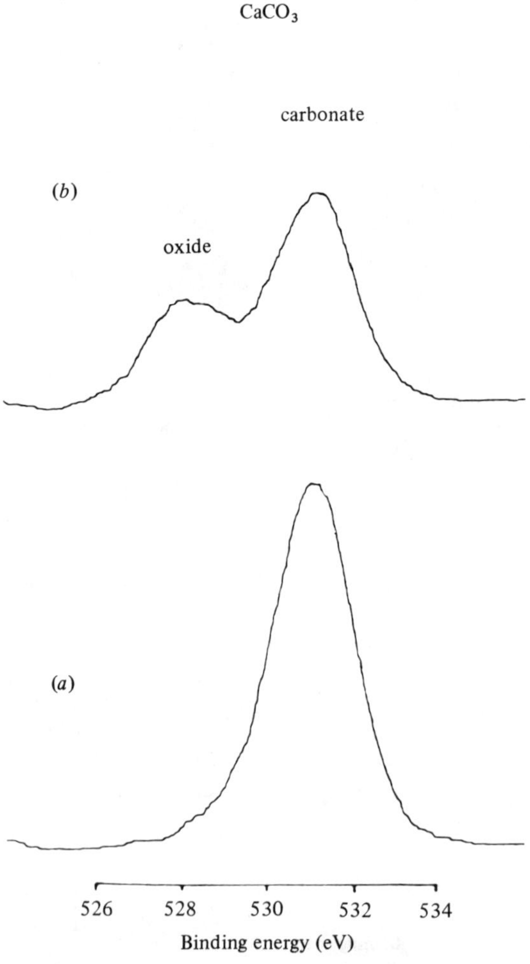

cases diffusion cannot be ignored. Both radiation-enhanced and thermal diffusion can occur if the sample is heated or bombardment takes place in the spike regime.

4.2.3 Ion-induced chemical decomposition

Another problem concerns the changes in composition brought about by ion-induced chemical decomposition. It is known that a range of metallic and non-metallic oxides are reduced under low-energy ion bombardment. Metal nitrates, carbonates and sulphates are often reduced to the oxide. Many polymers are reduced to elemental carbon. Although the extent of decomposition appears to be related to the free energy of formation, the precise mechanisms involved are not yet fully understood.

An example of ion-induced decomposition is shown in Fig. 2.23. This shows wide scan XPS spectra taken from a calcium carbonate surface before and after argon ion bombardment to a total dose of 7.5×10^{16} ions cm^{-2}. The spectra reveal a general reduction in the relative intensities of the carbon 1s and the oxygen 1s peaks. Examination of the high-resolution spectra of the 1-s photoelectron peaks in Fig. 2.24 reveals the presence of the oxide in the ion-etched surface (Christie et al., 1983).

Clearly this is a process which can cause serious uncertainties in the interpretation of chemical shift data obtained by XPS following ion cleaning or sputter-depth profiling. It is recommended practice to perform experiments on chemical standard materials to ensure that apparent changes in chemical structure are real and not ion-induced artefacts.

5 Conclusion

This chapter has discussed the principal ion sources used in surface analysis and the most important characteristics of the beam – target interactions when surfaces are subject to low-energy ion bombardment. Aspects of electron and ion electrostatic optics are treated separately in Chapter 3.

The chapter has shown the need for being extremely careful in interpreting results in surface analysis wherever ion beams are employed. Atomic mixing, surface topography formation, chemical effects are only just beginning to be fully understood. An experimental compendium of these effects can go some way to enable the investigator using surface analysis equipment to interpret his results but theory is also vitally important. The atomic mixing effects can only be fully explained and quantified using a theoretical approach.

It is expected that the next few years will see rapid progress in all areas described here. This will enable accurate quantification and reproducibility of results, independent of the particular surface analytical instrument and achievable over a wide range of conditions.

References

Almen, O. & Bruce, G. (1961). *Nucl. Inst. and Meth.*, **11**, 257.
Andersen, H.H. (1979). *Appl. Phys.*, **18**, 131.
Andersen, H.H. & Day, H.L. (1983). Chapter 4 in *Sputtering by Particle Bombardment 1*, Ed. R. Behrisch, Springer Verlag.
von Ardenne, M. (1961). *Expt. Tech. Physik*, **5**, 227.
Bayly, A.R., Waugh, A.R. & Vohralik, P. (1985). *Spectrochimica Acta*, **40**, 717.
Biersak, J.P. & Haggmark, L.G. (1980). *Nucl. Inst. and Mech.* 257.
Carter, G. (1983). Chapter in *Sputtering by Particle Bombardment 1*, Ed. R. Behrisch, Springer Verlag.
Christie, A.B., Lee, J., Sutherland, I. & Walls, J.M. (1983). *Appl. Surface Sci.*, **15**, 224.
Collins, R., Jimenez-Rodriquez, J.J., Wadsworth, M. & Badheka, R. (1985), Paper presented at 'Radiation Effects in Insulators. III', University of Surrey, Guildford, UK, 1985.
Duncan, S., Smith, R., Sykes, D.E. and Walls, J.M. (1984). *Vacuum*, **34**, 145.
Goldstein, H. (1980). *Classical Mechanics*, Addison Wesley.
Gurmin, B.M., Ryzhov, Y.A. & Skharaban, I. (1969). *Bull. Acad. Aci. USSR. Phys. Ser. (USA)*, **33**, 383.
Harrison, D.E. (1983). *Rad. Eff.*, **70**, 1.
Kaufman, H.R., Cuomo, J.J. & Harper, J.M.E. (1982). *J. Vac. Sci. Technol.*, **21**, 725.
Kingham, D.R. & Swanson, L.W. (1984). *Appl. Phys.*, **A34**, 123.
Lehmann, C. & Sigmund, P. (1966). *Phys. Stat. Sol.*, **16**, 507.
Liau, Z.L., Brown, W.L., Homer, R & Poate, J.M. (1977). *Appl. Phys. Lett.*, **30**, 626.
Lindhard, J. (1965). *K. Dan. Vidensk. Solksk. Mat. Fys. Medd.*, **34**, 14.
Littmark, U. & Hofer, W.O. (1980). *Nucl. Inst. Meth.*, **168**, 329.
Maddox, R.L. & Splinter, M.R. (1980). Chapter 4, p. 388 in *Fine Line Lithography*, Ed. R. Newman, North Holland.
Onderdelinden, D. (1968). *Can. J. Phys.*, **46**, 739.
Roth, J. (1983). Chapter 3 in *Sputtering by Particle Bombardment 2*, Ed. R. Behrisch, Springer Verlag.
Rousch, M.L., Andreadis, T.D. & Goktepe, O.F. (1981). *Rad. Eff.*, **55**, 119.
Sigmund, P. (1969). *Phys. Rev.*, **184**, 383.
Sigmund, P. (1983). Chapter 2 in *Sputtering by Particle Bombardment 1*, Ed. R. Behrisch, Springer Verlag.
Smith, R., Carter, G. & Nobes, M.J. (1986). *Proc. Roy. Soc. A.*, **407**, 405.
Smith, R., Valkering, T.P. & Walls, J.M. (1981). *Phil. Mag.*, **A4**, 879.
Thompson, M. W. (1969). *Defects and Radiation Damage in Metals*. Cambridge University Press.
Waugh, A.R., Bayly, A.R. & Anderson, K. (1984). *Vacuum*, **34**, 103.
Wittmaak, K. (1984). *Vacuum*, **34**, 119.

3

Electron and ion energy analysis

M.P. SEAH

1 Introduction

In Auger electron spectroscopy (AES) and X-ray photoelectron spectroscopy (XPS) the measurement of the energy spectrum of the emitted electrons is the whole basis of the technique. The way the spectrum is measured, however, varies considerably from application to application. In XPS and certain applications of AES, to determine chemical state data from precise lineshapes and energies, we require a good energy resolution, typically better than 0.5 eV over the kinetic energy range 0–2000 eV. The functioning of the spectrometer in this case is not the same as that used for mapping surfaces in AES where, to get high spatial resolution, the electron beam currents are low and the spectrometer energy resolution must be degraded in order to get a sufficient signal-to-noise ratio. For these and other reasons the electron spectrometers available today are quite complex and well instrumented. However, they are now sufficiently complex that the uninitiated user may fail to use them in their optimum mode. In this chapter we shall describe the basic principles of the important commercial spectrometers and attempt to show how they should best be used in various practical situations.

The electron spectrometers used today are all electrostatic and hence their focussing properties do not depend on the mass of the charged particle being energy analysed. This has allowed the electron spectrometers to be used directly for negative-ion energy analysis and, more importantly, with potentials of the opposite sign for positive-ion analysis in ion scattering spectroscopy (ISS). Here the requirements are simply for high sensitivity at moderate resolution. The higher energy form of ISS, Rutherford backscattering spectroscopy (RBS), uses a much simpler and very efficient form of energy analysis involving solid state detectors. These have many advantages over the electron optical devices, as will be discussed later but, sadly, are limited to ion energies above 100 keV. In the remaining technique, secondary ion mass spectroscopy (SIMS), the particles, which are ions, are mass analysed rather than energy analysed and energy analysis is only a smaller, but important, part of the technique. Most static SIMS instruments are based on quadrupole mass filters and with these it is difficult to separate particle clusters of nominally similar mass e.g. P and SiH at mass 31. The separation may be achieved by making use of the very different energy spectra of the monatomic and polyatomic clusters. For this purpose very simple, but often ill-optimised, energy filters may be added to the front of SIMS mass analysers.

Most of the recent developments in energy analysers have arisen through the developments in AES and XPS and it is here that we shall start the story in a somewhat historical order. We shall cover, in detail, the two most common spectrometers in use in these techniques and give references where the reader can find further details. Other spectrometers that have been successfully designed are outlined in reviews (Heddle, 1971; Steckelmacher, 1973; Wannberg, Gelius & Siegbahn, 1974; Roy & Carette, 1977; Seah, 1980) and the interested reader may wish to read of spectrometers such as the 127° cylindrical sector (Hughes & Rojansky, 1929; Marmet & Kerwin, 1960; Bennani, Duguet & Wellenstein, 1980) and parallel plate (Green & Proca, 1970; Proca & Green, 1970; Proca 1973) designs, which have found application outside the present fields, together with the variants of the grid-based retarding field analysers proposed for XPS use (Huchital & Rigden, 1972; Lee 1973; Staib & Dinklage, 1977; Dabbs, 1983). We shall not deal with these spectrometers here, although they are of considerable interest to those developing the techniques, but, as mentioned before, will concentrate on the instruments in common use to enable the analyst to understand how best to use them.

2 Auger electron spectroscopy

In Auger electron spectroscopy a focused electron beam, generally in the energy range 1–10 keV, is used to irradiate the sample, as discussed in Chapter 4. Electrons in the surface layers become excited and decay, emitting characteristic Auger electrons whose intensity and energy are measured with the spectrometer. The Auger electrons, with energies in the range 0–3000 eV, have typical intensities of 10^{-8} to 10^{-5} of the primary beam current (Bishop & Riviere 1969). Historically, the first popular spectrometer for measuring these electron energy spectra was the retarding field analyser (RFA) (Taylor, 1969). These analysers evolved from the construction used for low energy electron diffraction (LEED) instruments and involve only minor modifications and some modest electronics. The

Electron and ion energy analysis

RFA is important historically as it was used to establish the basic AES technique; however, today it is only used as an adjunct to LEED experiments on single crystals. The principle of its operation involves the electronic removal of large unwanted signals whose associated noise, unfortunately, cannot be removed. The analyser thus suffers from a poor signal-to-noise ratio and hence poor sensitivity. The sensitivity problem was overcome by the introduction of the cylindrical mirror analyser (CMA) by Palmberg, Bohn & Tracy (1969) as discussed below.

2.1 The cylindrical mirror analyser (CMA)
2.1.1 Principle of operation

The CMA is now marketed by a number of manufacturers. The basic form is shown in Fig. 3.1. The focal point of the exciting electron beam on the sample surface is arranged to be at the analyser focus, S. Electrons of energy E then move out radially in a field-free space to pass through a mesh-covered slit in the wall of the inner earthed cylinder. The

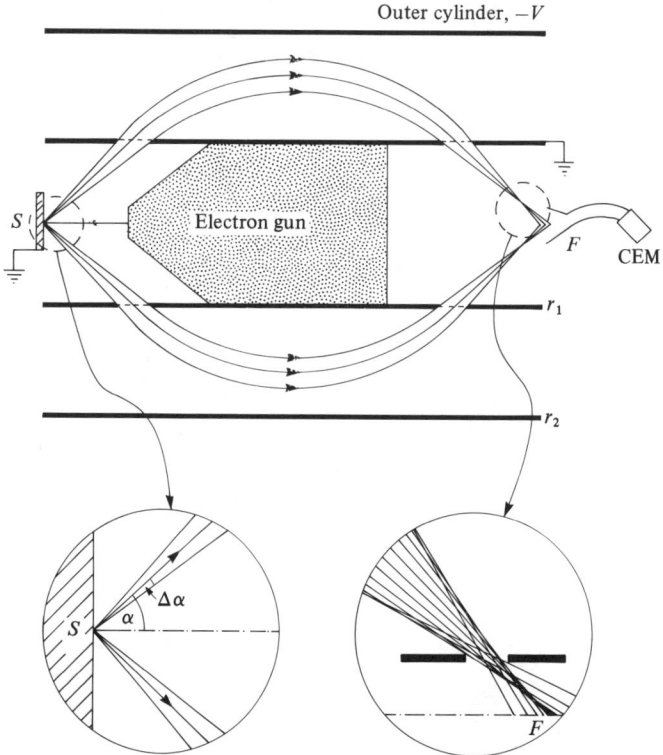

Fig. 3.1. Cross-sectional view of a typical CMA arrangement with enlarged views of the electron trajectories at the source and focal points.

electrons then experience a repulsive field from the outer cylinder at potential $-V$, and are deflected to pass through a second mesh slit in the earthed inner cylinder, eventually coming to a focus on the axis at F where they pass into a channel electron multiplier (CEM) for detection purposes. The channel multiplier may either be used at high gain to give one output pulse per electron for counting at rates below 10^6 per second, or at lower gain to give current amplification for higher intensity signals. The CMA has rotational symmetry about its axis and so has a large entrance aperture and high sensitivity.

The electron optical behaviour of this analyser, in the non-relativistic approximation, is considered in detail by Sar-El (1967). The field between the cylinders causes the potential at a radius r to behave as $\ln(r/r_1)$, hence the focal condition is expressed by the relation

$$E = \frac{K_0 e}{\ln(r_2/r_1)} V, \tag{1}$$

where e is the electron charge and r_1 and r_2 are the radii of the inner and

Fig. 3.2. The variation of the axis crossing distance, L, with the angle of emission, α, of the electron for three values of K_0. A low value of K_0 is caused by a high voltage V on the outer cylinder, and vice versa, for any given energy of the electrons. Thus by setting the window in α as shown and detecting at $L = 6.13 r_1$ as the voltage is increased, no electrons are detected until K_0 has fallen to 1.31 where a sharp intense peak will occur.

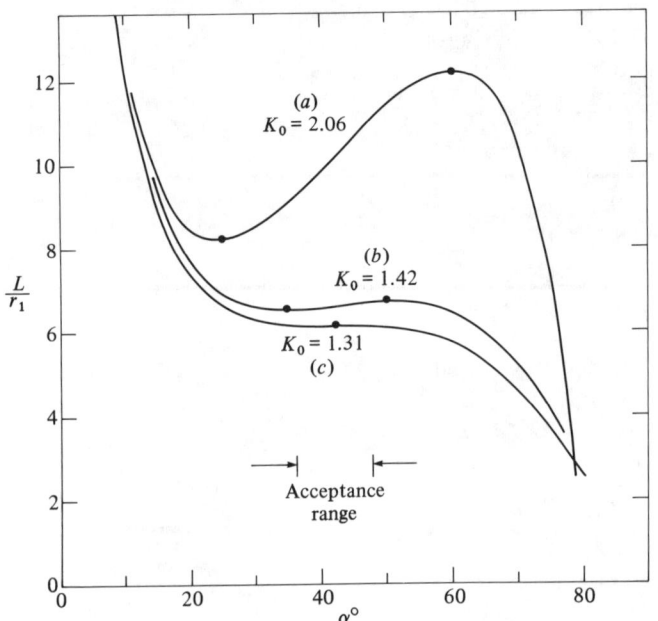

outer cylinders respectively. To illustrate the behaviour, if we fix V at a low value such as $-E\ln(r_2/r_1)/2.06e$, it is found that first order focal conditions occur for electrons centred at α values of 25° and 60°. By this we mean that at low α values the distance between S and F in Fig. 3.1 is very long. As α increases SF decreases, approaching a minimum at 25°. As α increases further SF now increases but reaches a maximum at 60°. Again, at higher values of α, SF decreases as shown by curve (a) in Fig. 3.2. As the potential V increases in magnitude the two first order foci at the above minimum and maximum move together until, at a K_0 value of 1.3099, the two first order foci merge at an α value of 42° 18.5' and form a second order focus, as shown by curve (c). At this focus electrons from a very wide range of α's meet at the focal point F where the distance SF is given by L in the relation (Hafner, Simpson & Kuyatt, 1968)

$$L = 6.130\, r_1. \tag{2}$$

This means that, for a given resolution, we can accept a very high solid angle of emission from the sample and therefore have high sensitivity. For small values of the semiangle, $\Delta\alpha$, either side of α, and for a small energy increases, ΔE, it can be shown that the shift in the axis crossing point, ΔL, is given by

$$\Delta L = 5.6\, r_1(\Delta E/E) - 15.4\, r_1(\Delta\alpha)^3 + 10.3\, r_1(\Delta E/E)\Delta\alpha. \tag{3}$$

For a small $\Delta\alpha$ and very small ΔL we find approximately

$$\Delta E/E = 2.75\,(\Delta\alpha)^3. \tag{4}$$

Thus the base resolution, ΔE_b, for $\pm\Delta\alpha$ is given by

$$\Delta E_b/E = 5.5\,(\Delta\alpha)^3. \tag{5}$$

2.1.2 Transmission of the CMA

The fraction T of the half-space in front of the sample intercepted by the analyser is simply $2\sin\alpha\Delta\alpha$, thus:

$$T = 1.346\,\Delta\alpha \tag{6}$$

and

$$\Delta E_b/E = 2.255\, T^3. \tag{7}$$

As an example, a CMA with $\Delta\alpha = 5°$ has a transmission of 12% and a resolution, $\Delta E_b/E$, of 0.37%. If $\Delta\alpha$ is halved the transmission also is halved but $\Delta E_b/E$ reduces to well below 0.1%.

The above analysis is valid for the source at S and the (Gaussian) image at F. It is clear in equation (2) that the image is one with classical spherical aberration and that a smaller image is possible, analogous with the circle of least confusion, just short of the image plane. Use of a ring slit at a radius $5.28 r_1 (\Delta\alpha)^2$ is associated with a reduction of the coefficient of $(\Delta\alpha)^3$ in equations (3) and (5) by a factor of 4. This is shown schematically in the right-hand insert of Fig. 3.1. In this case, with a ring slit of width W, the base resolution is given approximately by Sar-El (1970):

$$\frac{\Delta E_b}{E} = \frac{0.18 W}{r_1} + \frac{5.5}{4}(\Delta\alpha)^3. \tag{8}$$

Many detailed analyses of CMAs have been published but the commercial CMA cannot be made as efficient as the ideal calculations suggest. The transmission is always less than described above due to the meshes covering the slits which are obliquely traversed at 48° from the surface normal. Additionally, the meshes introduce aberrations due to the difficulty of keeping them rigidly cylindrical. For this reason commercial CMAs are rarely made with resolutions better than 0.25%.

Equations (7) and (8) show that for a given slit, T is independent of energy and ΔE_b is proportional to E. The area of peaks in the energy spectrum is proportional to the product $T \Delta E_b$ which is, in turn, proportional to E. This product gives the analyser intensity–energy response function and, here, changes a true energy spectrum, $n(E)$, into a measured spectrum, $En(E)$.

2.1.3 Positional sensitivity of the CMA

The main disadvantages of the CMA are the fall in signal as the sample moves away from the focal point and the concomitant shift in the apparent energy of the peaks. As the sample is moved from its correct position, the apparent peak shift for an axial movement ΔL may be determined from equation (3) to be:

$$\Delta E / E = \Delta L / 5.6 r_1, \tag{9}$$

which, for $r_1 = 14.605$ mm, gives a shift of 12 eV per 1000 eV per mm error in position, as observed experimentally by Sickafus & Holloway (1975). In more recent CMAs of larger size this shift may be halved. A further disadvantage which is to some extent overcome in commercial systems by careful design of the ancillary components, is the very restricted working space around the sample.

2.1.4 Operation of the CMA

In summary, commercial spectrometers are generally made with $r_2/r_1 = 2.15$ to 2.30 so that

$$E = k\,\text{eV}, \tag{10}$$

where k, the spectrometer constant, is in the range 1.6–1.7. The energy spectrum thus is defined by the potential on one electrode. The analysers are simple, robust and very efficient and are a main workhorse of AES studies. A major development of convenience has been the insertion of high spatial resolution electron guns down the centre of the analyser, as discussed in greater detail in the next chapter.

A typical spectrum from a CMA is shown in Fig. 3.3. As shown in equation (5) the resolution or energy acceptance of the analyser is proportional to E. Thus the analyser measures not the energy spectrum, $n(E)$, but the product $En(E)$, as noted in Section 2.1.2. In applications where the primary electron beam current is below about 10 nA the detection is by pulse counting single electrons and $En(E)$ is obtained directly. Above this current the counting system may become non-linear and the electron multiplier mode is changed from pulse counting to current amplification. In this mode $En(E)$ may still be obtained using a voltage to frequency converter to extract the signal from the high potential of the multiplier output but, more traditionally, the spectrum derivative method has been used to give the spectrum shown in Fig. 3.4 which represents $d(En(E))/dE$, i.e. the derivative of the earlier spectrum with respect to E. This differentiation removes much of the background and gives enhanced visibility to all of the peaks. It is in this mode that *Handbooks of Spectra*

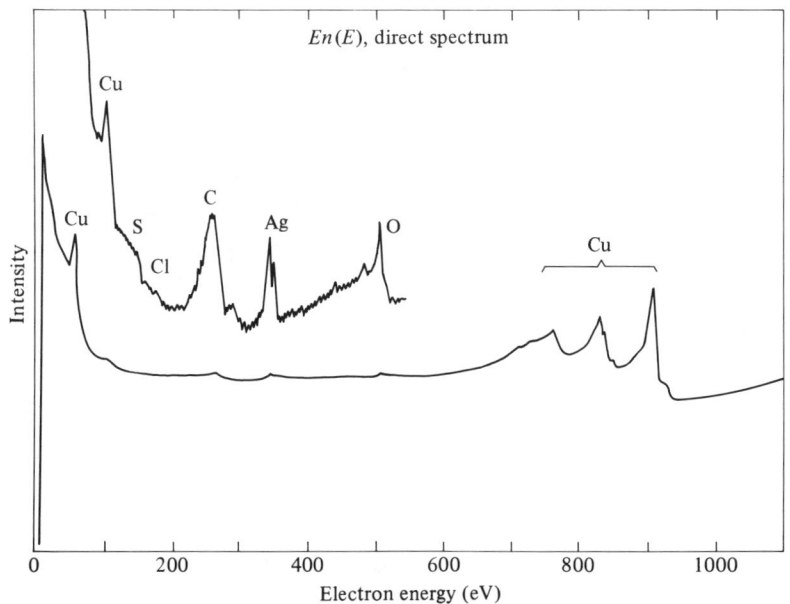

Fig. 3.3. A directly measured electron energy spectrum from a contaminated copper sample excited by a 5 keV electron beam and measured with a CMA.

have been presented (Davis et al., 1976; McGuire, 1979) although, more recently, the direct spectra have also become available (Sekine et al., 1982).

The derivative spectrum is obtained historically by applying a small sinusoidal potential modulation v to the mirror voltage V. This produces a small AC signal at the multiplier output in addition to the DC signal proportional to $En(E)$. The AC signal for small values of v is simply proportional to the derivative $d(En(E))/dE$ shown in Fig. 3.4, and is measured using phase sensitive detection methods to maintain a high signal-to-noise ratio. A schematic of the circuit is shown in Fig. 3.5. As v increases the derivative signal increases but the noise level is left unaltered; v is usually expressed as the term $v_e = E(v/V)$, in electron volts, and typical values are 2 eV peak-to-peak and 5 eV peak-to-peak. For quantitative analysis the non-linear response of the signal intensity to changes in v_e must be appreciated (Anthony and Seah 1983). This non-linearity may be predicted theoretically and for many singlet peaks, as shown in Fig. 3.6, the calculations are accurate. Note that the abscissa is not v_e but the ratio of v_e to the width of the peak measured at low values of v_e. Similar, but slightly different functions occur if the derivative is obtained by computer differentiation of the direct spectra (Seah, Anthony & Dench, 1983).

Fig. 3.4. The differential of Fig. 3.3 obtained by applying a 5 eV peak-to-peak modulation signal (v_e) to the mirror electrode.

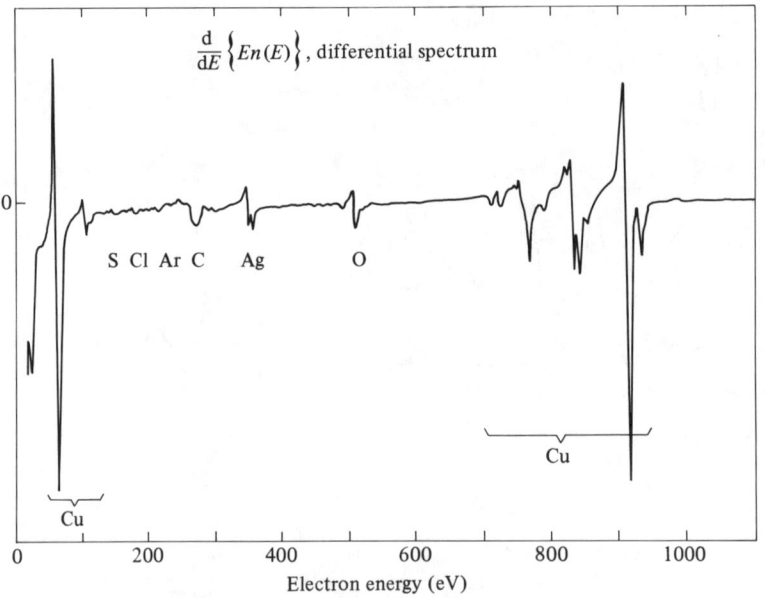

Electron and ion energy analysis

The reason for the choice of the CMA for AES is its high acceptance solid angle. This leads to high signal-to-noise ratios in the spectra obtained. The noise is random shot noise and so the signal-to-noise ratio becomes proportional to $(It)^{1/2}$ where I is the primary electron beam current (μA) and t is the detector integration time (ms). For the main peaks of the transition metals the signal-to-noise ratio, S/N, is typically given by

$$S/N = 10v_e(It) \tag{11}$$

up to a maximum value of $v_e = 5\,\text{eV}$ for the modulated differential case, and with $v_e = 5$ for the pulse counting mode.

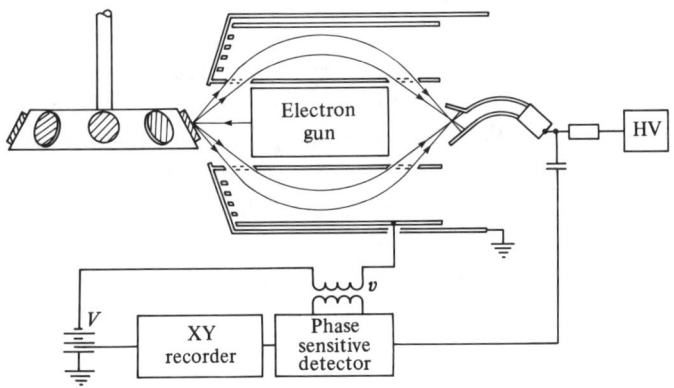

Fig. 3.5. Schematic electronic circuit for obtaining the differential spectrum from a CMA.

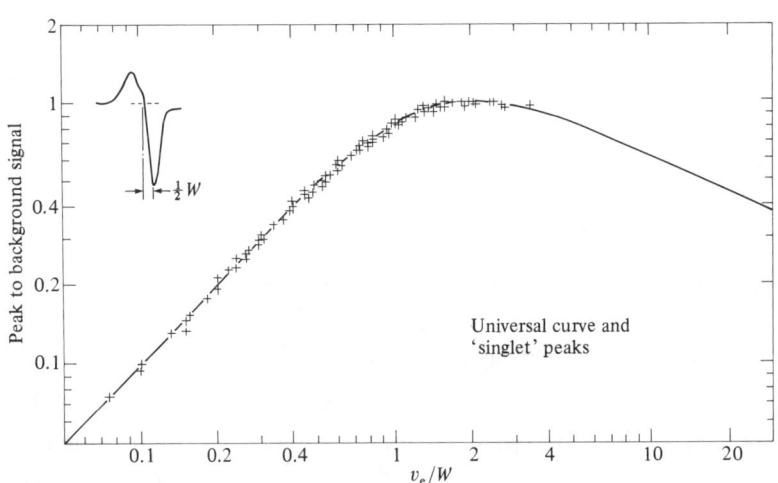

Fig. 3.6. The variation in measured intensity of a peak as a function of the reduced modulation, v_e/W, after Seah et al. (1983).

Another version of the CMA has been applied to XPS studies but, before considering this, another AES analyser will be evaluated. Evolving from its applications in XPS we find, today, concentric spherical sector analysers increasing in popularity. These are now discussed.

2.2 The concentric spherical sector analyser
2.2.1 Principle of operation

In the $1/r$ potential between two spherical sector electrodes electrons describe circular or elliptical orbits that are precise analogues of planetary motion. This leads to an instrument giving a first order focus in the radial direction and a higher order focus in the direction at right-angles to this, as shown schematically in Fig. 3.7. Early XPS analysers were of this form with or without pre-retardation. In order to improve the working area in complex instruments many of the manufacturers remove the spherical analyser from the specimen area and convey the electrons from one point to the other by an electron lens, as shown in Fig. 3.8. We shall first deal with the electron optics of the spherical sector analyser and then consider, briefly, some of the lens systems.

In its simplest form the analyser consists of two hemispherical concentric shells of radius R_1 and R_2 with potentials $-V_1$ and $-V_2$ respectively. Electrons are injected with energy E tangentially mid-way between the hemispheres at the radius R_0. The potential $-V_0$ at R_0 is simply given

$$V_0 = V_1 \frac{R_1}{2R_0} + V_2 \frac{R_2}{2R_0}. \tag{12}$$

If we inject electrons of energy eV_0 on the circumference at R_0, they will describe the circular orbit of radius R_0 if (Purcell, 1938; Kuyatt & Simpson, 1967)

$$V_1 = V_0(3 - 2R_0/R_1) \tag{13}$$

and

$$V_2 = V_0(3 - 2R_0/R_2), \tag{14}$$

hence the voltage difference between the hemispheres is given by

$$V_2 - V_1 = V_0 \left(\frac{R_2}{R_1} - \frac{R_1}{R_2} \right). \tag{15}$$

Electrons of the correct energy, emitted from a point source at S, tangentially to the radius, all focus at F, irrespective of the plane out of the page. If we reverse equation (15) in the form

$$V_0 = k(V_2 - V_1), \tag{16}$$

then, as before, k is known as the spectrometer constant. For the VG Scientific ESCALAB Mk II with $R_1 = 115$ and $R_2 = 185$ mm, k may be

Electron and ion energy analysis 67

Fig. 3.7. Cross-sectional view of a hemispherical electron spectrometer.

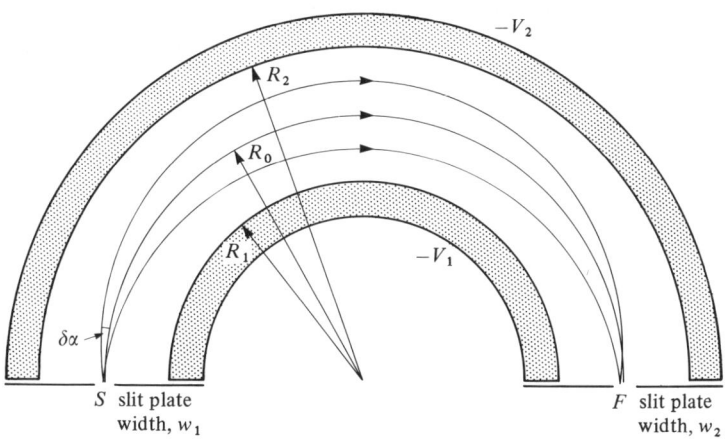

Fig. 3.8. Schematic cross-sectional view of a spherical sector analyser with input lens.

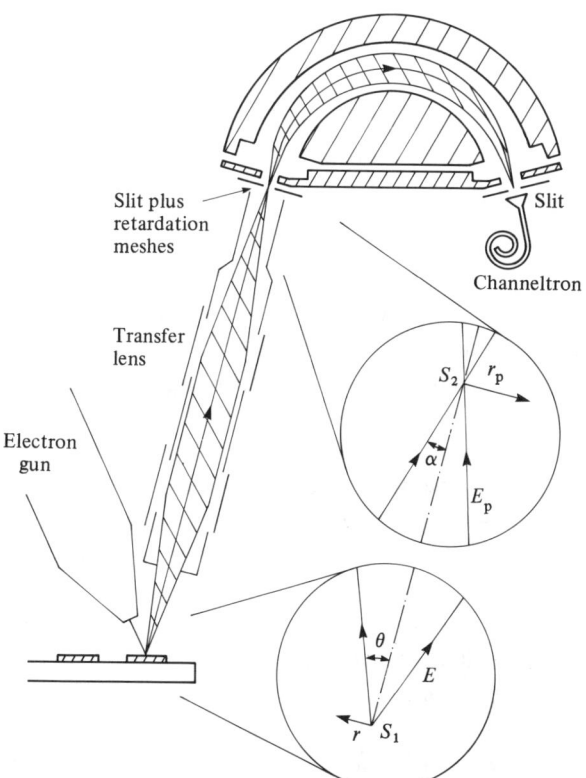

shown to be 1.013. For older analysers with a smaller gap, k is appropriately larger. If we now consider electrons making an angle $\delta\alpha$ to the tangential direction and at an energy ΔE above the correct energy E, then the shift in radial position after traversing the 180° is given by

$$\Delta R = 2R_0(\Delta E/E) - 2R_0(\delta\alpha)^2. \tag{17}$$

In a similar manner to the analysis for the CMA it may thus be shown that the base resolution is given by

$$\frac{\Delta E_b}{E} = \frac{W_1 + W_2}{2R_0} + (\delta\alpha)^2, \tag{18}$$

where W_1 and W_2 are the widths of the entrance and exit slits respectively.

2.2.2 Transmission of the spherical sector analyser

For AES we may consider a point source at the entrance to the analyser, in which case W_1 is effectively zero. If there is no internal restriction in the analyser and if internal stops are used to make

$$\alpha^2 = \frac{W_2}{2R_0}, \tag{19}$$

the transmission of the analyser is given by

$$T = \left(\frac{2W_2}{R_0}\right)^{1/2} \frac{\beta}{\pi}, \tag{20}$$

where β is the semiangular acceptance of the analyser in the plane normal to Fig. 3.7. Thus, in an analogous form to equation (7) for the CMA,

$$\frac{\Delta E_b}{E} = \frac{\pi^2 T^2}{2\beta^2}. \tag{21}$$

The maximum value for β with good design will be about 1 radian, therefore an analyser with a transmission of 2.7%, i.e. less than our CMA example given earlier, has the same resolution of 0.37%. In our example given here the spherical sector analyser gives a worse performance for all conditions of $\Delta E_b/E$; however, as we shall see later in Sections 2.2.3 and 2.2.5 this situation may be dramatically reversed and, with certain other advantages, this analyser probably has the greatest development potential.

As for the CMA, we see from equations (20) and (21), that for a given slit, T is independent of E and ΔE_b is proportional to E. Thus, again we find a spectrometer intensity-energy response function that is proportional to E so that, as described in Section 2.1.2, the measured spectrum is E times the true spectrum, $n(E)$.

2.2.3 The use of a lens with the spherical sector analyser

The hemispherical analyser, as discussed above, cannot be used on its own since the sample has to be in the entrance slit of the analyser.

Table 3.1. *Commercial electron spectrometers for AES and XPS*

Manufacturer	Model(s)	Application	Analyser configuration[a]				References
AEI (Kratos)	ES 100, 200, 300	XPS	S,	L+R,	, SSA,	S	Barrie, Drummond & Herd, 1974
Dupont	650	XPS	R followed by energy filter				Lee, 1973
Hewlett Packard	5950A	XPS	,	L+R,	, SSA,	M	Wannberg et al., 1974; Baird & Fadley, 1977
Kratos	X-SAM 800	AES/XPS	,	L+R,	S, SSA,	S or M	
Leybold	LHS 10	AES/XPS	,	L+R,	S, SSA,	S	Noller, Polaschegg & Schillalies 1974a, b; Polaschegg, 1974
Mc Pherson	ESCA 36	XPS			S, SSA,	S	Rendina, 1972
Perkin-Elmer	PHI 540, 545, 590, 595, 600	AES			, CMA,	S	Palmberg, 1969
	PHI, 548, 549 550, 560	AES/XPS	,	R,	, DPCMA,	S	Palmberg, 1974, 1975
	PHI 5000	XPS	,	L+R,	S, SSA,	S or M	
Riber	ASC 2000	AES			, CMA,	S	
Varian	IEE-15	XPS	,	R,	S, CMA,	S	Weichert & Helmer, 1970
	981-2807	AES			, CMA,	S	
	Automated Auger	AES			, CMA,	S	
VG Scientific	ESCA 3	XPS		R,	S, SSA,	S	Seah, 1980
	ESCALAB I	AES/XPS	L,	R,	S, SSA,	S	Seah, 1980; Hughes & Phillips, 1982
	ESCALAB II	AES/XPS	L,	R,	S, SSA,	S or M	

[a] S = slit, L = lens, R = retardation, SSA = spherical sector analyser, CMA = cylindrical mirror analyser, DPCMA = double pass cylindrical mirror analyser,
M = multichannel output. The order shows the order seen by an electron leaving the sample.

Working space around the sample may be created by reducing the angle of the analyser to, say, 150° instead of 180°. In this case the object point S is fixed on a line normal to the entrance slit and the image point F is on a line normal to the exit slit. The precise position of F is at the intersection of the above line with a line through S and the centres of the spherical sectors (Purcell, 1938). A second way of improving the working space is to use an electron lens to focus from the sample to the point S. This lens may simply transport the source away or it may also change the electron kinetic energy incident into the analyser. Some commercial spectrometers are listed in Table 3.1, showing which option is chosen in a particular case. For electrons in any electron optical system we may apply the Helmholtz-Lagrange relation and, for the points S_1 and S_2 in Fig. 3.8, this gives

$$r\theta E^{1/2} = r_p \alpha E_p^{1/2}, \tag{22}$$

where r is the radius of the electron source on the sample, r_p is the radius of the image at the entrance slit, θ is the electron cone semiangle at the sample and α that at the slit, E is the electron energy at the sample and E_p that at the slit. The ratio r_p/r is simply the magnification M of the lens and the ratio E/E_p is the retardation ratio R of the spectrometer. We may rewrite equations (20) and (21) to include the lens. For small values of W_2 and β, T would be increased by $M^2/R^{1/2}$ but, if β is 1 radian, few lens systems can fill greater than one radian and the limitations of the lens must be considered. If we assume the front of the lens can accept a cone of semiangle θ, there exist three regimes to consider:

(i) the lens defines the transmission: $\theta < MR^{-1/2}\beta, \theta < MR^{-1/2}\delta\alpha$;
(ii) the lens defines the transmission in the β direction and the spectrometer in the $\delta\alpha$ direction: $\theta < MR^{-1/2}\beta, \theta > MR^{-1/2}\delta\alpha$;
(iii) the spectrometer defines the transmission: $\theta > MR^{-1/2}\beta$, $\theta > MR^{-1/2}\delta\alpha$. With retardation at a fixed base resolution, $\delta\alpha$ and W_2 may be adjusted according to equation (19) and hence

$$\frac{\Delta E_b}{E} = \frac{W_2}{RR_0} = \frac{2(\delta\alpha)^2}{R}. \tag{23}$$

Thus, the analogue to equation (7) for the CMA is given by the equations

$$T = (1 - \cos\theta) \quad \text{case (i)}, \tag{24}$$

$$T = \left(\frac{2\Delta E_b}{E}\right)^{1/2} \frac{M\theta}{\pi} \quad \text{case (ii)}, \tag{25}$$

$$T = \left(\frac{2\Delta E_b}{E}\right)^{1/2} \frac{M^2\beta}{R^{1/2}\pi} \quad \text{case (iii)}. \tag{26}$$

To give an example, to maximise the transmission one should design a lens to accept as large a flux as possible and then to get all of this flux through the spectrometer. In this way it is possible to match or exceed the CMA's 12% transmission with $\theta \geqslant 28°, M \geqslant 7$ and $R \leqslant 2$. To do this and get all of the electrons through the analyser entrance slit requires the design of a very low aberration lens system (Harting & Read, 1976). Commercial manufacturers do go to great lengths to achieve this. From the above relations it is clear that the signal increases with M and may increase as R is reduced. The lenses are thus designed to magnify between S_1 and S_2 with M in the range 1–10 and, if options are available, a high M and low R will be found to give the best performance.

Unfortunately aberration coefficients for the lenses are not generally published. If the aberration, for the cone angle of electrons going through the analyser, becomes comparable to the entrance or exit slits of the hemispheres then the T values given above are not achieved. Indeed, depending on the lens behaviour, some of the parametric dependencies given in equations (25) and (26) may be changed. Also, in practice, few spectrometers are optimised according to equation (19), as assumed for simplicity here.

2.2.4 Positional sensitivity of the spherical sector analyser

The above considerations now enable modern spherical sector analysers to achieve similar signal-to-noise performances to the highly efficient CMA's, given in equation (11). Moreover, as we shall see later, for XPS studies, the spherical sector analysers have considerable advantages. Two additional advantages of the spherical systems are important. The first is their relative insensitivity to the sample positioning. Unlike the CMA, sample movements along the analyser axis have no effect on the peak positions in the energy spectrum. Again, movements at right-angles to this axis, out of the plane of the paper in Fig. 3.8, also have little effect. The only direction of sensitivity is in the third axis at right-angles to the lens and in the plane of the paper. In this case, the analogue to equation (9) is

$$\frac{\Delta E}{E} = \frac{\Delta L M}{2RR_0}, \qquad (27)$$

which, for the VG Scientific ESCALAB Mk II with $R_0 = 150$ mm, $M = 3$ and a typical R value of 4, gives a shift of 2.5 eV per 1000 eV per mm error, very much lower than the value for the CMA.

2.2.5 Multichannel detectors

The second of the advantages of the spherical sector is the prospect of adding multiple detection systems at the output slit of the hemispheres.

This may be in the form of a few multiple, parallel, equivalent detector chains or position sensitive detectors spread across the whole of the analyser output slit plane, as shown in Fig. 3.9. There are many ways of doing this, for instance using phosphor screens and TV cameras (Gelius, Basilier, Svensson, Bergmark & Siegbahn, 1974; Bertrand, Kalinowski, Tribble & Tolentino, 1983; Morris *et al.*, 1983), phosphor screens and charge-coupled devices (Hicks, Naviel, Wallbank & Comer, 1980; Delwiche, Hubin-Franskin, Furlan & Collin, 1982), resistive anode networks (Firmani *et al.*, 1982; Pollard, Trevor, Lee & Shirley, 1981), resistive filament (Hansson, Goldberg & Bachrach, 1981) and discrete anodes (van Hoof and van der Weil, 1980). These methods are briefly reviewed by Hicks *et al.* (1980). The maximum gain in efficiency is given by the ratio of the total collected flux in this mode to that in the normal mode. Whether or not this is achieved depends on the way the detectors are arranged and if the computer is correctly programmed to add the intensities appearing at different energies appropriately into the spectrum.

If the spectrometer is optimised as discussed above, a given resolution may be obtained without loss of intensity by using a low retard ratio and concomitantly low values of W_2 and $\delta\alpha$. For R values of 4 and 2 respectively, an overall resolution of 0.5% requires $W_2 = 0.02R_0$ and $0.01R_0$ and $\delta\alpha = 0.1$ and 0.7 respectively. It may be shown that, under these conditions with a spherical sector analyser of the geometry of the

Fig. 3.9. Two possible multichannel output systems for spherical sector spectrometers.

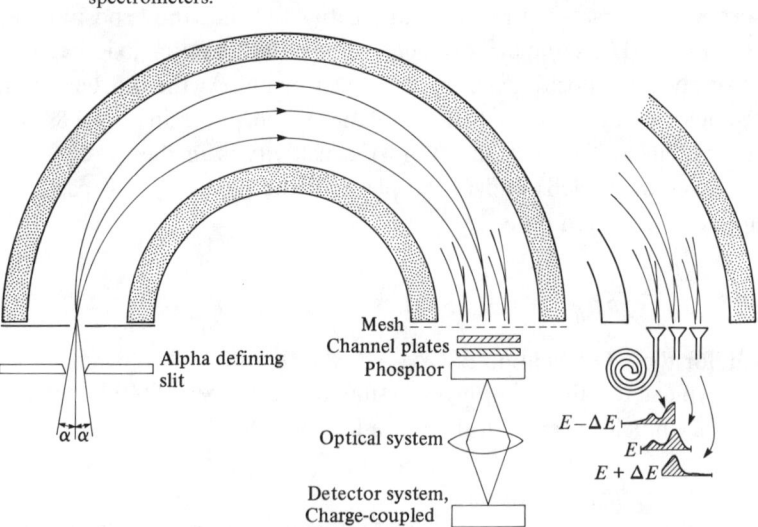

ESCALAB Mk II, if the useable gap between the hemispheres is 35 mm, and if R_0 is 150 mm, then the maximum gains would be of the order of 12 and 24 times, respectively, using as many channels as the gain, each of width W_2. In practice, it is unlikely that these full benefits will be achieved but clearly an improvement of over an order of magnitude could be attained in the spherical sector system. Thus, in an ideally arranged spherical sector lens system, the maximum transmission could well exceed 100%. It is most important when considering these multidetectors to ensure that they do not contribute to the noise in the spectrum or any real benefit may be lost. In the CMA the geometry makes multidetector arrangements difficult to visualise.

In the above we have attempted to present, in a simple and brief fashion, the essential behaviour of the CMA and spherical sector analysers in relation to their energy resolution, transmission and peak position sensitivity. For the measurement of intensities we must also consider the electron multiplier sensitivity as this is part of the measurement chain detailed in Figs. 3.1, 3.7 and 3.8. The electrons may strike the front of the multiplier at energy E or E_p. The efficiency of the multipliers in pulse counting varies with this energy, as shown in Fig. 3.10, and appears to be rather variable. Also a difference occurs between pulse counting and analogue applications. This is currently an unsolved problem. We now consider the behaviour of the above spectrometers when used in XPS.

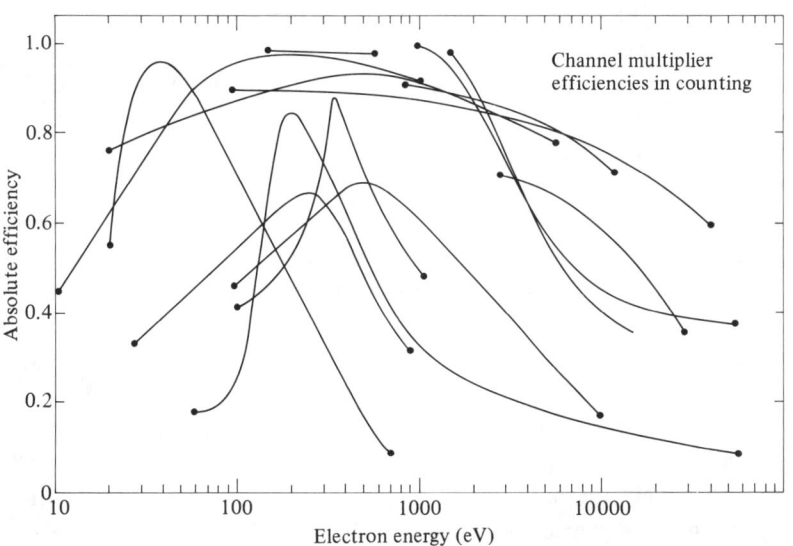

Fig. 3.10. A compilation of channel multiplier efficiencies from the literature.

3 X-ray photoelectron spectroscopy

In conventional XPS an X-ray source emitting a strong characteristic X-ray line is used to illuminate the whole surface of a sample, typically of the order to 1 cm square. The emission from this whole area is then energy analysed but with higher energy resolution than is customarily used for AES. The energy range and the intensities are similar to those for AES and consequently the same spectrometer can be used for the two techniques. However, as we shall see below, the way the spectrometer should be used is very different for the two techniques. In addition, with the current interest in small area analysis and the use of imaged X-ray sources such as monochromators, the optimum settings are again altered.

Unlike the AES case, where the analyser is scanned with all voltages on electrodes in proportion to one another and in which therefore ΔE is proportional to E, here in XPS we use a mode in which ΔE is kept constant. These two modes are common to all commercial electron spectrometers and are variously called the fixed retard ratio (FRR), constant retard ratio (CRR) or constant $\Delta E/E$ mode and the fixed analyser transmission (FAT), constant analyser energy (CAE) or constant ΔE mode. Also, unlike the AES case the electron detection is always by counting and is in the undifferentiated form.

Although historically the spherical sector electron spectrometers were the first to be used for XPS, we shall follow the order in Section 3.2 and start with the CMA.

3.1 The cylindrical mirror analyser

For XPS it is very important to achieve energy resolutions of 0.5 eV or better for kinetic energies up to 1500 eV and for the peak energy calibration to be accurate to better than this figure. With the CMA described in Section 2.1 these are both very difficult. The CMA can easily achieve 0.5 eV for 100 eV electrons and so the high resolution is maintained over the whole energy range by retarding the electrons to enter the CMA at fixed energies, usually 25, 50 and 100 eV. This retardation is made between two hemispherical meshes (Gerlach 1973) centred on the sample, as shown in Fig. 3.11. The first mesh is at the sample potential, usually earthed and the second at the same potential as the CMA inner cylinder. The electrons leave the sample at their kinetic energy E and are retarded by $E - E_p$ to enter the CMA at the pass energy E_p. In this way the ratio $\Delta E/E_p$ is defined by the equation (5) in Section 2.1.1 and ΔE remains constant through the spectrum at the chosen value.

In order to ensure that the peak positions remain constant, the volume of space that can project electrons into the analyser is limited by the use of a second CMA in line with the first (Palmberg, 1974, 1975). The single stage CMA has only one slit and the energy is defined by that slit and the source position. In the double pass CMA, on the other hand, the energy is defined by the two CMA slits and, if the source position is incorrect, the electrons fail to be transmitted. Thus, the system shown in Fig. 3.11 overcomes the main problems of the single stage CMA.

In XPS the output signal for a fully illuminated sample is proportional to the product of the transmission efficiency T, discussed in Section 2.1.2, and the area, A, of the sample from which the electrons are detected. This product, AT, has been termed the étendue, G, of the analyser. It is difficult to estimate the precise values of A or T since both are dependent on the quality of the meshes, their sphericity and wire mesh size. Some points may, however, be made.

If we consider refraction at the meshes it is easy to show, using the symbols defined in Fig. 3.11, that

$$\gamma E^{1/2} = \alpha E_p^{1/2}. \tag{28}$$

Since there is a limitation on the total solid angle of electrons emerging

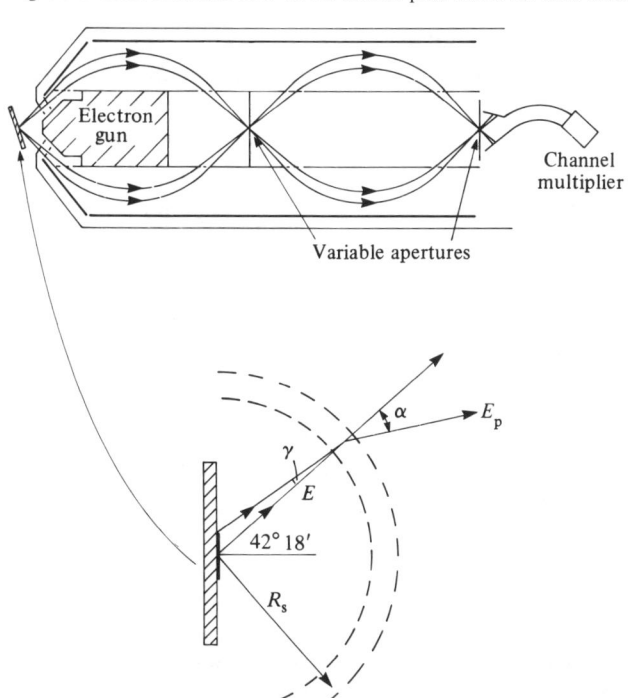

Fig. 3.11. Cross-sectional view of the double pass CMA for XPS studies.

from the second mesh that can be transmitted by the double pass CMA it is clear that the solid angle from the sample falls as the energy increases according to

$$T = T_0 \frac{E_p}{E},\qquad(29)$$

where $T = T_0$ in the absence of retardation. The above refraction has an additional effect of increasing the area of analysis at low energies. At high energies the area of analysis is defined by the image of the output slits of the CMA. However, as the energy is reduced this image becomes blurred by an additional term of the order of r_s where

$$r_s = R_s \alpha \left(\frac{E_p}{E}\right)^{1/2}.\qquad(30)$$

Here R_s is the mean radius of the retarding spheres, of the order of 30 mm. For 500 eV electrons retarded to 25 eV, with $\alpha = 0.1$, r_s is 0.67 mm, increasing to 1 mm for 250 eV electrons. In practice the blurring is greater than this because of the aberrations of both the spherical retarding meshes and the cylindrical inner cylinder slit meshes.

3.2 The concentric spherical sector analyser
3.2.1 Principle of operation

The analysis for the concentric spherical sector analyser is more complex than that for the CMA as there is a greater range of parameters which may be used to optimise the performance and give instrument flexibility. The basic optics are described in Section 2.2; however, for XPS, the emitting source area is no longer negligible and the front slit in the analyser becomes important.

If the spectrometer has no lens we may consider the resolution directly in terms of equation (18). For monochromatic sources of electrons evenly spread over the sample, the intensity measured is proportional to W_1 but only provided $W_2 \geqslant W_1$. For this reason it is customary to make $W_2 = W_1 = W$ so that, from equation (18)

$$\frac{\Delta E_b}{E} = \frac{W}{R_0} + (\delta\alpha)^2.\qquad(31)$$

In this case an optimum condition is that

$$\alpha^2 = \frac{W}{R_0},\qquad(32)$$

slightly different from equation (19). Generally some compromise will be made so that $\delta\alpha$ in a commercial instrument is coupled to the slits to optimise the overall instrument performance. The intensity measured is controlled by the étendue G and not simply by T. Thus the analogue to equation (20) is approximately

$$G = lW\left(\frac{W}{R_0}\right)^{1/2}\frac{\beta}{\pi}, \tag{33}$$

where l is the length of the input and output slits. Equation (33) is valid for some of the earlier spectrometers operated in the constant $\Delta E/E$ mode. The resolution is set by W and $\delta\alpha$ and the étendue is then fully defined.

In order to operate in the constant ΔE mode, electrons are retarded into the analyser which is set at the pass energy E_p. In this case, if the spectrometer is optimised as above,

$$\frac{\Delta E_b}{E_p} = \frac{2W}{R_0}. \tag{34}$$

In the simplest case, the retardation occurs between flat parallel meshes with the first mesh at the sample earth potential and the second mesh retarding the electron to a kinetic energy E_p for injection into the hemispheres. Now, using equation (28) we find

$$G = \frac{lW^{3/2}\beta}{R_0\pi}\left(\frac{E_p}{E}\right), \tag{35}$$

so that, for a given ΔE,

$$G = \frac{l\beta\Delta E_b^{3/2}}{2\pi E}\left(\frac{R_0}{2E_p}\right)^{1/2}, \tag{36}$$

showing that for a given resolution a greater intensity will be obtained with a larger spectrometer (allowing W to increase) and at low pass energies (also allowing W to increase). Note that the intensity now falls as the electron energy increases and the analyser intensity–energy response function, described previously in Section 2.1.2, is proportional to E^{-1}. Thus, the measured energy spectrum is now proportional, not to $n(E)$ the true spectrum, but to $n(E)/E$.

3.2.2 The use of a lens with the spherical sector analyser

The more complex analysers for combined AES and XPS studies, as described in Section 2.2.3, use a lens to take electrons from the sample and inject them into the analyser. As before, we may define the three operating regimes for deciding which part of the electron optics limits the electron flux

(i) the lens defines the transmission: $\theta < M\beta(E_p/E)^{1/2}$, $\theta < M\delta\alpha(E_p/E)^{1/2}$;

(ii) the lens defines the transmission in the β direction and the spectrometer in the $\delta\alpha$ direction: $\theta < M\beta(E_p/E)^{1/2}$, $\theta > M\delta\alpha(E_p/E)^{1/2}$

(iii) the spectrometer defines the transmission: $\theta > M\beta(E_p/E)^{1/2}$, $\theta > M\delta\alpha(E_p/E)^{1/2}$.

Following the previous analysis for a given resolution ΔE_b,

$$G = \frac{(1 - \cos\theta)lR_0\Delta E_b}{2M^2 E_p} \quad \text{case (i)}, \tag{37}$$

$$G = \left(\frac{\Delta E_b}{2E}\right)^{1/2} \frac{\theta l R_0 \Delta E_b}{\pi M E_p} \quad \text{case (ii)}, \tag{38}$$

$$G = \left(\frac{\Delta E_b}{2E_p}\right)^{1/2} \frac{\beta l R_0 \Delta E_b}{\pi E} \quad \text{case (iii)}. \tag{39}$$

We see that, for XPS, low pass energies and low magnifications together with a larger analyser size are extremely beneficial. These are the reverse operating modes from those for AES and, to achieve an optimum for both, the instrument must be very flexible.

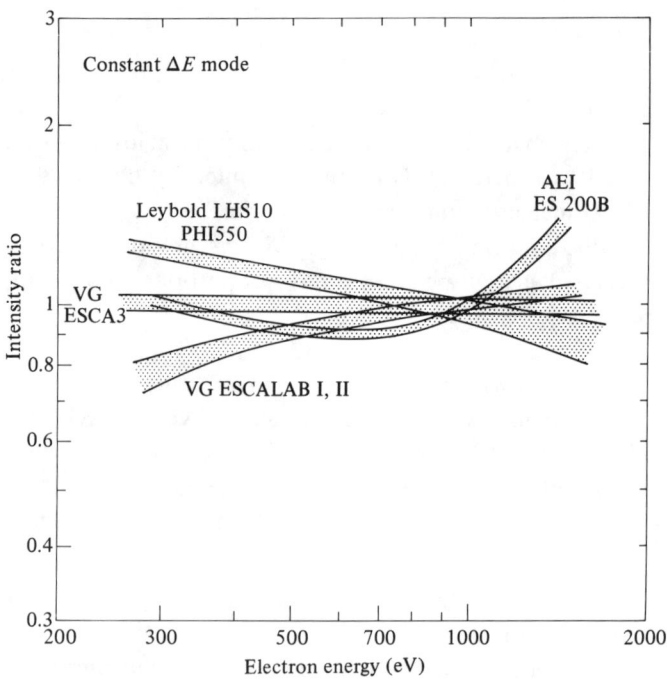

Fig. 3.12. The relative intensity energy response functions of commercial XPS analysers in the constant ΔE mode, normalised to the VG Scientific ESCA 3 Mk II, after Seah, Jones & Anthony (1984).

We see also that, as the energy E increases, all spectrometers operated in the constant ΔE mode satisfy case (i) at low energies, move into case (ii) as the energy rises and eventually fulfil case (iii). Thus, the intensity-energy response function, described earlier in Section 2.1.2 and given by the product $G\Delta E_b$, is simply proportional to G. To transfer data from one spectrometer to another we need the ratios of the intensity–energy response functions and this is defined as the relative intensity energy response function. The functions for a number of commercial XPS spectrometers, relative to that of the VG Scientific ESCA 3 Mk II, are shown in Fig. 3.12.

3.2.3 XPS of small areas

Today we see a growing interest in small area XPS analysis (Gurker, Ebel & Ebel, 1983; Yates & West, 1983; Smith and Seah, 1987). This may be achieved by imaging an aperture at the analyser entrance back on to the sample. If the lens spherical aberration is sufficiently low and the aperture diameter is d, the diameter analysed on the sample is d/M. It is interesting to see if the above conclusions for R_0, M and E_p also hold for such a fixed analytical area on the sample. If the $\delta\alpha$ in the analyser is set according to equation (32), where W is replaced by the variable output slit size, W_2, we return to equations (24)–(26) with a slight change in the numerical constant and with R written as (E/E_p). We have now reverted to T since we no longer wish to increase the area of analysis to increase the signal. Contrary to the standard XPS case, the higher magnifications and pass energies become the optimum operating modes. This difference merely arises since, in this case as in AES, we consider a fixed analysis area whereas for the standard XPS mode we wish to incorporate electrons from as large an area as possible.

3.2.4 X-ray monochromators

A last case to consider is the use of the spherical sector analyser with X-ray monochromators where there is a line source of width s projected onto the plane at right-angles to the lens entrance axis. In these cases, if the signal is maximised by matching Ms with the input slit then

$$G = \frac{(1 - \cos\theta)ls}{M} \qquad \text{case (i)}, \tag{40}$$

$$G = \left(\frac{\Delta E_B}{2E}\right)^{1/2} \frac{2\theta ls}{\pi} \qquad \text{case (ii)}, \tag{41}$$

$$G = \left(\frac{\Delta E_B}{2E}\right)^{1/2} \frac{2M\beta ls}{\pi}\left(\frac{E_p}{E}\right)^{1/2} \qquad \text{case (iii)}, \tag{42}$$

and again, the optimisation has changed so that M and E_p become less important.

3.2.5 Multichannel detectors

Above we have seen three common uses of XPS each with different optimisation conditions. Before leaving the analyser we should consider the use of multidetectors. For the normal use of XPS the multidetector system discussed in Section 2.2.5 will not give such high gains as for AES. For an energy resolution of 0.5 eV, for instance in the VG Scientific ESCALAB II basic geometry, an ideally optimised spectrometer operated at 10 eV pass energy would require 3.75 mm slits and, if the useable space between the hemispheres is 35 mm as discussed before, a gain of nine times in efficiency is achieved. If the pass energy is increased by a factor of n, the étendue falls by a factor of n but, for the multidetector output, the increase in efficiency exactly offsets this. Thus, for all pass energies in the spectrometer design discussed above the ideal multidetector gives nine times improvement in detection efficiency over the basic instrument operated at low pass energies. For spatially resolved XPS or for use with monochromators the gain in yield is of course much higher. Current versions of some commercial instruments listed in Table 3.1 are achieving these order of magnitude improvements.

4 Ion scattering spectroscopy

In ion scattering spectroscopy, as discussed in Chapter 8, the energy of the scattered ion, E, is related to both the mass of the target atom, M_t, and the scattering angle, θ, between the spectrometer and the incident ion beam according to

$$E = E_0 \frac{M_i^2}{(M_i + M_t)^2} \left\{ \cos\theta \pm \left[\left(\frac{M_t}{M_i}\right)^2 - \sin^2\theta \right]^{1/2} \right\}^2, \qquad (43)$$

where M_i is the mass of the bombarding ion of energy E_0. Thus, since θ is fixed by the instrument geometry, M_t is uniquely determined by E. The energy spectrum thus gives a mass analysis of the atoms in the target layer. The energy analysis, usually of primary ions of He^+ or Ne^+ in the energy range up to 2 keV can be efficiently performed with both the CMA and spherical sector analysers discussed for AES studies, with a simple reversal of electrode polarities. With the CMA the ion beam must be concentric with the analyser to ensure that θ has only one value, in this case 138°.

The mass spectrum from equation (43) is not linear with energy and the separation of adjacent higher masses is difficult. The behaviour of equation (43) is best discussed at $\theta = 90°$ where the result simplifies to

$$E = E_0 \left(\frac{M_t - M_i}{M_t + M_i} \right). \tag{44}$$

For high masses the fractional energy loss follows $2M_i/M_t$ so that for a He^+ ion beam, unit mass resolution is lost after Na. For an Ne^+ ion beam this limit occurs around Fe, or slightly higher masses if $\theta > 90°$. This loss of discrimination arises because the peaks at energies near E_0 rarely have widths less than 2% of E_0 due to the physics of the scattering problem, even when studied with very high resolution spectrometers (Bertrand, Delannay & Streydio, 1977; Niehus & Bauer, 1975). Thus typical spectrometers are operated at about 2% resolution and the range of input scattering angles is limited to about $\pm 3°$. The range of input angles, it should be remembered, gives a loss of mass resolution due to equation (43) in addition to the energy resolution terms discussed for AES and XPS.

Thus the requirements for ISS appear to have been contained fully within the instrumental developments required for AES and XPS. If the highest sensitivity is required the multi-channel detection systems are applicable although they do not, as yet, appear to have been tried for ISS.

5 Rutherford backscattering spectroscopy

Rutherford backscattering spectroscopy (RBS), as described in Chapter 9, is based on the same principles as ISS but involves the use of higher energy ions. Most RBS studies use helium ion beams in one of two energy ranges either 1–4 MeV or 100–500 keV. These may be termed the high- and medium-energy ranges, respectively where ISS covers the low-energy range. The technique is excellently reviewed by Chu, Mayer & Nicolet (1978) and also in Chapter 9 but here we consider solely the measurement of the energy spectrum of the back-reflected ions and the constraints and limitations of the solid state detectors used. Electrostatic or electromagnetic spectrometers can be used but the solid state detectors offer far greater simplicity and in some ways much greater efficiency. Electrostatic spectrometers are normally used for the low energy range of ISS but only occasionally in the medium energy range.

The principle of the solid state detector is that the energetic particle enters a semiconductor, either Si or Ge, and creates electron–hole pairs over the whole track length. This is typically a cylinder of 1 μm diameter and 5 μm length for 1 MeV helium ions, increasing to 60 μm length at 10 MeV. If the energy of creation of a pair is ε eV and if the particle energy is E, the number of carriers generated is E/ε. These carriers are collected by a biased gold electrode on the semiconductor surface to give an output pulse. The pulse strength gives a measure of the particle energy, irrespective of its charge state, and the number of pulses at this strength gives the number of particles at that energy and hence the energy spectrum. The energy resolution of the device depends on the system's ability to define the pulse strength, and hence particle energy, accurately. If the number of carriers varies as a Poisson distribution, the standard deviation of the pulse strength is proportional to $(E/\varepsilon)^{1/2}$ so that the full width at half maximum of a peak in the spectrum, ΔE, is $2.35\,(E\varepsilon)^{1/2}$ and the resolution $\Delta E/E$ is $2.35\,(E/\varepsilon)^{-1/2}$. Values of ε for Si and Ge are 3.55 and 2.94 eV, respectively, so that the energy measurement of 4 MeV α particles by Si and Ge detectors may be limited to resolutions of 0.22 and 0.20% each, respectively. Resolutions of 0.18% in Si at 6 MeV, equivalent to 0.22% at 4 MeV, have been attained for many years, as shown in Fig. 3.13, and current resolutions with slightly improved detectors obtain marginally better values. To achieve this resolution the detector must only collect particles over a narrow range of angles $\Delta\theta$. For example, from equation (43) it is easy to show that for helium ions scattered through 90° from a relatively heavy atom of mass M_t,

$$\frac{\Delta E}{E} \simeq -\frac{4\Delta\theta}{M_t}.$$

Fig. 3.13. The α spectrum of Th—C observed with a silicon surface barrier detector, after Watt and Ramsden (1964).

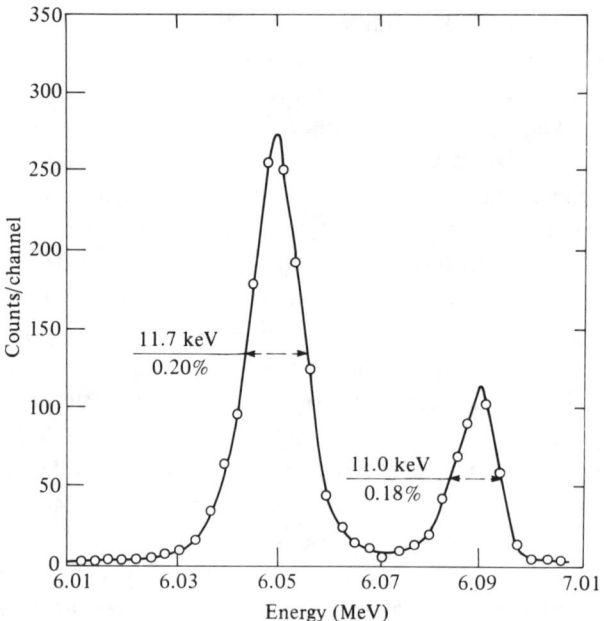

To achieve 0.25% resolution for ions reflected from Si the collector should subtend only 1.15° at the target and of course the primary beam must be collimated to better than 1°. More typical resolutions are around 0.75% and these may be achieved with a detector of 5 mm by 5 mm at 80 mm from the sample.

High resolution is important since the peak width and hence peak height in studies of surface atom sites is limited by the detector resolution. For instance Feldman *et al.* (1977), as shown in Fig. 3.14, in studies of the scattering of 1.83 MeV helium ions from a $W(001)$ surface, see the surface height step in the spectrum for a randomly orientated beam, but a peak of only 14 keV (0.75% resolution) width for the beam aligned along a $\langle 100 \rangle$ direction. This peak arises from scattering by the surface atoms only. The limit of 5 keV resolution for 2 MeV He ions is only achieved with cooled detectors. For the highest resolution the pulse preamplifier is placed adjacent to the detector and both are cooled. The shortness of the lead reduces input capacitance and allows high count rates to be observed. The cooling reduces noise in the detector, feedback resistor and input FET.

The overall energy resolution of the detector is governed by the straggling of the incoming ions in the gold metallisation and the oxide on the surface of the detector, the evenness of these layers, the statistical

Fig. 3.14. RBS spectra for 2 MeV He incident on W along a $\langle 100 \rangle$ axial direction (○) and in a non-channelling direction (●) using a surface barrier detector of 14 keV resolution, after Feldman *et al.* (1977).

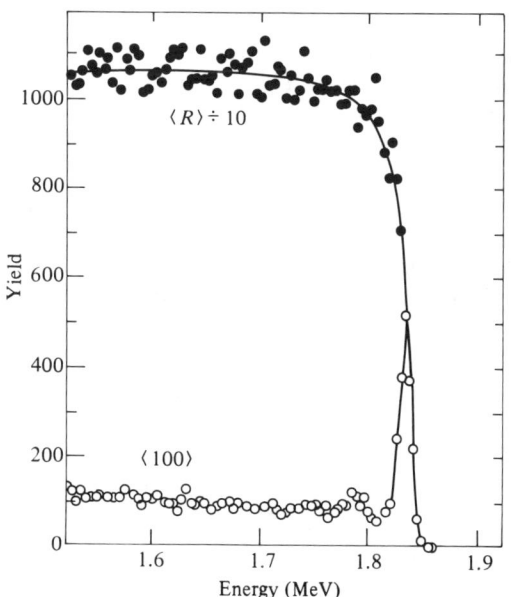

problem of the Poisson distribution discussed earlier, the elimination of unwanted traps in the detector and the detector capacitance. The last may be reduced by using a small collection area.

In addition to the resolution terms given above is the inherent noise level of the detectors, currently equivalent to around 1–2 keV in the energy spectrum for He ions. Thus, for optimised detectors

$$\Delta E \simeq (2.10^6 + 5.5\, E\varepsilon)^{1/2} \tag{45}$$

whereas a typical figure for an electrostatic spectrometer would be $0.005E$. At this figure, the electrostatic spectrometer gives better energy resolution only for beam energies below 800 keV. However, most medium energy ion beam studies use solid state detectors rather than electrostatic analysers for three main reasons (i) solid state detectors measure the intensity of all the backscattered particles whereas electrostatic analysers only deflect ions and ignore the large neutral content of the yield, (ii) solid state detectors measure the energy of all of the ions in a given solid angle of emission all of the time whereas electrostatic spectrometers only measure those within the resolution setting ΔE at the energy E, at any given time and (iii) the solid state detector is much simpler, cheaper and easier to use in practice.

References

Anthony, M.T. & Seah, M.P. (1983). *Journal of Electron Spectroscopy and Related Phenomena*, **32**, 73.

Baired, R.J. & Fadley, C.S. (1977). *Journal of Electron Spectroscopy and Related Phenomena*, **11**, 39.

Barrie, A., Drummond, I.W. & Herd, Q.C. (1974). *Journal of Electron Spectroscopy and Related Phenomena*, **5**, 217.

Bennani, A.L., Duguet, A. & Wellenstein, H.F. (1980). *Journal of Electron Spectroscopy and Related Phenomena*, **20**, 75.

Bertrand, P., Delannay, F. & Streydio, J.M. (1977). *Journal of Physics E: Scientific Instruments*, **10**, 403.

Bertrand, P.A., Kalinowski, W.J. Tribble, L.E. & Tolentino, L.U. (1983). *Rev. Sci. Instr.*, **54**, 387.

Bishop, H.E. & Riviere, J.C. (1969). *J. App. Physics*, **40**, 1740.

Chu, W.K., Mayer, J.W. & Nicolet, M.A. (1978). *Backscattering Spectrometry*, Academic Press, New York.

Dabbs, J.E. (1983). *Journal of physics E: Scientific Instruments*, **16**, 57.

Davis, L.E., Macdonald, N.C., Palmberg, P.W., Riach, G.E. & Weber, R.E. (1976). *Handbook of Auger Electron Spectroscopy*, 2nd edition, Physical Electronics Division of Perkin Elmer Corp, Minnesota.

Delwiche, J., Hubin-Franskin, M.J., Furlan, M., Collin, J.E., Roy, D., Leclerc, B., Roy, P., Poulin, A. & Carette, J.D. (1982). *Journal of Electron Spectroscopy and Related Phenomena*, **28**, 123.

Feldman, L.C., Kauffmann, R.L., Silverman, P.J., Zuhr, R.A. & Barrett, J.H. (1977). *Phys. Rev. Lett.*, **39**, 38.

Fellner-Feldegg, H., Gelius, U., Wannberg, B., Nilsson, A.G., Basilier, E. & Siegbahn, K. (1974). *Journal of Electron Spectroscopy and Related Phenomena*, **5**, 643.
Firmani, C., Ruiz, E., Carlson, C.W., Lampton, M. & Paresce, F. (1982). *Rev. Sci. Instr.*, **53**, 570.
Gelius, U., Basilier, E., Svensson, S., Bergmark, T. & Siegbahn K. (1974). *Journal of Electron Spectroscopy and Related Phenomena*, **2**, 405.
Gerlach, R.L. (1973). *J. Vac. Sci. and Tech.*, **10**, 122.
Green, T.S. & Proca, G.A (1970). *Rev. Sci. Instr.*, **41**, 1409.
Gurker, N., Ebel, M.F. & Ebel, H. (1983). *Surface and Interface Analysis*, **5**, 13.
Hafner, H., Simpson, J.A. & Kuyatt, C.E. (1968). *Rev. Sci. Instr.*, **39**, 33.
Hansson, G.V., Goldberg, B. & Bachrach, R.Z. (1981). *Rev. Sci. Instr.*, **52**, 517.
Harting, E. & Read, F.H. (1976). *Electrostatic Lenses*, Elsevier, New York.
Heddle, D.W.O. (1971). *Journal of Physics E: Scientific Instruments*, **4**, 589.
Hicks, P.J., Daviel, S., Wallbank, B. & Comer, J. (1980). *Journal of Physics E: Scientific Instruments*, **13**, 713.
Huchital, D.A. & Rigden, J.D. (1972). *J. Appl. Phys.*, **43**, 2291.
Hughes, A.E. & Phillips, C.C. (1982). *Surface and Interface Analysis*, **3**, 220.
Hughes A. Li. & Rojansky, V. (1929). *Phys. Rev.*, **34**, 284.
Kuyatt, C.E. & Simpson, J.A. (1967). *Rev. Sci. Instr.*, **38**, 103.
Lee, J.D. (1973). *Rev. Sci. Instr.*, **44**, 893.
Marmet, P. & Kerwin, L. (1960). *Canadian Journal of Physics*, **38**, 787.
McGuire, G.E. (1979). *Auger Electron Spectroscopy Reference Manual*, Plenum Press, New York.
Morris, A., Johnathan, N., Dyke, J.M., Francis, P.D., Keddar, N. & Mills, J.D. (1983). *Rev. Sci. Instr.*, **55**, 172.
Niehus, H. & Bauer, E. (1975). *Rev. Sci. Instr.*, **46**, 1275.
Noller, H., Polaschegg, H.D. & Schillalies, K. (1974a). *Journal of Electron Spectroscopy and Related Phenomena*, **5**, 705.
Noller, H.G., Polaschegg, H.D. & Schillalies, H. (1974b). *Japanese Journal of Applied Physics*, Supplement 2, Part 1, 343.
Palmberg, P.W., Bohn, G.K. & Tracy, J.C. (1969). *Appl. Phys. Lett.*, **15**, 254.
Palmberg, P.W. (1974). *Journal of Electron Spectroscopy and Related Phenomena*, **5**, 691.
Palmberg, P.W. (1975). *J. Vac. Sci. Tech.*, **12**, 379.
Pollard, J.E., Trevor, D.J., Lee, Y.T. & Shirley, D.A. (1981). *Rev. Sci. Instr.*, **52**, 1837.
Polaschegg, H.D. (1974). *Appl. Phys.*, **4**, 63.
Proca, G.A. & Green, T.S. (1970). *Rev. Sci. Instr.*, **41**, 1778.
Proca, G.A. (1973). *Rev. Sci. Instr.*, **44**, 1365.
Purcell, E.M. (1938). *Phys. Rev.*, **54**, 818.
Rendina, J.F. (1972). *International Laboratory*, May/June issue, pp. 17–23.
Roy, D. & Carette, J.D. (1977). In *Electron Spectroscopy for Suface Analysis Topics in Current Physics*, Vol. 4, pp. 13–58, Springer Verlag, Berlin.
Sar-El, H.Z. (1967). *Rev. Sci. Instr.*, **38**, 1210.
Sar-El, H.Z. (1970). *Rev. Sci. Instr.*, **41**, 561.
Seah, M.P. (1980). *Surface and Interface Analysis*, **2**, 222.
Seah, M.P., Anthony, M.T. & Dench, W.A. (1983). *Journal of Phys. E: Scientific Instruments*, **16**, 848.
Seah, M.P., Jones, M.E. & Anthony, M.T. (1984). *Surface and Interface Analysis*, **6**, 242.
Sekine, T., Nagasawa, Y., Kudoh, M., Sakai, Y., Parkes, A.S., Geller, A.D.,

Mogami, A. & Hirata, K. (1982). *Handbook of Auger Electron Spectroscopy*, JEOL, Tokyo.
Sickafus, E.N. & Holloway, D.M. (1975). *Surface Science*, **51**, 131.
Smith, G.C. & Seah, M.P. (1987). *Journal of Electron Spectroscopy and Related Phenomena*, **42**, 359.
Staib, P. & Dinklage, U. (1977). *Journal of Physics E: Scientific Instruments*, **10**, 914.
Steckelmacher, W. (1973). *Journal of Physics E: Scientific Instruments*, **6**, 1061.
Taylor, N.J. (1969). *Rev. Sci. Instr.*, **40**, 792.
van Hoof, H.A. & van der Wiel, M.J. (1980). *Journal of Physics E: Scientific Instruments*, **13**, 409.
Wannberg, B., Gelius, U & Siegbahn, K. (1974). *Journal of Physics E: Scientific Instruments*, **7**, 149.
Watt, D.E. & Ramsden, D. (1964). *High Sensitivity Counting Techniques*, Pergamon Press, London.
Weichert, N. & Helmer, J.C. (1970). *Advances in X-ray Analysis*, **13**, 406.
Yates, K. & West, R.H. (1983). *Surface and Interface Analysis*, **5**, 217.

4

Auger electron spectroscopy

H.E. BISHOP

1 Introduction

Auger electron spectroscopy (AES) together with X-ray photoelectron spectroscopy (XPS) are the two major surface analytical techniques. They were both developed in the late 1960s at a time when the ultra-high vacuum technology, necessary for most surface studies, became commercially available and was no longer a province for the specialist. The two techniques are largely complementary; in general, XPS is rather more sensitive and gives more useful chemical information, whilst AES has the advantages of greater speed and the potential for high spatial resolution. In particular, it is the capability of high spatial resolution that has made it such a popular technique in technological applications, such as in the study of semiconductor structures and in the investigation of the role of trace impurities in the intergranular fracture of metals. The high spatial resolution is achieved because, in AES, the specimen is excited by an electron beam that can be focussed into a fine probe. It should be remembered, however, that Auger electrons also appear in XPS spectra and are frequently used in conjunction with the photoelectron peaks, although here we will be concerned exclusively with electron excited Auger spectra. In this chapter we will discuss the underlying principles of AES and its associated instrumentation, how to evaluate the spectra and the various modes of operation, illustrated by practical examples.

2 Basic principles
2.1 *The Auger effect*

The Auger effect is the de-excitation of an ionised atom by a non-radiative process. When an electron is ejected from an inner shell of an atom the resultant vacancy is soon filled by an electron from one of the outer shells (Fig. 4.1). The energy released may appear either as an

X-ray photon, the characteristic X-rays used in electron probe microanalysis and in X-ray fluorescence spectroscopy, or be transferred to another electron which is ejected from the atom with an energy, E_A, determined by the three energy levels concerned, where

$$E_A = E_1 - E_2 - E_3^*. \tag{1}$$

E_1 and E_2 are the binding energies of the atom in the singly ionised state while E_3^* is that for the doubly ionised state. For lower energy events leading to a photon or electron with energy less than 2 keV the Auger process is dominant with more than 95% of ionisations leading to the ejection of an Auger electron.

The Auger electron moves through the solid and soon loses its energy through inelastic collisions with bound electrons. However, if the Auger electron is released sufficiently close to the surface, it may escape from the surface with little or no energy loss and be detected by an electron spectrometer. When the primary ionisation has been produced by an electron beam, peaks due to the Auger electrons appear superimposed on a background of back-scattered electrons. Fig. 4.2 shows the energy spectrum of electrons below 2 keV from a number of elements excited by a 10 keV electron beam, the distinctive Auger peaks for each element stand out clearly above the smooth background. On a sample of unknown composition the elements present at the surface may be determined from the presence of their characteristic Auger peaks.

Fig. 4.1. Characteristic X-ray and Auger electron production.

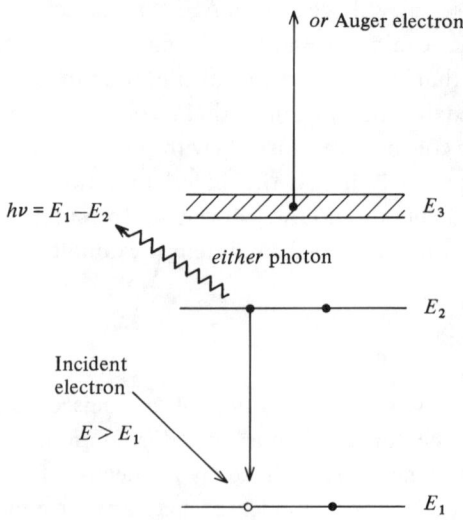

Auger electron spectroscopy

Standard Auger spectra for the majority of elements have been assembled in the handbooks referred to in Chapter 3 and a summary chart from one such compilation is given in Fig. 4.3. There is usually little problem in identifying the major elements present, but small peaks may be difficult to assign unequivocally, particularly at lower energies, where many elements have peaks. Interpretation may be further complicated by peak shifts due to chemical bonding effects, as will be discussed later.

2.2 Electron scattering

The exciting electron beam plays a central role in AES. It must be chosen to give an adequate excitation of the Auger peaks and also to give the desired spatial resolution. To achieve this it is usual to employ a primary beam of energy between 5 and 10 keV although it is sometimes advantageous to move outside this range.

On entering a solid the electron beam is scattered, both elastically out of its initial direction and inelastically, until it either escapes back through the surface or reaches thermal energies. In this process the primary electrons produce ionisations in the solid and excite secondary electrons,

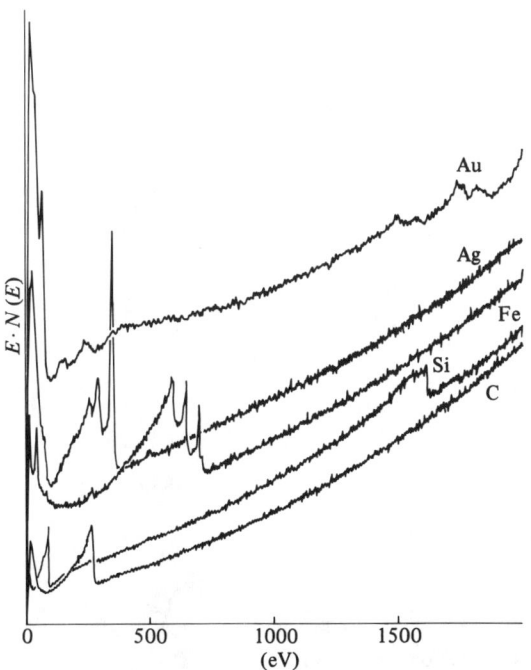

Fig. 4.2. $E \cdot N(E)$ spectra for a range of elements recorded under identical conditions, Bishop (1984).

mainly with a few eV energy but some with energies of up to 1 keV or more. Because of the many interactions taking place, calculations of electron scattering in solids are complicated and normally entail many approximations. However, an empirical relation for the penetration, R, of an electron, energy E_0 (keV) into a solid density, ρ, is given by

$$\rho R = 8.6 E_0^{1.5} \, \mu\text{g cm}^2. \tag{2}$$

Although very approximate this expression brings out two important features of the electron range; first the range depends mainly on the density of the specimen and second the energy dependence is through a power law with an exponent of about 1.5.

The overall spectrum of back-scattered electrons from a boron sample is shown in Fig. 4.4. Although a low primary energy has been chosen in

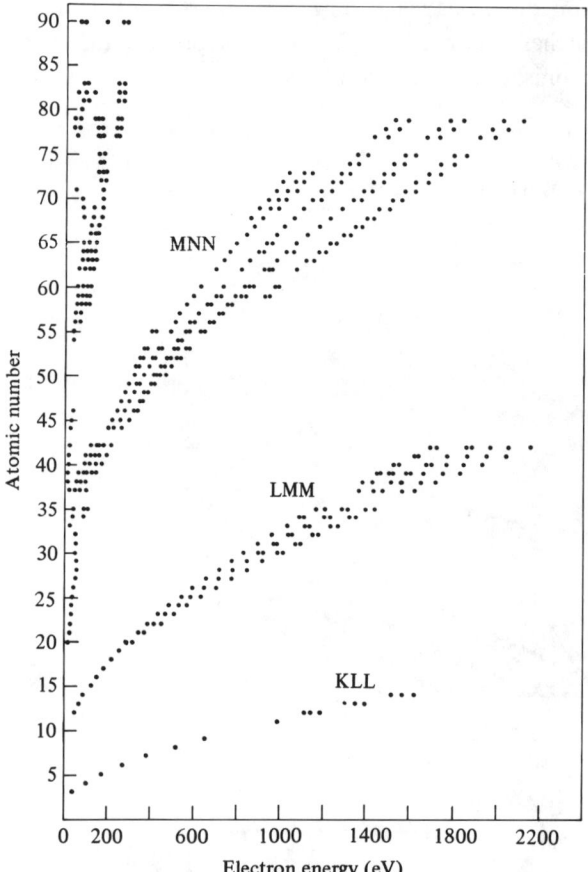

Fig. 4.3. Principal Auger electron energies, after Davis *et al.* (1976).

order to allow the whole spectrum to be recorded, the same general principles apply for higher primary energies. At the primary beam energy there is a sharp peak due to electrons that have been elastically scattered back out of the specimen, without energy loss. If the specimen is crystalline, these electrons will carry the diffraction information, which is exploited in techniques such as low energy electron diffraction and reflection high-energy electron diffraction. At slightly lower energies there are smaller peaks due to electrons that have suffered characteristic losses, for instance through the excitation of a plasmon. These peaks soon merge into a continuum of inelastically back-scattered primary electrons. The information contained in this region of the spectrum is exploited in the technique of low energy electron loss spectroscopy. At the other end of the spectrum is the large peak of low-energy secondary electrons. At a few hundred eV

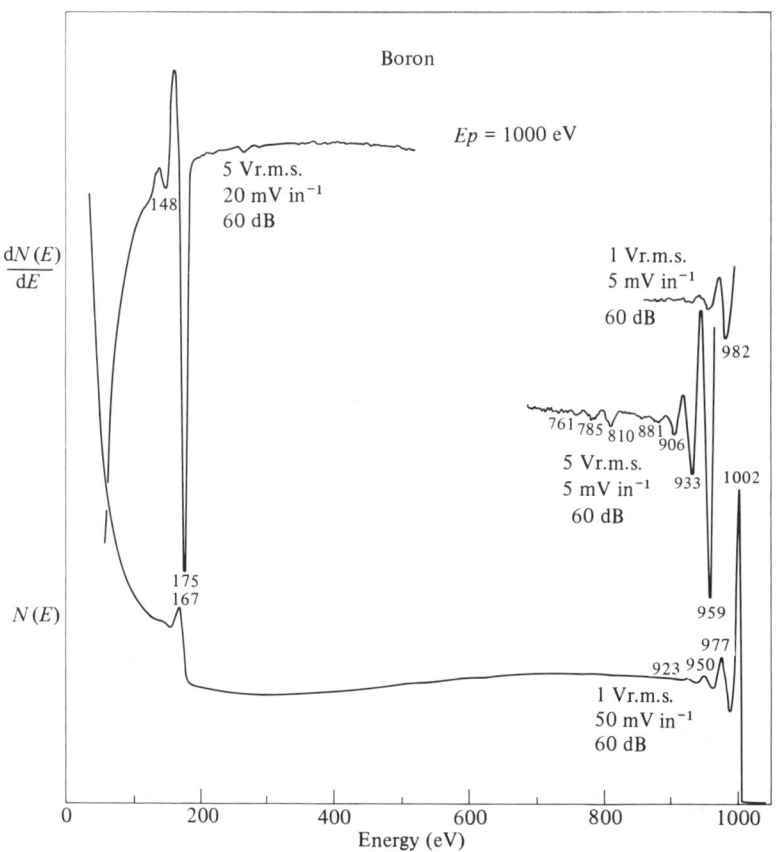

Fig. 4.4. Direct (lower) and differential spectrum from boron for a primary beam energy of 1 keV, recorded with a retarding field analyser.

there is a poorly defined cross-over point above which the distribution is dominated by back-scattered primary electrons and below which secondary electrons form the major component. This cross-over point moves to higher energies as the primary beam energy is increased. Superimposed on this distribution are the peaks due to Auger electrons which have their own inelastic tails due to Auger electrons excited some way below the surface that have lost energy passing through the solid before escaping from the surface.

Both the major components of the secondary electron spectrum, the back-scattered primaries and the low-energy secondaries may be used to form the image in a scanning electron microscope (SEM) although they carry rather different information. The secondary electrons, which are the ones most commonly used, are very sensitive to the state of the surface itself and to topography. In the poor vacuum of a conventional SEM the surface is covered by contamination of one sort or another which largely obscures any variations in emission due to the surface composition of the sample itself so that image contrast is largely determined by specimen topography. However, in the UHV conditions of an Auger instrument one often sees markedly different secondary electron contrast. Perhaps the most striking effect is the apparent darkening of the image obtained

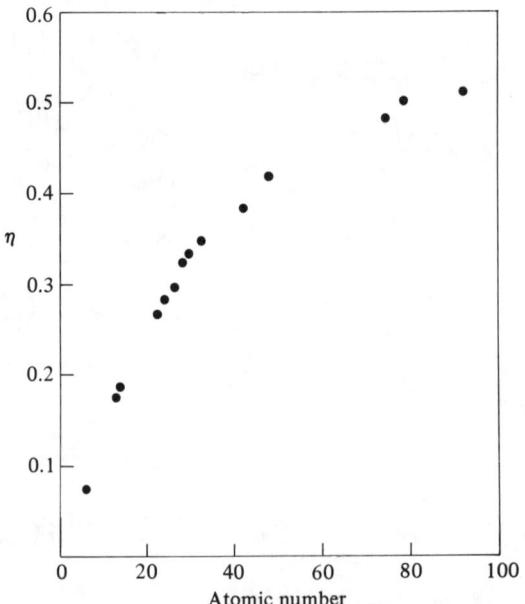

Fig. 4.5. Fraction of electrons back-scattered as a function of atomic number for normal beam incidence and a primary beam energy of 10 keV, Bishop (1965).

Auger electron spectroscopy

from a specimen just introduced into the system through an airlock. This is caused by the desorption of contaminants such as water from the surface by the electron beam. The back-scattered primaries are less sensitive to the immediate surface but depend strongly on the atomic number of the substrate. The variation of the back-scattering coefficient, η, with atomic number is shown in Fig. 4.5 (this is for electrons with energy greater than 50 eV which excludes most secondary electrons). An interesting property of η is that it is almost independent of primary beam energy. Images formed with the high-energy back-scattered electrons are sensitive to the mean atomic number of the specimen and are very useful in picking out different phases on a polished specimen. Unfortunately this mode of operation is not yet commonly available in UHV instruments.

2.3 Spatial resolution

The intensity of the Auger signal and the area of the surface from which it comes is determined by the near-surface ionisations produced by

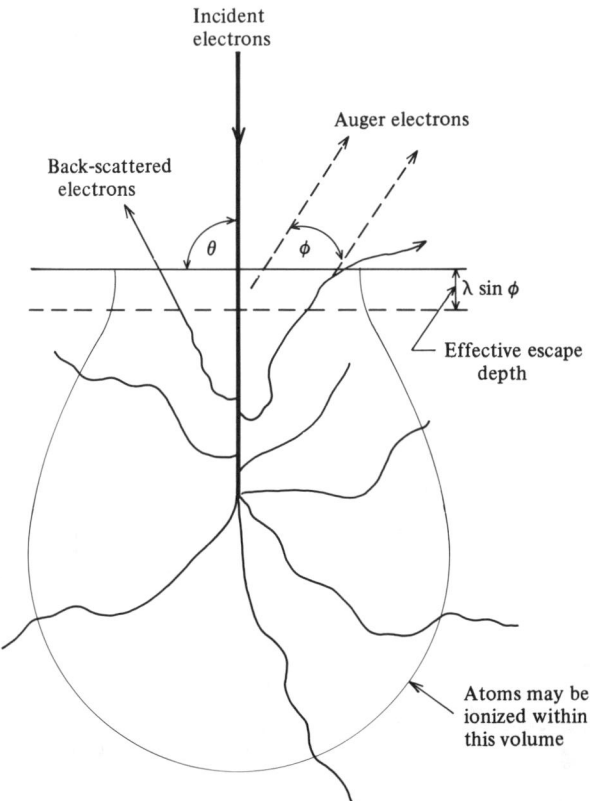

Fig. 4.6. Spatial distribution of Auger production.

the exciting electron beam (Fig. 4.6). The observed Auger signal I_A from an element A is determined by the expression

$$I_A = I_0 \cdot N_A \cdot T(E_A) \cdot \sigma(E_0)(1 + r(E_c, E_0)) \cdot \gamma \cdot \lambda(E_A) \sin\phi \cdot \operatorname{cosec}\theta, \quad (3)$$

where I_0 = incident beam intensity, E_0 = primary beam energy, E_A = Auger electron energy, E_c = critical ionisation potential, N_A = atom density of element A, $T(E_A)$ = detection efficiency of the spectrometer and detector, $\sigma(E_0)$ = ionisation cross-section for relevant shell of A, $r(E_c, E_0)$ = fractional contribution to total ionisation by flux of back-scattered electrons, γ = probability of ionisation giving rise to an Auger electron (including the effect of Coster–Kronig transitions), $\lambda(E_A)$ = inelastic mean free path for electrons of energy E_A, θ = angle of incidence of primary beam, ϕ = effective exit angle of Auger electrons entering spectrometer.

The parameter that gives AES its surface sensitivity is λ, often referred to as the escape depth. A compilation of experimental measurements by Seah & Dench (1979) is given in Fig. 4.7. The large scatter in the results reflects both experimental errors due to the difficulty in making such

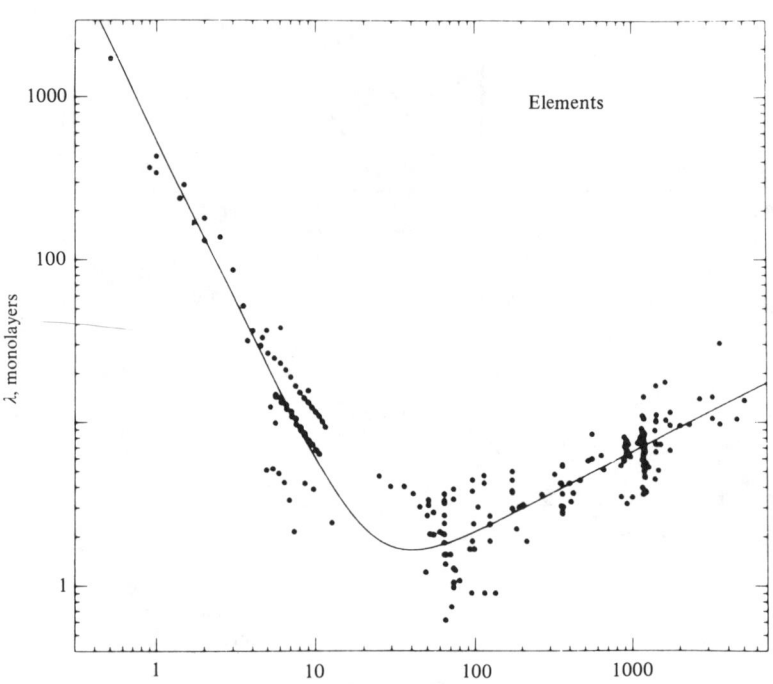

Fig. 4.7. Variation of elastic mean free path with energy, Seah and Dench (1979).

Auger electron spectroscopy 95

measurements and also some variation from element to element. For Auger peaks with energies in the region of 100 eV, λ is close to a minimum value of about two monolayers with the result that the Auger signal is dominated by atoms at the outer surface. Above 100 eV, the higher the energy of an Auger peak the greater λ, with the result that the signal becomes less surface specific. An upper energy limit to AES is usually set from 2.5 to 3.0 keV by which energy the escape depth has risen to ten monolayers and also the intensity of the Auger peaks has become too small to be of much practical use.

The spatial resolution of AES is determined by the area of the specimen surface from which Auger electrons originate. Those Auger electrons excited by the incident beam as it enters the specimen will leave the specimen at a distance from the point of incidence comparable to λ. The additional Auger electrons produced by the back-scattered electrons as they pass through the surface may leave at any distance from the impact point up to approaching the electron range. Thus for an infinitely fine electron probe the source distribution of Auger electrons will be a narrow peak a few nm wide surrounded by a much broader diffuse ring caused by the back-scattered electrons, a very similar situation to that for secondary electrons in the scanning electron microscope (SEM). One may therefore argue that the ultimate spatial resolution that may be achieved in AES is similar to that found in the SEM, that is a few nm. However, one must remember that resolution measurements in the SEM ignore the low spatial resolution contribution to the signal from the back-scattered component. In the majority of AES instruments where probe sizes are $\sim 0.5\,\mu$m or more the electron probe diameter will dominate the effective spatial resolution, but in more advanced instruments where probe sizes may be pushed down towards 10 nm or less the relative contributions of the incident and back-scattered electron flux must be considered carefully when interpreting Auger maps and linescans.

2.4 Choice of primary beam energy

Before moving on to describing Auger instrumentation it is useful to discuss the choice of primary beam energy E_0. The dependence of the Auger signal itself on E_0 is mainly through the energy dependence of the ionisation cross-section, σ, which rises from zero at the critical ionisation potential, E_c, to a maximum at about three times E_c and then falls off with increasing E_0. The effect of the back-scattering contribution to the signal is to broaden the maximum in the yield and to shift it to a somewhat higher energy. In practice we are not so much concerned with the absolute magnitude of the Auger signal as with choosing a primary beam energy that would give the best signal to noise for our measurements. As the noise is related to the background on which the peaks are superimposed the variation of background with energy must also be taken into account. Table 4.1 shows the results of experimental measurements made for a number of beam energies on Auger peaks over the range of energies used in AES (Bishop, 1984). In all cases the background shows a monotonic decrease with increasing energy while the higher energy Auger peaks show the expected maximum in intensity. The signal-to-noise is determined by the ratio of signal to the square root of the background which is shown in the lowest section of Table 4.1. If we were simply concerned with low energy peaks a primary beam energy of 2 keV would be perfectly adequate, but, if the higher energy peaks are to be excited efficiently, a beam energy of 5–10 keV is indicated. Above 10 keV any additional gain in sensitivity for the high-energy peaks is offset by a corresponding loss at the low-energy end of the spectrum.

Table 4.1. *Variation of Auger peak height and background with incident beam energy*

Beam energy keV	Cu 57 eV	Cu 912 eV	Ta 164 eV	Ta 1670 eV
2	7 500	3 150	7 500	—
5	5 050	7 500	5 400	1 950
10	3 000	6 100	3 950	5 150
20	1 600	4 150	2 400	3 800
30	1 300	3 350	1 350	3 500
	Corresponding background $(cs)^{-1}$			
2	60 000	93 000	46 500	—
5	31 500	24 000	29 000	60 000
10	18 000	10 500	18 500	23 500
20	10 200	5 200	10 500	11 300
30	7 700	3 550	8 000	8 000
	Peak background			
2	0.125	0.03	0.16	—
5	0.16	0.32	0.19	0.03
10	0.17	0.58	0.21	0.22
20	0.16	0.80	0.23	0.34
30	0.16	0.93	0.23	0.44
	Peak/(background)$^{\frac{1}{2}}$			
2	31	109	35	—
5	28	49	32	7.9
10	22	59	29	33.0
20	16	57	24	36.0
30	15	56	21	39.0

On the basis of signal-to-noise arguments we may conclude that a good working compromise for the primary beam energy lies in the range 5–10 keV and many commercial instruments are routinely used in this range of energies. There are two main considerations that may lead one to operate outside this range. These are spatial resolution and specimen charging. For a given probe size the maximum current is proportional to the beam brightness which in turn is proportional to beam energy; therefore if a particular spatial resolution is required it is easier to achieve it at a higher beam voltage. However, high-beam voltages have two drawbacks: first the electron range and hence the area from which the back-scattered component of the Auger signal comes is greater, and second more energy is deposited in the specimen which may give rise to undesirable heating effects.

Specimen charging is a major limitation to the use of AES on insulating specimens. If the total current reaching the specimen is greater than that leaving through back-scattering and secondary emission the specimen will charge until the incoming and outgoing electron fluxes are equal. In an extreme case the surface may charge to the full incident beam potential. The simplest way to overcome this problem is to increase the secondary emission by rotating the sample so the incident beam is at glancing incidence. This approach may not work or may not be practicable in the case of a rough sample. A second alternative is to reduce the primary beam energy. Fig. 4.8 shows schematically how the total secondary emission changes with energy. For most insulators there will be an energy range over which the secondary yield is greater than unity so that the

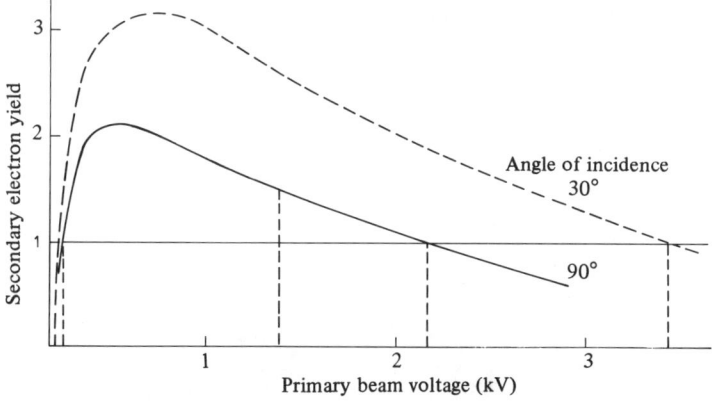

Fig. 4.8. Secondary electron yield curve as a function of primary beam energy and angle of incidence.

specimen will only charge a few volts positive before equilibrium is established and a reasonable Auger spectrum may be recorded, although the primary beam energy may be too low to excite many of the more energetic peaks efficiently. Although charging can be suppressed in the region of the specimen under the incident beam, surrounding areas of the specimen may be charged by scattered electrons. This additional charging may distort the recorded Auger spectrum but the effect can be minimised by reducing the exposed insulating area by evaporation of a conducting layer or by masking the specimen.

3   Instrumentation

The basic components of an AES system are shown schematically in Fig. 4.9, these are an electron gun, an electron spectrometer, an electron detector, to record secondary electron images, and an ion gun for specimen cleaning and depth profiling. The choice of the individual components will depend on the purpose of the system. If the aim is to provide a convenient monitor of surface cleanliness in a low-energy electron diffraction system it may be quite sufficient to provide a very simple electron gun and use the LEED optics as a retarding field analyser (RFA). Most early AES systems were of this type. On the other hand a dedicated scanning AES system will require a high-performance electron gun to give as fine a probe as possible, and a high-transmission electron spectrometer, together with good ion profiling and efficient specimen loading and processing facilities.

In the previous chapter the two types of electron spectrometer in common use, the CMA and CHA are described. Apart from use in conjunction with LEED the RFA is now rarely used because of its poor sensitivity. Until recently the most commonly used analyser for AES was the CMA because of its higher transmission whilst the CHA tended to

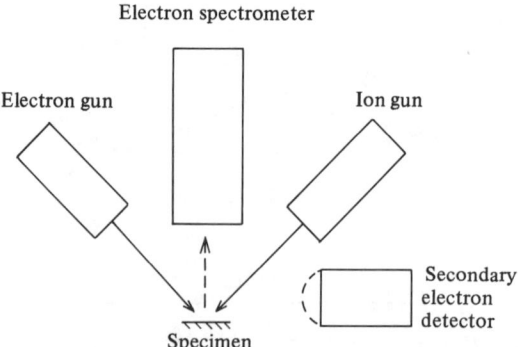

Fig. 4.9. Basic components of an AES system.

be used mainly in combined AES/XPS instruments. However, as the transmission of the modern CHA with an optimised input lens and multichannel detectors can approach or even exceed that of the CMA, the CHA is being increasingly used for AES.

3.1 The cylindrical mirror analyser

A typical CMA system is shown in Fig. 3.5, which illustrates a number of the attractive features of this type of spectrometer. The analyser has a very large acceptance angle comprising a full cone about the spectrometer axis. This gives both a high sensitivity and reduces the dependence of the signal on specimen topography. From the practical point of view the electron gun may be incorporated into the inner cylinder making a convenient compact unit. The physical constraints set on such an integral electron gun limits its performance and filament replacement is difficult. These problems are avoided in a modified form of the CMA shown in Fig. 4.10 where the analysed electrons are collected off axis; only the final probe forming lens is located in the inner cylinder with the electron source and condenser lenses moved back behind the analyser.

The large solid angle of the CMA does have some drawbacks as it severely restricts access to the sample. To gain full advantage of the CMA, the specimen surface should be normal to the analyser axis. However, for ion etching or if one wishes to use a non-integral electron gun the specimen must be rotated to give a reasonable incident angle thus losing some of the advantages of the CMA.

3.2 Concentric hemispherical analyser

This type of analyser was initially optimised for XPS but careful design of the input lens has greatly increased the effective solid angle so

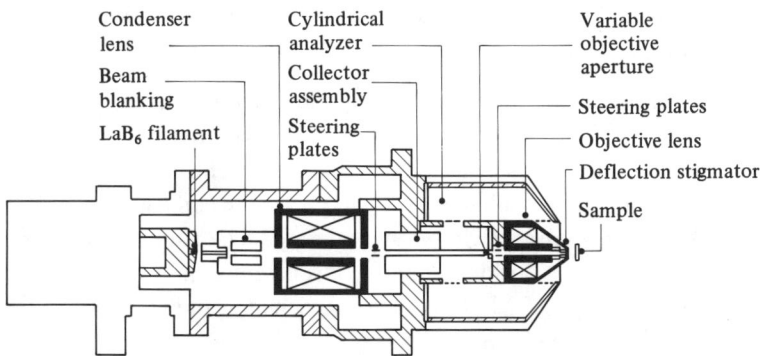

Fig. 4.10. CMA modified to incorporate a high-performance magnetic probe forming lens, Gerlach (1980).

that its transmission may now be comparable to that of a CMA. The main advantages the CHA has over the CMA is much better access to the sample and the ability to vary analyser resolution electrostatically without changing physical apertures. Although an arrangement similar to that shown in Fig. 4.10 using a CMA is very good for high spatial resolution AES, such a system requires a specially designed electron optical system and analyser. If AES together with other analytical techniques are to be combined with a high-resolution scanning electron microscope, a CHA is often chosen as it occupies less space near the specimen and is less sensitive to specimen position than a CMA.

3.3 Electron probe formation

The nature of the electron probe required in an AES instrument is determined by a number of considerations:
— the speed of analysis, which argues for a high beam current;
— the desired spatial resolution, which sets an upper limit to the probe current;
— beam-induced changes to the specimen, which sets a limit to current density.

The range of probe currents normally used in AES is between 10^{-9} and 5×10^{-6} A. The lowest current is used for high spatial resolution although the resolution achieved will depend on the quality of the probe forming system. The highest currents may be used to give speed and high sensitivity where spatial resolution is of little concern. However, such a high current carries a very great risk of beam induced charges to the specimen and should normally be avoided.

The electron optical system forming the probe has two critical components, the electron source and the probe forming lens. In most cases the electron source is thermionic but for the highest spatial resolution the brighter field emission source may be used. However, the cost and stability problems associated with field emission sources have limited their use to a few specialised systems. The commonly used thermionic sources are

— a strip filament, which has a long life but poor brightness. This source is used in situations where high spatial resolution is not essential and filament failure is very undesirable.
— a tungsten hairpin filament of the type used in conventional SEM. This is reasonably bright but has a limited life of about 100 hours.
— lanthanum hexaboride. This source is brighter than tungsten and has a longer life but is more expensive and a little more complicated to operate.

The electron lenses used to form the probe may be either magnetic or electrostatic. The best performance is obtained from magnetic lenses as they have low aberrations. On the other hand, it is much easier to fit an electrostatic lens into the UHV system required for AES. Submicron resolution can be readily obtained with a 10 keV electrostatic gun. The more complicated and expensive electromagnetic lenses are only used where an SEM resolution below 100 nm is required.

3.4 Specimen stage

In instruments with high spatial resolution the design of the specimen stage can become crucial. For resolutions down to about 1 μm the type of stage conventionally used in surface analysis equipment is acceptable but for higher resolution applications a more sophisticated stage design similar to the type used in the SEM must be adopted to limit vibration and specimen drift to acceptable levels.

3.5 Ion gun

An important use of AES is in depth profiling, for which an ion gun is required. Details of ion erosion are discussed in Chapter 2. For AES we need to be able to erode a small area of the specimen and record the variations in the Auger signal from the centre (or in some cases selected points within) this area. The current density must be such that the specimen is eroded sufficiently fast to allow a profile to be recorded within an acceptable time scale. It is usual to have a focussed ion beam rastered over a few mm square to give a flat-bottomed pit. Typical conditions are for a 1 μA 5 kV argon beam to be focussed into a probe of around 50 μm and rastered over 3 mm square, which will give an erosion rate of about 1000 Å per hour. Lining up the electron beam to coincide with the centre of the raster can present problems and the system should be designed to simplify this procedure.

In automatic depth profiling it is usual for the computer to be able to blank off both the ion gun and the electron gun. In the case of the ion gun it may be necessary to remove the ion beam from the sample during measurement of the Auger spectra if measurements of a number of elements at fixed depths is required. In addition the ion beam itself produces secondary electrons and even low-energy (< 100 eV) Auger peaks. It is therefore desirable to switch off the ion beam for measurements on low-energy peaks. If the electron beam is left on some types of sample (for example oxides such as silicon oxide) the area irradiated by both ions and electrons is eroded more rapidly than areas bombarded by ions alone. In such cases it is recommended that the electron beam should be either

4 Data recording

4.1 The direct and differential spectrum

As may be seen in Figs. 4.2 and 4.4, the Auger electrons appear as peaks on a smooth background of back-scattered elctrons. In the case of a clean surface the main peaks are readily seen and identified; however, smaller peaks and those due to minor impurities on a contaminated surface

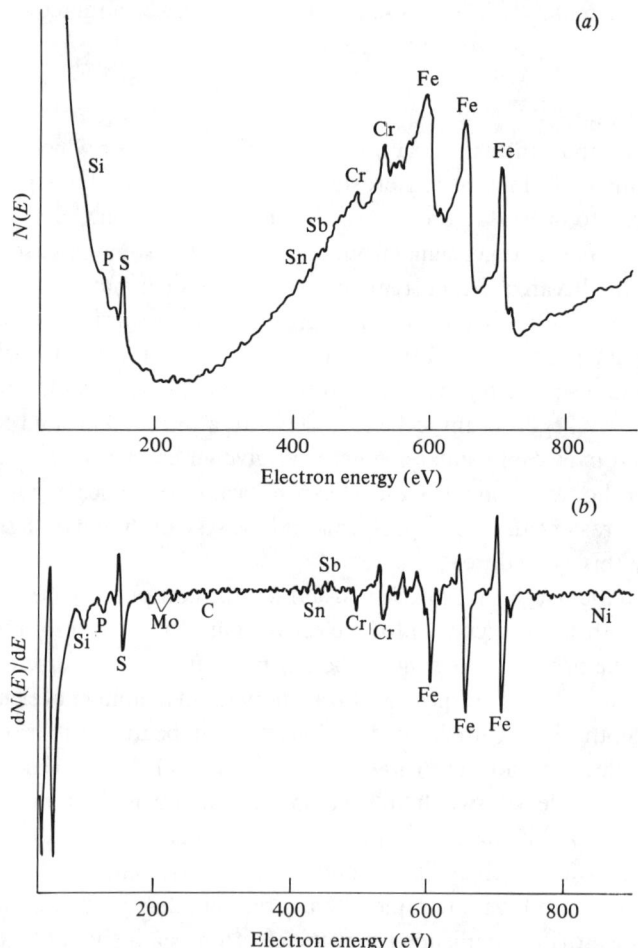

Fig. 4.11. Auger spectrum from a grain boundary on a fractured steel specimen, (a) direct $E \cdot N(E)$ spectrum, (b) differential spectrum.

are often difficult to pick out from the background. Because the background is usually sloping, even increasing the gain of the electron detection system and applying a zero offset is frequently not a great help. As a result it has been usual to record Auger spectra in a differential form. The spectrum in Fig. 4.4 was recorded in both direct and differential forms. In the differential mode it is easy to increase the system gains to reveal detail structure not visible in the direct spectrum. A more complicated spectrum from a steel sample is shown in Fig. 4.11a. Here there is an overlap of the L Auger peaks of the transition elements which makes it difficult to recognise what elements are present in the direct spectrum, but the sharper differential peaks reduce this interference and greatly simplify interpretation. Notice also that the low energy phosphorus, sulphur and iron M peaks are much more easily distinguished in the differential spectrum.

The most distinctive feature of an Auger peak in the differential spectrum is the negative going feature at the high-energy side of the peak. By convention, the minimum of this feature, which represents the point of maximum slope of the parent peak, is used to define the energy of the transition. In summary, the differential Auger spectrum has been found to have a number of practical advantages over the direct spectrum:

- away from the peaks the background is close to zero except at low energies so that amplifier gain can easily be changed;
- the differential peaks are sharper than the parent peaks reducing overlap problems and providing a reproducible energy reference point;
- the peak-to-peak height provides a convenient measure of peak intensity;
- in the CMA and CHA which use electron multipliers the output is at high potential, it is experimentally simpler to measure the high frequency differential rather than the DC direct signal.

These advantages are relevant to systems where data are recorded directly on an analogue chart recorder. Over the last few years it has become increasingly common to use a computer to control data acquisition and display. Although at first the computer systems simply mimicked the analogue recording systems they replaced, it is now becoming increasingly common to record data in the direct mode and differentiate numerically if so desired. This is particularly the case when working at high spatial resolution where the signal is low enough to use direct electron counting.

A typical analogue recording system is outlined in Fig. 3.5. An oscillator in the lock-in amplifier superimposes a sinusoidal modulation, typically

10–20 kHz, on the potential applied to the outer cylinder of the analyser. The AC component of the signal from the electron multiplier passes through a capacitor to decouple it from the multiplier HT and the magnitude at the modulation frequency is detected in the lock-in amplifier and fed to an XY recorder. The amplitude of the modulation is chosen to give a compromise between sensitivity and resolution. The variation in amplitude and shape of a two-component Auger peak with modulation voltage is shown in Fig. 4.12. At modulation voltages lower than the analyser resolution, the amplitude is proportional to modulation voltage but the peak shape is almost invariant. As the modulation is increased beyond the analyser resolution further increases in amplitude are matched by a progressive loss in spectral resolution. The peak amplitude reaches a maximum when the modulation is about equal to the total width of the Auger feature. The operator has to choose a modulation appropriate to his experiment. In the example given in Fig. 4.12 a modulation voltage of

Fig. 4.12. Variation of Ag MNN peak with modulation voltage.

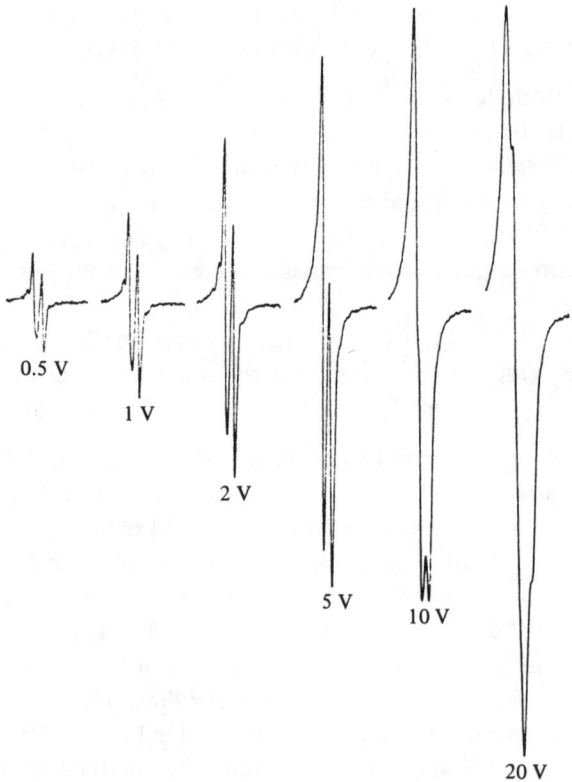

Auger electron spectroscopy 105

5 V is a reasonable compromise between resolution and sensitivity for routine applications.

The resolution of the CMA and the CHA in the Constant Retard Ratio (CRR) mode normally used in AES, is proportional to the energy of the analysed electrons. Thus when making a scan through a spectrum the optimum modulation voltage will change. A much higher modulation voltage is required for the high-energy peaks which also tend to be small and therefore need to be recorded efficiently. Although it is simplest to use a constant modulation voltage when recording a spectrum some instruments have the facility to make the modulation proportional to the analyser energy so that the ratio of modulation to analyser resolution is constant. This alternative mode of operation must be borne in mind when comparing Auger spectra from different instruments.

We have seen above that before recording a spectrum in the differential mode an important decision has to be made between sensitivity and resolution. Data recorded in the direct mode do not suffer from this

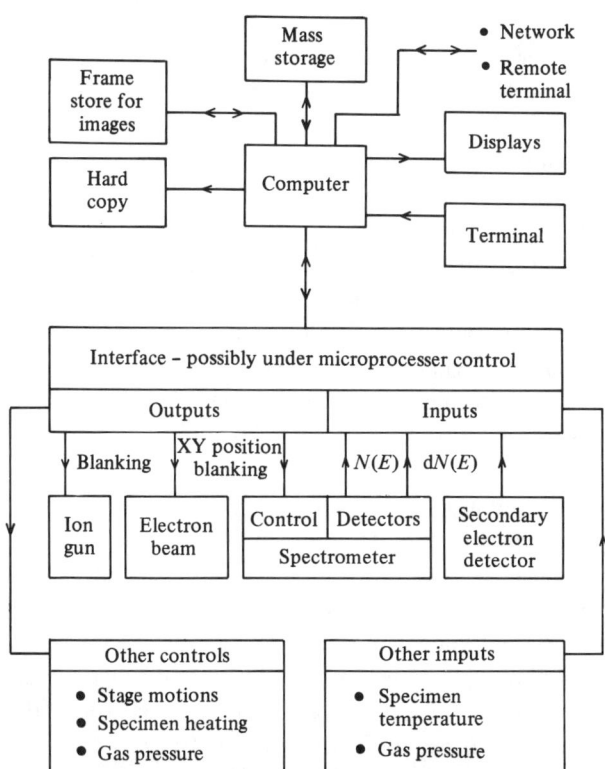

Fig. 4.13. Schematic diagram of computer control system.

restriction. The spectrum is recorded with the full resolution of the spectrometer, provided sufficient data points are recorded. By a suitable combination of differentiation and smoothing high-resolution or high-sensitivity differential spectra may be obtained at will from a single spectrum recorded in the direct mode.

Another advantage the direct spectrum has over the differential is that it retains any information carried in the background. It can be seen from Fig. 4.2 that the magnitude of the background increases with the atomic number of the sample and may therefore give an indication of the mean atomic number of the substrate. However, the depth of the surface layer from which electrons contributing to the background originate in the 100–1000 eV energy range has not yet been clearly established. The most important practical use of the background is in correcting the Auger signal for topography (Jansen et al., 1977). Although the measured Auger signal has a strong dependence on the angle of incidence of the primary beam the ratio of peak intensity to the background is almost independent of incident angle. This property is used in correcting Auger maps or line profiles for the effects of specimen topography.

4.2   Computer control and data system

The majority of AES instruments now operate under some degree of computer control. A schematic diagram of a comprehensive control system is shown in Fig. 4.13, although a given system is unlikely to have all these features implemented. A computer control system has two main functions, to acquire and store data efficiently giving the operator real time information on the progress of the measurements, and to provide processing and display facilities for subsequent data evaluation. As data evaluation can often be a lengthy process it is convenient to be able to run the instrument while earlier data are being evaluated, either through some time sharing arrangement or on a separate data processing system.

5   Quantitative aspects of AES
5.1   *Quantitative analysis*

A vitally important aspect of any analytical technique is how to relate the observed peak intensities to the actual composition of the specimen. The X-ray analogue of AES, electron probe X-ray microanalysis, is capable of giving a good (better than 1%) quantitative measure of the average concentrations of a wide range of elements within the volume excited by the electron beam. Quantitative analysis in AES, or XPS for that matter, is unfortunately not so straightforward because the observed signal from an element is strongly dependent not only on its average concentration but also on how it is distributed within the first few atomic layers of the surface. The sensitivity to Auger electrons in the outer surface layer is greater than that to electrons originating from subsurface layers because of strong inelastic scattering. Thus any quantitative analysis in AES carries with it some assumptions about the distribution of the component elements in the surface and near-surface layers, consequently the accuracy of the results will depend on how good these assumptions are. In spite of this reservation it is important to have a quantitative scheme routinely available to give the user some idea of the concentrations involved, and to allow comparisons to be made between samples. The approach that is usually adopted is to assume that the composition of the sample in the near surface region is homogeneous.

A good first approximation to the atomic concentration C_A of an element A in a sample is given by

$$C_A = \frac{I_A/I_A^\infty}{\sum_i I_i/I_i^\infty}, \qquad (4)$$

where I_A is the intensity of the Auger signal from the specimen and I_A^∞ is that from a pure standard recorded under identical conditions. The summation is for the corresponding intensity ratios from all other elements present in the sample. It is only necessary to know the relative values of I_A^∞ and the other I_i^∞ because of the form of equation (4) and handbooks of Auger spectroscopy give relative intensity factors that may be used under defined experimental conditions.

Equation (4) gives a reasonable estimate of composition but ignores a number of matrix effects that may lead to a systematic error of up to 50% in extreme cases. In many situations, particularly for technological samples where the state of the surface is not well known, such a semiquantitative analysis is quite acceptable. An improved analysis may be obtained by including a set of matrix correction factors that allow for atomic density, electron back-scattering and variations in escape depth. These corrections are described in detail in the handbooks and in Briggs & Seah (1983).

5.2   *Measurement of peak intensities*

The conventional and most convenient measure of Auger peak intensity is the peak-to-peak height in the differential spectrum. Even when recording spectra in the direct mode on a computer system, differentiating and measuring the peak-to-peak height is a good way of determining peak intensities. More sophisticated peak area measurements and peak resolution techniques similar to those employed in XPS may be used when chemical shifts or peak interferences make simple peak-to-peak height measurements impracticable.

In many AES measurements spatial variations in composition are explored either as depth profiles, line scans or images. For these measurements it is necessary to measure data as rapidly and as efficiently as possible. Early AES analogue systems used a multiplexer that scanned cyclically through a few volts about nominated peaks and recorded the peak-to-peak height for each region on a chart recorder as the specimen was eroded by ion bombardment to give a depth profile. A similar approach is often used with computer controlled systems. The main problem with this method is that an automatically determined peak-to-peak height is systematically too high unless some correction is made for the noise amplitude (Fig. 4.14(a)). On the other hand, any small peak shifts that may occur are accommodated in this approach.

Auger mapping is normally carried out in the direct counting mode as low primary beam currents are necessary in order to achieve good spatial resolution. Two scans are made for each line of the image with the spectrometer set first to the peak and then to the background just above the peak (Fig. 4.14(b)). The image constructed from the background intensity (B) is similar to the secondary electron image and reflects a combination of topography and the mean atomic number of the surface. The image from the difference ($P-B$) shows the variation of Auger intensity across the imaged area but the normalised intensity ($P-B$)/B gives a better representation of variation in elemental concentration as topographical effects are largely removed. For lower energy peaks, where there is an

Fig. 4.14. Peak height measurement, (a) differential peak, h true peak height, $h+n$ measured by automatic peak to peak system, (b) direct peak.

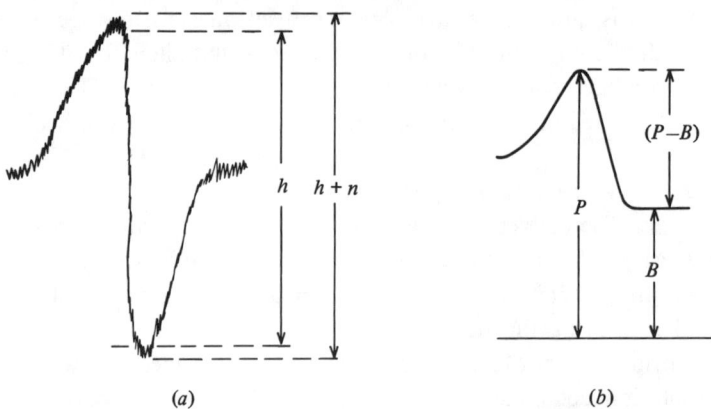

Auger electron spectroscopy

appreciable curvature of the background under the peak, two background measurements on either side of the peak are required.

5.3 Crystallographic effects

The intensity of Auger peaks can be strongly influenced by the crystalline nature of the sample (Bishop et al., 1984). The crystalline order of the specimen may affect the Auger signal through three mechanisms:

(a) differences in the composition or density of atomic planes parallel to the surface;
(b) anisotropic emission of Auger electrons, either through anisotropy in the emission process itself or by subsequent diffraction of Auger electrons as they leave the specimen;
(c) effects related to the diffraction of the exciting electron beam, so-called channelling effects.

The first mechanism is observed for layer compounds such as MoS_2 where the composition of the atomic planes alternates between the two constituent elements. The Auger signal from the element in the surface plane is relatively high and that from the other element low compared to a comparable amorphous compound. Anisotropic emission effects can be large, particularly for low energy Auger peaks. However, the high input solid angle of most spectrometers tend to smear out these effects in practice. The most striking crystallographic effects are produced by electron channelling. Fig. 4.15 shows an example of the potential magnitude of the effect for a single crystal of Al covered with a superficial oxide layer, formed

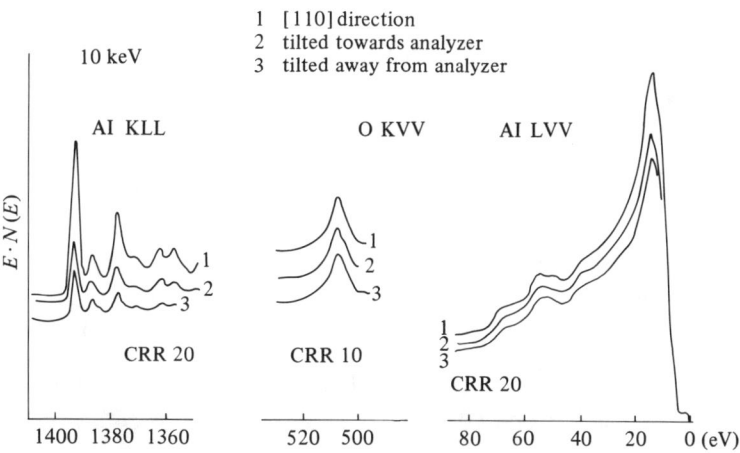

Fig. 4.15. Crystallographic contrast on an Al single crystal, Bishop et al. (1984).

by exposure of the clean surface to oxygen at room temperature. The three spectra were obtained first with the incident electron beam directed in the 110 direction and then with the specimen rotated $\pm 5°$ from this direction. The oxygen signal and the low-energy Al LVV peak show little or no dependence on the incident beam direction but the intensity of the Al KLL peak varies by a factor of almost 3 as the incident angle is rotated through the 110 direction.

In practical AES work these effects are not frequently met because the surface is not sufficiently well ordered. However, the possibility of crystalline contrast affecting quantitative measurements must always be considered, particularly in the case of intergranular fracture and of samples

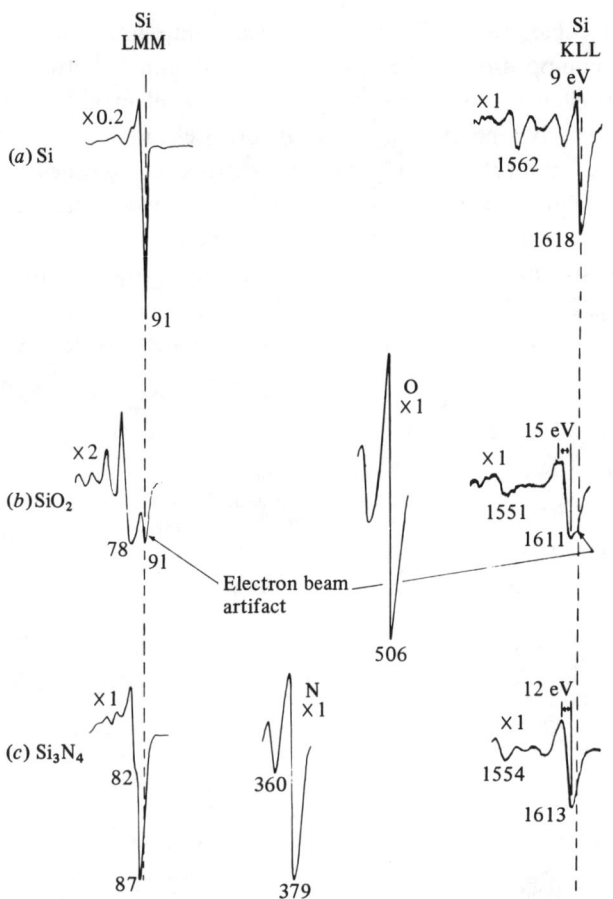

Fig. 4.16. Chemical shifts in Si KLL and LMM peaks, Holloway (1976), courtesy North Holland Physics Publishing.

that have been heated in vacuum so that the surface region may have become well ordered.

6 Chemical effects

The primary use of AES peaks is to identify the elements present on the surface but many peaks, particularly those involving valence electrons also carry information about the chemical state of the surface. Usually XPS is the preferred technique for investigating the chemistry of the surface as photoelectron peaks are sharper and chemical effects are more easily interpreted than is the case for the corresponding Auger peaks.

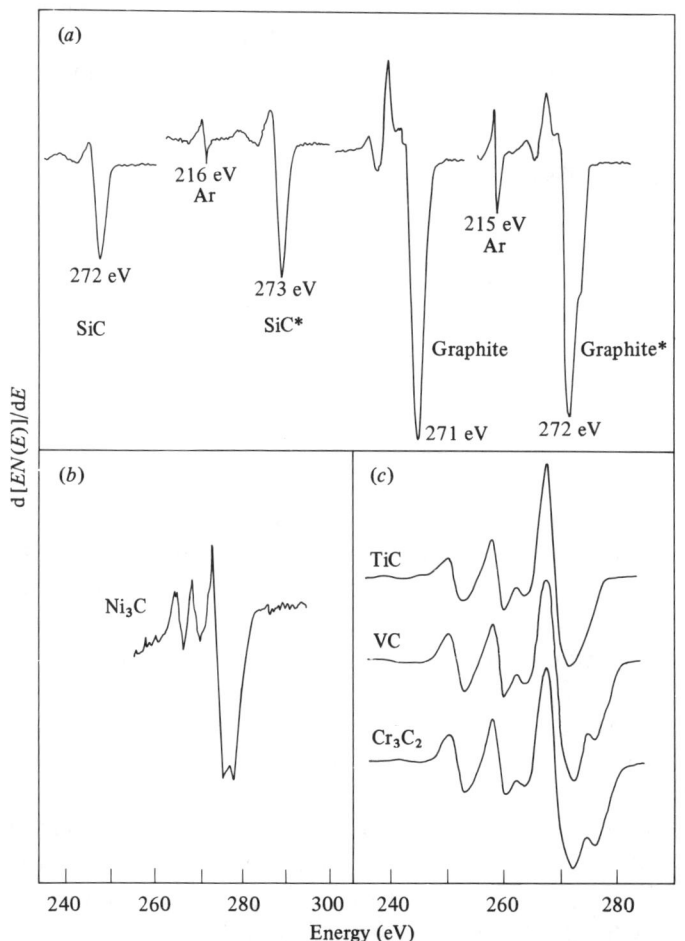

Fig. 4.17. Chemical dependence of the carbon KVV Auger peak, * indicates ion bombardment of specimen, Rivière (1983).

There are, however, numerous occasions when useful chemical information can be deduced from Auger spectra. The chemical effect may appear as a simple peak shift, a change in peak shape or, as is frequently the case, as both. A good example of a peak shift is shown by silicon (Fig. 4.16). Both the KLL and LMM peaks show shifts in position between pure silicon and the oxide or the nitride; for the LMM peak there is a shift of 12 eV between silicon and silicon dioxide. There is also a change in peak shape and width. These chemical effects severely complicate quantitative measurements involving peak-to-peak heights and peaks showing such effects should be avoided if at all possible for quantitative AES unless special steps are taken to correct for them.

Carbon shows only small chemical shifts but the peak shape is very characteristic of the particular carbon compound responsible for the peak. Fig. 4.17 shows a variety of carbon peaks demonstrating this effect. The shape of the carbon peak is of considerable practical value for instance when studying metal surfaces as it is easy to distinguish between carbon contamination and carbon as carbide.

Fig. 4.16 incidentally illustrates a very important problem in AES, i.e. beam-induced effects. Many compounds, particularly oxides, are readily decomposed under electron irradiation. The Auger spectra from silicon oxide in Fig. 4.16(b) show distinct peaks in the elemental position because oxygen has been lost from the surface under the action of the electron beam. In many cases the susceptibility of the specimen to beam damage is the factor limiting spatial resolution. The current density must be limited so that the specimen damage is acceptable in the time taken to make the measurement.

7   Practical analysis

AES may be used in a number of ways to solve materials problems. In this section we will consider various modes of operation for AES with practical examples. AES investigations can normally be divided into two classes, those where the surface itself is of interest and those where variations of composition with depth is required. The simplest investigation would involve a series of point analyses from areas of interest selected from the SEM image. A light ion bombardment may be used in conjunction with such measurements to remove any superficial contamination and to establish which elements are present only at the outer surface and which persist into the near-surface region. A more detailed study may involve the identification of elements of interest and then the determination of their spatial distribution over the surface using a line scan or mapping procedure. If a depth profile is required the elements of interest must first

Auger electron spectroscopy 113

be determined and then their Auger intensities recorded as the surface is eroded. If it is necessary to follow the composition of the specimen over about 2 μm into the surface sputter depth profiling becomes a lengthy procedure and suffers from loss of depth resolution because of uneven sputtering effects discussed in Chapter 2. For such samples it is worth considering producing a taper section so that the subsurface region is exposed for analysis.

7.1 Point analysis

Almost invariably the first step in any AES investigation is to record wide scan spectra from various regions of the sample to identify what elements are present. For many investigations this is all that is required as the materials problem is adequately explained. A good example of such an investigation is given by Lowry & Hogrefe (1980) who were investigating poor bonding to an aluminium pad following a dry plasma etching process. The spectra, (a) and (b), in Fig. 4.18 were recorded from the contaminated region of a bond pad and from an adjacent area. The contaminated region showed significant peaks from C, N and F while the

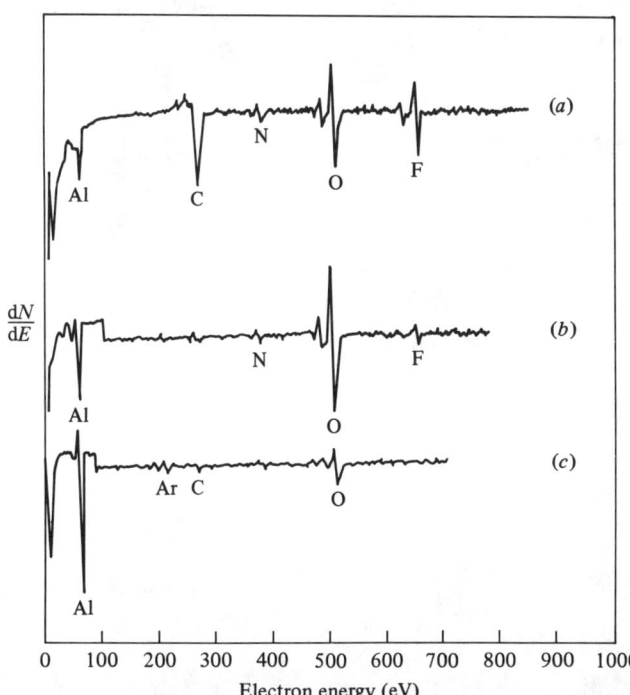

Fig. 4.18. AES characterisation of contaminated bond pad.

Fig. 4.19. Grain boundary topography associated with intergranular facets showing different segregation behaviour, (a) group of facets identified by segregation type, (b) type A facet with associated precipitates, (c) type B facet showing S segregation, predominantly smooth, Wall (1986).

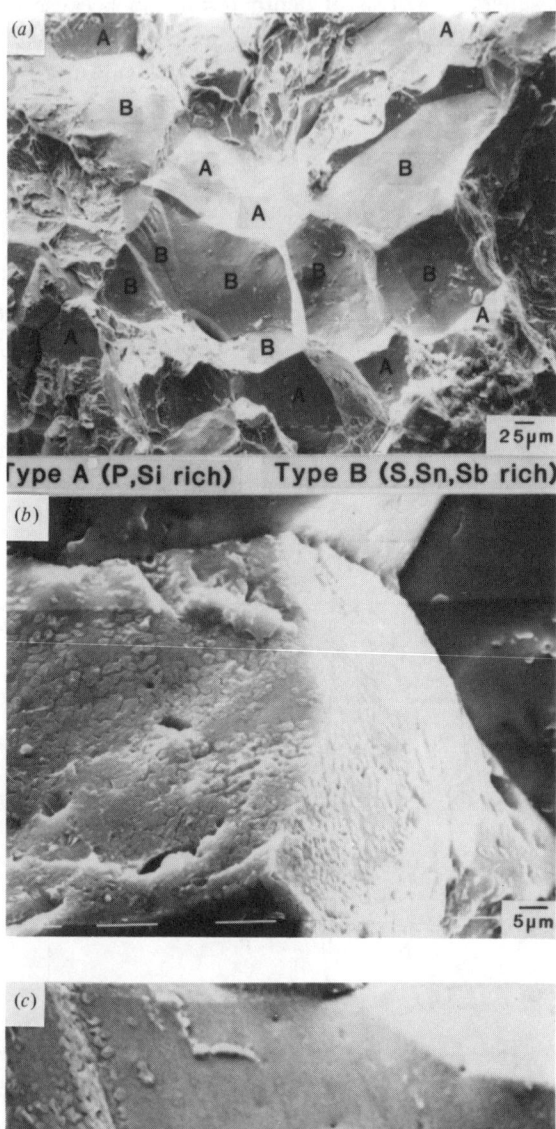

adjacent region showed only small concentrations of these elements. After a light ion bombardment to remove 10 nm from the surface spectrum III was recorded from the contaminated area showing that the contaminant layer was less than 10 nm thick.

Embrittled metals often fail in an intergranular manner. An important contribution of AES to the study of this phenomenon is to determine the composition of grain boundaries on the fracture surface. As atmospheric contamination would obscure the surface of interest specimens must be prepared by fracturing test pieces within the AES system. Fig. 4.19 shows the scanning electron image of the fracture face of a commercial 9Cr 1Mo steel obtained in our laboratory (Wall, 1986). At high magnification the intergranular facets show two different types of topography. (These images

Fig. 4.20. Representative spectra from (*a*) type A facets, (*b*) type B facets.

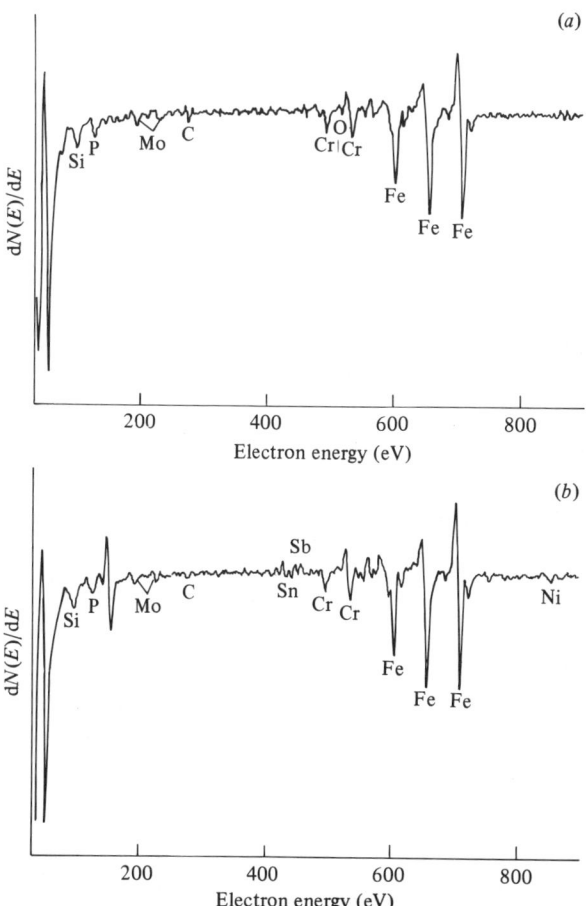

Table 4.2. *Auger analysis depth profile of type B (S, Sn, Sb) grain boundary facets in steel CA1 (Sandvik HT7 9Cr 1Mo Q + T). The surface layers are removed in situ by argon ion bombardment: 2 μA min cm^{-2} corresponds to the removal of approximately 1 monolayer from the grain boundary*

Ageing treatment	Ion bombardment μA min cm^{-2}	Monolayers removed (approx.)	Surface composition atomic %												
			Fe	Cr	Mo	P	Si	S	Sn	Sb	N	Ni	Cu	Mn	C	O
5000h at 500°C	0	0	52.1	11.4	4.8	2.5	1.6	7.8	1.4	2.3	2.2	2.3	2.0	2.7	6.0	1.1
	1.1	0.5	53.9	11.3	5.2	3.0	3.1	4.6	0.5	1.4	4.4	2.0	1.5	0.8	7.5	0.7
	2.2	1.0	54.2	10.9	5.3	3.2	3.5	3.4	0.6	1.3	3.8	2.1	1.8	1.0	7.7	1.2
	4.5	2.0	58.7	11.3	7.5	2.8	5.3	2.0	—	—	3.4	—	—	—	8.0	1.1
	12.0	6.0	58.6	11.1	7.9	3.1	5.4	0.9	—	—	4.1	—	—	—	7.5	1.5
	32.0	16.0	60.0	11.9	8.7	2.8	6.5	0.6	—	—	2.7	—	—	—	5.9	0.8
10 000h at 500°C	0	0	45.9	12.0	3.5	1.7	0.9	7.8	2.2	4.2	4.2	3.5	3.0	6.6	2.5	2.0
	4	2	54.2	12.3	5.9	1.4	3.9	2.5	1.2	1.4	4.2	1.5	3.1	2.0	3.3	3.1
	10.6	5	56.9	11.7	6.0	1.2	5.1	1.0	—	—	5.4	2.0	2.8	—	4.5	3.4
	17.6	9	58.0	11.5	6.9	1.0	5.1	—	—	—	4.0	2.6	2.6	—	4.3	4.0
	45.8	23	61.3	12.4	7.6	1.1	4.4	—	—	—	3.8	2.0	—	—	4.6	2.8
	88.0	44	63.6	13.0	7.5	1.0	2.8	—	—	—	4.9	—	—	—	4.5	2.7
	172	86	67.7	12.6	7.2	1.1	3.3	—	—	—	2.6	—	—	—	2.7	2.8

Auger electron spectroscopy

were obtained from a conventional SEM as the resolution in the Auger instrument was not adequate to show up the fine structure on the fracture faces clearly.) Representative AES spectra from the two types of facet are shown in Fig. 4.20. The type B facets show a strong S-peak together with Sn and Sb peaks which are not present in the spectra from type A facets. A semiquantitative analysis of the type B facets, of specimens subjected to different heat treatments, as fractured and after a sequence of ion bombardments is shown in Table 4.2. The elements that have segregated to the surface such as S, Sn, Sb and Mn are rapidly removed by ion bombardment. P, which in many instances will segregate to grain boundaries, is unaffected by ion bombardment showing that in this case it is incorporated in precipitates formed at the grain boundaries.

7.2 Depth profiling

Depth profiling is one of the most important applications of AES as it provides a convenient way of analysing the composition of thin surface layers. Table 4.2 shows an example of its use in surface or grain boundary segregation. The technique is particularly suited to many microelectronics problems. A typical example of its use (Lindfors, 1979) is shown in Fig. 4.21. A layer of TiW had been deposited on a silicon substrate to improve bonding between an outer gold layer and the substrate. The depth profile however revealed that gold had diffused through the WTi layer and accumulated at the silicon interface. The presence of this gold at the silicon surface would lead to poor adhesion and defeat the object of the TiW layer.

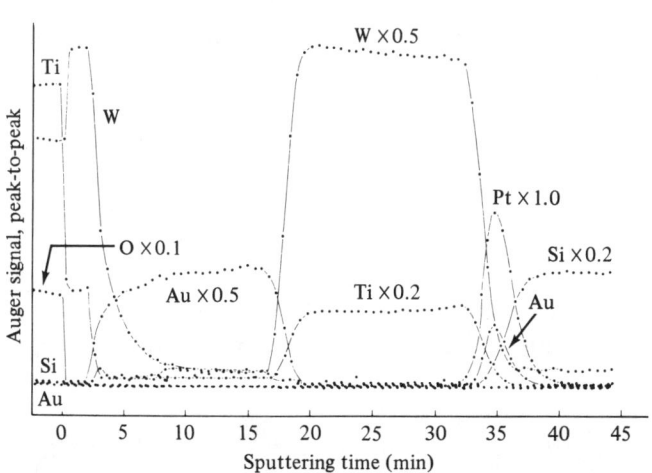

Fig. 4.21. Depth profile through a multilayer device (courtesy Plenum Press).

The data in Fig. 4.21 are recorded as raw peak-to-peak heights plotted directly against sputtering time. A more sophisticated data processing system would allow atomic concentrations to be plotted for each depth, either by using relative sensitivity factors or applying more complex matrix correction formuli. One point to watch in evaluating such profiles is that artefacts may appear at interfaces due to the time necessary to build up equilibrium sputtering conditions and to changes in the back-scattering contribution to the signal, where there is a significant change in mean atomic number between adjacent layers.

7.3 Taper sections

For information beyond about $2\,\mu$m below the surface, sputter depth profiling is slow and involves a loss resolution. An alternative approach is to produce a taper section thus exposing the subsurface regions that can be analysed after cleaning off superficial contamination by ion bombardment. The taper can be produced either by conventional lapping or by a ball-cratering technique (Walls *et al.*, 1979). A taper angle between 0.1 and 1° is required if useful depth resolution is to be achieved. The effective depth resolution, Δz, for a taper angle, α, for an electron probe of diameter, b, assuming a perfect polish is

$$\Delta z = b \tan \alpha.$$

Mounting and polishing samples by the lapping technique requires considerable skill and is very time consuming. It is, however, capable of giving very high-quality finishes and can cope with fragile surface layers that may suffer from spalling when subjected to ball cratering.

Ball cratering is a quicker and self-calibrating technique that has been developed over the past few years. In its original form a steel ball coated with diamond paste is rotated in contact with the specimen to form a well-defined spherical crater in the specimen surface. A more easily controlled crater can be produced by the dimpling polishing machine originally developed for the preparation of transmission microscope specimens. In this case the crater is formed by rotating the specimen in contact with the edge of a rotating polishing wheel.

By measuring the diameter of the crater, D, and knowing the radius, R, of the ball or polishing wheel it is a simple matter to calculate the depth, z, corresponding to a radial distance, x, from the edge of the crater, i.e.

$$z = \frac{x}{2R}(D - x), \tag{5}$$

Fig. 4.22. Crater profiling, (a) schematic diagram of crater, (b) ball crater in a chrome coating, Walls *et al.* (1979).

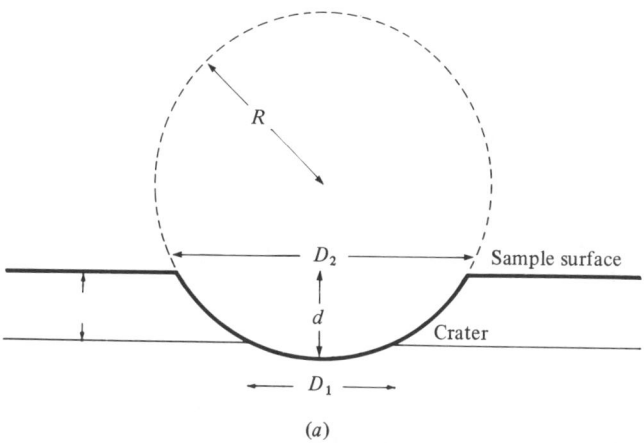

(a)

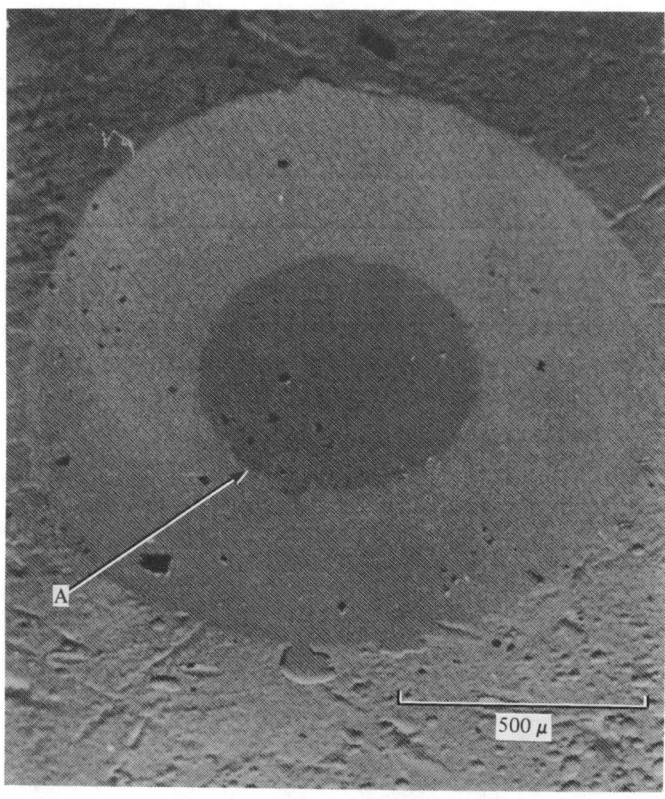

(b)

and that

$$\Delta z = \frac{b}{2R}(D - 2x) + m, \qquad (6)$$

where m is a factor determined by the quality of the polish.

On many specimens it can be difficult to define precisely the edge of the crater. This problem can be overcome by coating the specimen with a layer of gold or some other convenient material. The gold interface then defines the position of the specimen surface.

An example of the appearance of a ball crater is shown in Fig. 4.22, where a crater has been made through a hard chrome coating (6 μm thick) on a mild steel substrate. An example of the use of ball cratering to produce a compositional depth profile through an electrodeposited zinc coating on steel is shown in Fig. 4.23. The surface chromate treatment that had been applied to improve paint adhesion can be seen to have penetrated up to 5 μm into the surface. The Auger data for such a line scan can be acquired by positioning the electron probe manually but more usually data would be collected under computer control in exactly the same way as for a depth profile with the exception that the probe is moved rather than the ion gun switched on.

Ion depth profiling uses a rastered ion probe to produce a well-defined pit in the specimen, the edge of which, incidentally, is a shallow taper profile, although its shape is not as well defined as in the case of a mechanically produced profile. This pit edge taper does, however, offer the opportunity of examining an interface in more detail than may have been possible during the depth profile itself. Fig. 4.24(a) is a conventional depth profile through an SiO$_2$ layer on InP which shows a sharp interface. Higher depth resolution was obtained by taking a line scan across the SiO$_2$/InP interface on the pit edge. The taper angle was measured by

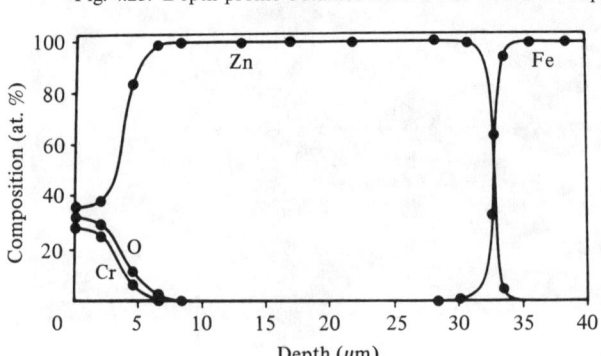

Fig. 4.23. Depth profile obtained from a ball cratered sample.

Fig. 4.24. (a) Conventional sputter depth profile through an SiO$_2$ layer on InP. (b) Line profile across interface at the crater edge. The 25 μm corresponds to a depth of 50 Å.

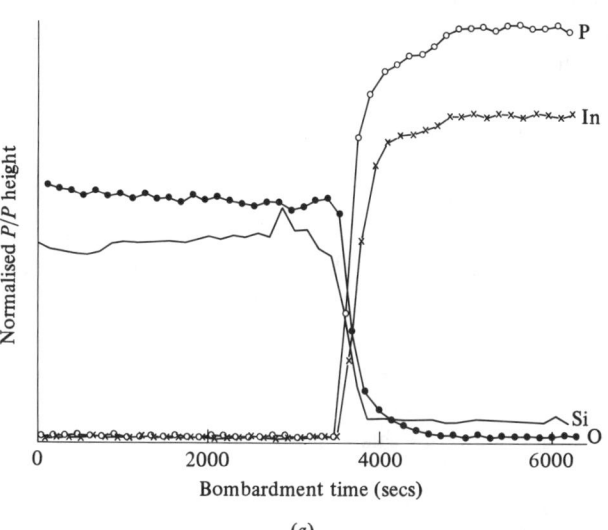

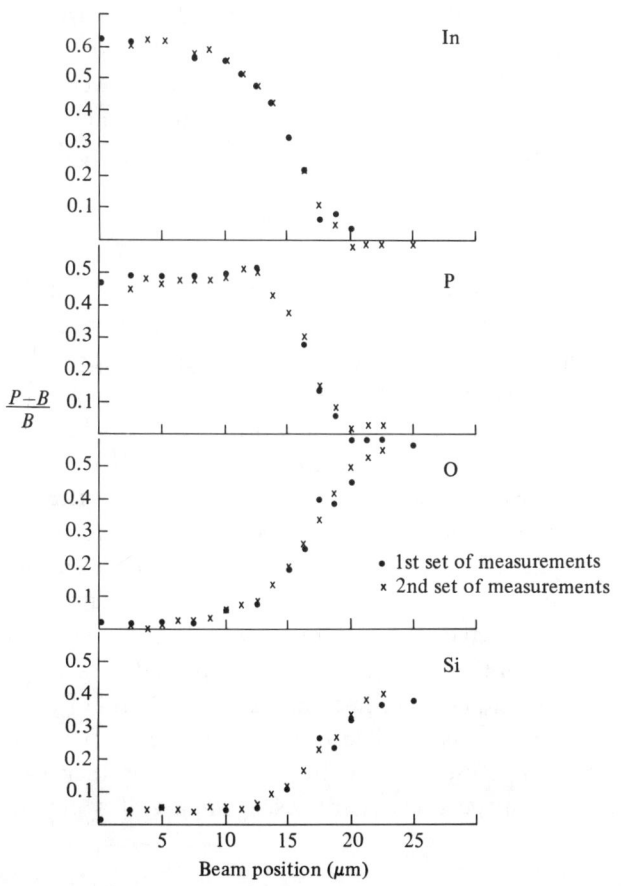

taking an EPMA line scan for oxygen. Knowing the total film thickness the taper gradient was measured $2\,\text{Å}/\mu\text{m}$ across the sample. Fig. 4.24(b) shows an interesting example of the type of effect that can be met at an interface. One would expect the In and P to fall off together if the interface were simple, but the In starts to fall off first while the P shows a maximum at the interface. There are two possible explanations of this effect. The obvious one is that the native oxide on the InP, that would remain at the interface, is P rich but this would have involved a chemical shift in the P peak that was not observed. An alternative explanation is that P is preferentially sputtered from the surface of InP so that the P signal from the InP is less than would be expected. However, close to the interface, where the InP has just been exposed, or indeed is still partially covered by SiO_2, the equilibrium depletion of surface P has not been established leading to a higher P signal in this region.

7.4 Auger images

The possibility of using the Auger signal to produce compositional maps of the surface similar to the X-ray maps produced in EPMA was realised relatively early in the development of AES (MacDonald & Waldrop, 1971). In practice, imaging was found to be of limited use because images took a long time to record and topological effects made interpretation difficult. The latter problem has been largely overcome by normalising the peak intensity to the background, as described in section 5.2. On a modern instrument it may still take 5–10 min to record the image for a relatively small Auger peak, but the ability to store the image in a computer with image-processing capabilities, allowing for data smoothing and the choice of optimum contrast conditions, means that good-quality images can be obtained much more easily than on older analogue instruments.

Because Auger peaks appear on a high background, Auger images are inherently noisy, so that images with more than two or three useful grey levels are rarely obtained. If more quantitative information is required a line scan, where more time can be spent on each data point, is preferable. Auger maps are only part of an investigation. Before any map is recorded a number of point analyses will have been made to establish what elements are present. From these analyses and the SEM image, the operator will have some idea of the elemental distribution over the surface. Auger maps can then be used to confirm these ideas. For quantitative data, the maps can be used to select the best places for line scans across the field of view. Even though in many cases it is possible to solve a materials problem without a map, it can still be useful to record Auger maps together with the corresponding SEM image because the visual impact of a picture

Auger electron spectroscopy 123

can greatly simplify the communication of results, particularly to non-specialists.

As an example we can return to the fractured steel sample shown in Fig. 4.9. A sulphur map from a facet showing both types of topography is shown in Fig. 4.25(c) (dark contrast signifies high S). The normalised $(P-B)/B$ signal has been used to produce this map. Fig. 4.25(b) is the topographical image produced from the background to the S signal while Fig. 4.25(a) is a conventional scanning electron micrograph of the same facet. Because of time considerations the Auger image was only recorded

Fig 4.25. Sulphur map for grain boundary showing mixed topography. (a) SEM image, (b) secondary electron image constructed from background B, (c) normalised S image $(P-B)/B$ (dark contrast is high sulphur).

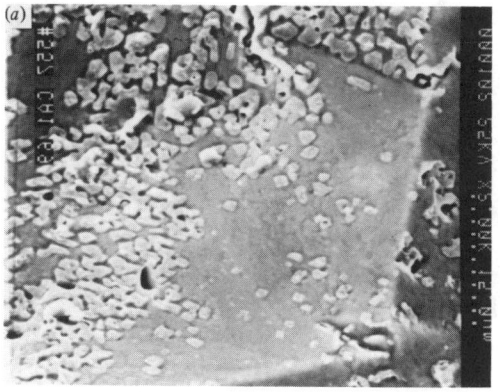

Grain boundary

Secondary electrons

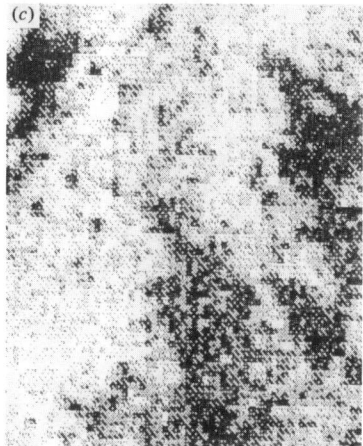

Sulphur
$(P-B/B)$

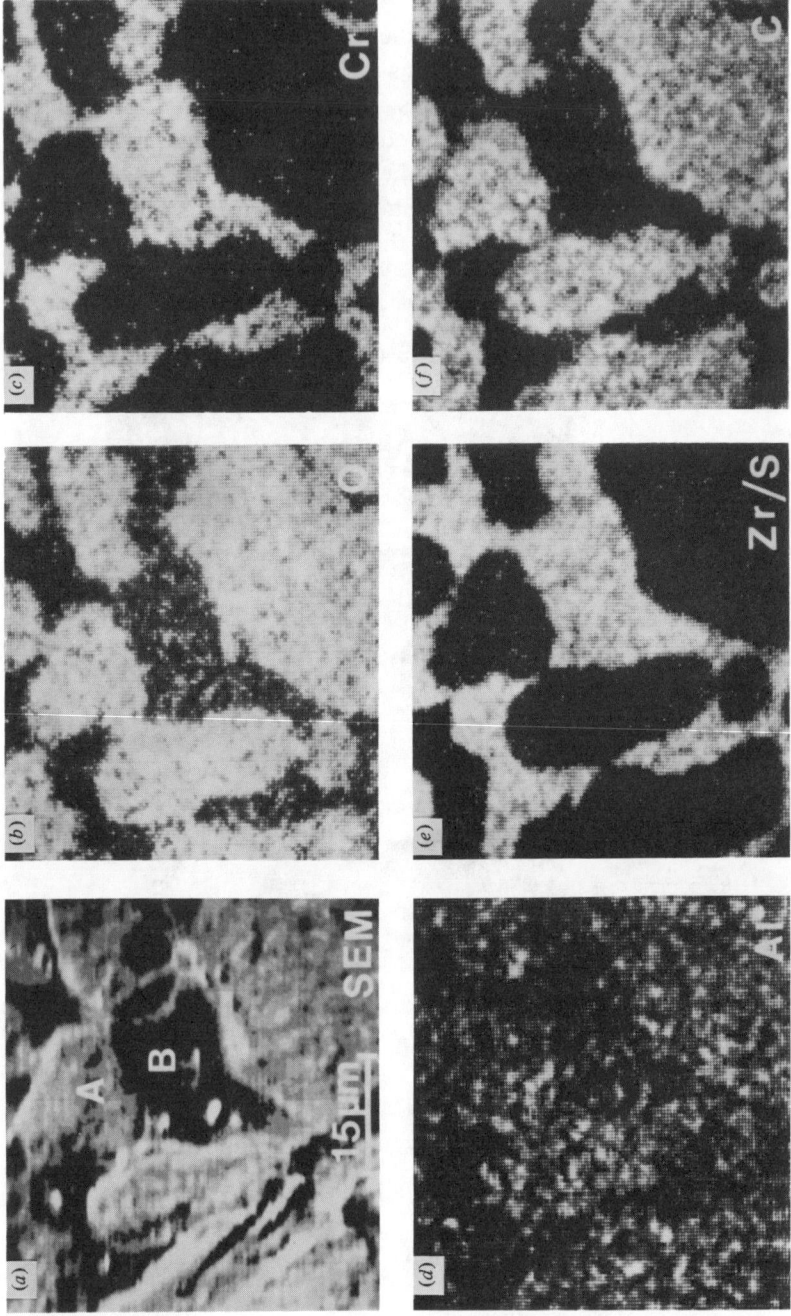

Fig. 4.26. Auger maps from superalloy, heated in vacuum, (a) SEM image, (b) O, (c) Cr, (d) Al, (e) Zr/S, (f) C.

at low resolution using 64 × 64 pixels. In spite of this a strong correlation can be seen between the S signal and the smooth areas of the grain boundary confirming the deductions made from the point analyses.

Mapping can be particularly useful in studies of multiphase material where the oxidation or segregation behaviour of the phases may differ. Fig. 4.26 shows a set of images obtained from the surface of a superalloy that had been ion cleaned and then annealed at 350 °C (El Gomati *et al.*, 1985). The Auger images show the surface of one phase to be rich in Zr and Cr while the other has high O and C. Although point analysis shows that there is a 50% difference between the Al concentration in the two phases the Al KLL Auger signal was too weak to give a convincing image.

8 Conclusions

Auger emission spectroscopy is a very powerful surface analytical technique that has found applications in many fields of materials science. However, no single technique can solve all problems and an important aspect of any analyst's task is to select which technique, out of those available to him, are most appropriate to his problems. The use of an inappropriate technique can be both expensive and frustrating. When faced with a materials problem it is important to discover as much as possible about the nature of the specimen before embarking on an investigation using sophisticated surface analytical techniques. In our laboratory we have found by experience that a preliminary investigation using optical microscopy and if appropriate scanning electron microscopy, with energy dispersive X-ray analysis, greatly assists in defining the problem more precisely, helps in deciding what further techniques need to be used and assists in interpreting the results from these techniques.

Before selecting AES one must weigh up its relative merits and demerits with respect to allied techniques. The main advantages of AES are:

— surface or near surface analysis;
— high spatial resolution;
— relatively rapid analysis;
— sensitive to light elements (other than H and He);
— chemical information is sometimes available;
— gives reliable semiquantitative analysis.

Against this must be set the following drawbacks:

— the electron beam can damage the surface;
— surface charging can make the analysis of insulators difficult;
— precise quantitative analysis presents problems;
— sensitivity is only modest, 0.1–1 atom %;
— depth analysis by ion profiling or sectioning is destructive.

In this chapter we have only been able to take a brief overview of Auger spectroscopy and many fascinating aspects of the subject have only been touched on, or omitted completely. It is hoped, however, that the reader will have gained sufficient insight to grasp the great value of this technique.

References

Bishop, H.E. (1965). *X-ray Optics and Microanalysis*, eds. Castaing *et al.* (Paris: Hermann), p. 153.

Bishop, H.E. (1984). In *Electron Beam Interactions with Solids*, ed. D.F. Kyser (SEM Inc., A.M.F. O'Hare, Chicago), p. 259.

Bishop, H.E., Chornik, B., Le Gressus, C. & Le Moel, A. (1984). *Surf. Interface Anal.*, **6**, 116.

Briggs, D. & Seah, M.P. (1983) eds., *Practical Surface Analysis* (John Wiley: Chichester, New York).

Davis, L.E., MacDonald, N.C., Palmberg, P.W., Riach, G.E. & Weber, R.E. (1976). *Handbook of Auger Electron Spectroscopy*, 2nd edition, Physical Electronics Division of Perkins Elmer Corp., Minesota, USA.

El Gomati, M.M., Walker, C.G.H., Peacock, D.C., Prutton, M., Bishop, H.E., Hawes, R.M.H. & Smialek, J. (1985). *Surface Science*, **152/153**, 917.

Gerlach, R.L. (1980). *Electron Microscopy* eds. P. Brederoo & V.E. Cosslett (7th European Congress on Electron Microscopy Foundation: Leiden), **3**, 210.

Holloway, P.H. (1976). *Surface Science*, **54**, 506.

Janssen, A.P., Harland, C.J. & Venables, J.A. (1977). *Surface Science*, **62**, 277.

Lindfors, P.A. (1979). *Surface Contamination, Genesis, Detection and Control*, ed. K.L. Mittall (Plenum Press, New York), Vol. 2, 587.

Lowry, R.K. & Hogrefe, A.W. (1980). *Solid State Technol.*, **71**.

MacDonald, N.C. & Waldrop, J.R. (1971). *Appl. Phys. Lett.*, **19**, 315.

Rivière, J.C. (1983). *The Analyst*, **108**, 649.

Seah, M.P. & Dench, W.A. (1979). *Surf. Interface Anal.*, **1**, 2.

Wall, M. (1986). Thesis, Imperial College, London.

Walls, J.M., Hall, D.D. & Sykes, D.E. (1979). *Surf. Interface Anal.*, **1**, 204.

5
X-ray photoelectron spectroscopy

A.B. CHRISTIE

1 Introduction

From the early pioneering work of Kai Siegbahn (Nobel prize winner, 1981) and his colleagues at Uppsala, Sweden, through intensive and exciting years of instrumental, interpretational and applicational development over the last decade, X-ray photoelectron spectroscopy (usually referred to by its acronym of XPS) has matured into one of the most powerful and valuable methods of surface analysis currently available. The universal applicability, high throughput, ease of interpretation and wealth of information provided by the technique place XPS above most other methods of surface analysis in a large number of areas of application, such as polymers, organics, biological specimens, fibres, films, powders and particles.

Although familiar to the majority of the scientific and industrial community concerned with the analysis of solid surfaces, XPS is a specialist's tool, and is likely to remain so for some years to come. In addition, the technique is undergoing continuous improvement and development as greater demands are put upon instrument manufacturers – some instruments are already into their fifth generation within 15 years of introduction. The purpose of this chapter, therefore, is to provide an introduction to XPS for those readers who are unfamiliar with the technique, a useful source of XPS references for those interested in surface analysis in general, and a review of recent developments in the technique for those who were brought up on the earlier texts on this subject (Carlson, 1975; Briggs, 1977; Brundle & Baker, 1981).

Rapid developments in XPS over the last decade have resulted in many of these early texts becoming prematurely dated. Indeed, a great number of earlier publications on photoelectron spectroscopy were concerned mainly with UV excited spectra (a large field of study which falls outside the scope of this chapter), and more recent texts often reflect a state-of-the-art in XPS which has since been superseded (Walls & Christie, 1982; Briggs & Seah, 1983). Small-area XPS, photoelectron imaging, multichannel detection and automated operation are but four examples which are discussed in more detail in Section 6.

References to applications are deliberately restricted. Complete volumes may be written on applications to, for example, corrosion science or polymer technology, and excellent sources of such references may be found by skimming through the current literature (Hofmann, 1986).

Applications data in this chapter has therefore been selected to illustrate particular aspects of instrumentation or interpretation, and is not intended to serve as any form of review or bibliography.

2   Basic principles and theory

XPS is one of a large number of instrumental *in-situ* surface analytical techniques that have been developed over the past 20 years. In each of these techniques, the specimen of interest is excited by some form of controllable energy (the excitation source) and its subsequent response to that excitation, in the form of an emission of some species, is observed by some type of spectroscopy or microscopy. In XPS, the primary excitation is accomplished by irradiating the specimen by a source of (more or less) monochromatic X-rays. The X-rays cause photoionisation of atoms in the specimen and the response of the specimen (photoemission) is observed by measuring the energy spectrum of the emitted photoelectrons.

2.1   The photoelectric effect

In addition to the valence electrons, which provide the bonding for the system, each atom present in the surface (except hydrogen) possesses core electrons not directly involved in the bonding. The so-called 'binding energy' (E_b) of each core electron (conceptually, but not strictly, equivalent to the ionisation energy of that electron) is characteristic of the individual atom to which it is bound. Table 5.1 is a compilation of the core electron binding energies for most of the elements in the periodic table. Information on the binding energies of electrons within a sample allows qualitative elemental analysis. In the basic XPS experiment, the sample surface is irradiated by a source of low-energy X-rays under ultra-high vacuum (UHV) conditions. *Photoionisation* then takes place in the sample surface, the resultant *photoelectrons* having a kinetic energy (E_k) which is related to the X-ray energy (hv) and E_b by the Einstein relation (Einstein, 1905):

$$E_k = hv - E_b. \tag{1}$$

Table 5.1. Core-level electron binding energies. (Electron binding energies in electron volts of elemental core-levels commonly encountered in XPS ($20\,eV < E_b < 1400\,eV$)/(after Siegbahn et al., 1967)

	$1s_{1/2}$	$2s_{1/2}$	$2p_{1/2}$	$2p_{3/2}$	$3s_{1/2}$	$3p_{1/2}$	$3p_{3/2}$	$3d_{3/2}$	$3d_{5/2}$
1 H	14								
2 He	25								
3 Li	55								
4 Be	111								
5 B	188								
6 C	284								
7 N	399								
8 O	532	24							
9 F	686	31							
10 Ne	867	45							
11 Na	1072	63	31						
12 Mg	1305	89	52						
13 Al		118	74	73					
14 Si		149	100	99					
15 P		189	136	135					
16 S		229	165	164					
17 Cl		270	202	200					
18 A		320	247	245	25				
19 K		377	297	294	34				
20 Ca		438	350	347	44	26			
21 Sc		500	407	402	54	32			
22 Ti		564	461	455	59	34			
23 V		628	520	513	66	38			
24 Cr		695	584	575	74	43			
25 Mn		769	652	641	84	49			

Table 5.1 (Cont.)

	$1s_{1/2}$	$2s_{1/2}$	$2p_{1/2}$	$2p_{3/2}$	$3s_{1/2}$	$3p_{1/2}$	$3p_{3/2}$	$3d_{3/2}$	$3d_{5/2}$
26 Fe		846	723	710	95				
27 Co		926	794	779	101				
28 Ni		1008	872	855	112				
29 Cu		1096	951	931	120				
30 Zn		1194	1044	1021	137				
31 Ga		1298	1143	1116	158	107	103		
32 Ge			1249	1217	181	129	122		29
33 As			1359	1323	204	147	141		41

	$3s_{1/2}$	$3p_{1/2}$	$3p_{3/2}$	$3d_{3/2}$	$3d_{5/2}$	$4s_{1/2}$	$4p_{1/2}$	$4p_{3/2}$	$4d_{3/2}$	$4d_{5/2}$
34 Se	232	168	162		57	27				
35 Br	257	189	182	70	69					
36 Kr	289	223	214		89	24				
37 Rb	322	248	239	112	111	30				
38 Sr	358	280	269	135	133	38		20		
39 Y	395	313	301	160	158	46		26		
40 Zr	431	345	331	183	180	52		29		
41 Nb	469	379	363	208	205	58		34		
42 Mo	505	410	393	230	227	62		35		
43 Tc	544	445	425	257	253	68		39		
44 Ru	585	483	461	284	279	75		43		
45 Rh	627	521	496	312	307	81		48		
46 Pd	670	559	531	340	335	86		51		
47 Ag	717	602	571	373	367	95	62	56		
48 Cd	770	651	617	411	404	108		67		
49 In	826	702	664	451	443	122		77		24
50 Sn	884	757	715	494	485	137		89		32
51 Sb	944	812	766	537	528	152		99		40
52 Te	1006	870	819	582	572	168		110		50
53 I	1072	931	875	631	620	186		123		63
54 Xe	1145	999	937	685	672	208		147		
55 Cs	1217	1065	998	740	726	231	172	162	79	77
56 Ba	1293	1137	1063	796	781	253	192	180	93	90

	$4s_{1/2}$	$4p_{1/2}$	$4p_{3/2}$	$4d_{3/2}$	$4d_{5/2}$	$4f_{5/2}$	$4f_{7/2}$	$5s_{1/2}$	$5p_{1/2}$	$5p_{3/2}$	$5d_{3/2}$	$5d_{5/2}$
72 Hf	538	437	380	224	214			65	38	31		
73 Ta	566	465	405	242	230	27	25	71	45	37		
74 W	595	492	426	259	246	37	34	77	47	37		
75 Re	625	518	445	274	260	47	45	83	46	35		
76 Os	655	547	469	290	273	52	50	84	58	46		
77 Ir	690	577	495	312	295	63	60	96	63	51		
78 Pt	724	608	519	331	314	74	70	102	66	51		
79 Au	759	644	546	352	334	87	83	108	72	54		
80 Hg	800	677	571	379	360	103	99	120	81	58		
81 Tl	846	722	609	407	386	122	118	137	100	76		
82 Pb	894	764	645	435	413	143	138	148	105	86	22	20
83 Bi	939	806	679	464	440	163	158	160	117	93	27	25
84 Po	995	851	705	500	473	184		177	132	104		31
85 At	1042	886	740	533	507	210		195	148	115		40
86 Rn	1097	929	768	567	541	238		214	164	127		48
87 Fr	1153	980	810	603	577	268		234	182	140		58
88 Ra	1208	1058	879	636	603	299		254	200	153		68
89 Ac	1269	1080	900	675	639	319		272	215	167		80
90 Th	1330	1168	968	714	677	344	355	290	229	182	95	88

If the photoelectrons have sufficient kinetic energy they are able to escape from the surface by overcoming the specimen work function, and *photoemission* is said to occur. The entire process is referred to as the *photoelectric effect*.

Photoemission may be observed from solids, liquids and gases (Siegbahn, 1985), in fact from any state of matter containing bound electrons, but it is the application of the photoelectric effect to solids that has aroused the most interest in recent years.

Since the energy levels are quantised (Table 5.1), the photoelectrons have a kinetic energy distribution, $N(E)$, consisting of a series of discrete bands that essentially reflects the 'shell' form of electronic structure of the atoms in the sample (Fig. 5.1). The experimental determination of $N(E)$ by a kinetic energy analysis of the photoelectrons produced by exposure to X-rays is termed X-ray photoelectron spectroscopy (XPS).

2.2 *Photoionisation cross-sections*

The number of photoelectrons produced from any given core-level of an element is determined by the photoionisation cross-section (σ) of that level for the photon energy ($h\nu$) concerned; σ is defined as the transition probability per unit time for excitation of a single photoelectron from the

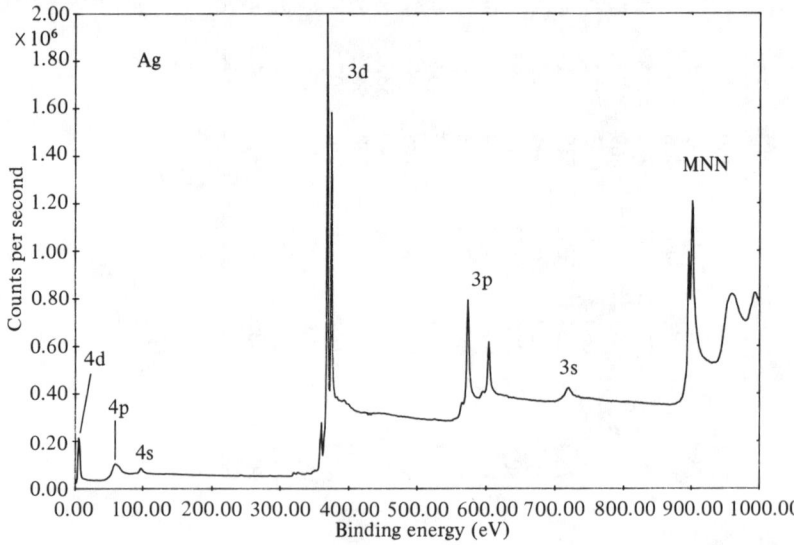

Fig. 5.1. MgK$_\alpha$ XPS survey spectrum from silver. Typical high sensitivity (2×10^6 cps), low-resolution XPS survey spectrum from a clean silver specimen. Signal-to-noise ratio of the 3d photoelectron peaks is 10^4:1. Features labelled MNN are due to silver MNN Auger transitions (data courtesy of VG Scientific).

core level of interest under an incident photon flux of $1\,\mathrm{cm^{-2}s^{-1}}$. In practice, σ is found to depend on $h\nu$, E_b atomic number (Z), and the relative directions of photon incidence and photoelectron emission (ϕ). The dependence of σ on ϕ is, by convention, characterised by an asymmetry factor, L, such that:

$$L(\phi) = 1 + \tfrac{1}{2}\beta(\tfrac{3}{2}\sin^2\phi - 1) \qquad (2)$$

where β is a constant for a given core level of a given atom and a given photon energy. Tabulations of σ and β may be found in the literature (Scofield, 1976; Reilman, Msezane & Manson, 1976). From equation (2), we see that no asymmetry correction is needed ($L(\phi) = 1$) for values of ϕ close to 54.7°, the so-called magic angle in XPS. Most of the current state-of-the-art commercial spectrometers fitted with non-monochromated X-ray sources employ ϕ values in the range 50°–60°. Although theoretical values of σ and β are available, quantitative interpretation of XPS spectra is most usually accomplished via empirical, experimentally derived atomic sensitivity factors. This topic is discussed in detail in Section 5.

2.3 Lineshapes and fine structure

Experimentally observed photoelectron lineshapes ($N(E)$ vs E) in XPS are determined by a convolution of the photon (X-ray) energy distribution, the initial state structure (electron energy distribution in the ground state), the final state structure (electron energy distribution in the photoionised state), lifetime broadening effects, electron energy loss structure acquired during transfer from the atom to the surface, and the spectrometer energy resolution function (Fig. 5.2). Each of these contributions is discussed briefly in turn.

The photon energy distribution from a conventional X-ray anode is dominated by characteristic X-rays resulting from core-level transitions of the type 2p–1s (for example). The individual components of the X-ray envelope are Lorentzian in shape, and their complexity and widths are determined by the detailed core-level structure (atomic number) of the element concerned. The energies and relative intensities of the characteristic X-rays produced from magnesium and aluminium anodes are presented in Table 5.2. In addition to the characteristic radiation, continuous Bremsstrahlung radiation is also produced, resulting in a contribution to the general photoelectron background in the XPS spectrum. Removal of Bremsstrahlung and unwanted minor X-ray features, in addition to a significant reduction in the characteristic X-ray linewidth, may be achieved by employing monochromatisation of the source

(Section 3.6). Alternatively, such features may be removed by subsequent data processing of the XPS spectrum (Section 4.5).

Initial-state structure is dominated by the phenomenon of spin–orbit coupling which, for all except the S-levels, causes a characteristic splitting of the photoelectron peaks. This phenomenon is well understood and discussed in detail in the literature (see Briggs & Seah, 1983). Final state structure may arise through relaxation of the core-hole following photo-

Table 5.2. *Characteristic X-ray satellite lines from magnesium and aluminium (data from Briggs, 1977)*

X-ray line	Kinetic energy (eV)		Relative intensity ($K_{\alpha 1,2}$ 100)	
	Mg	Al	Mg	Al
$K_{\alpha 1,2}$	1253.6	1486.6	100	100
$K_{\alpha'}$	1258.1	1492.2	1.0	1.0
$K_{\alpha 3}$	1262.0	1495.2	9.2	7.8
$K_{\alpha 4}$	1263.6	1497.1	5.1	3.3
$K_{\alpha 5}$	1270.9	1505.4	0.8	0.4
$K_{\alpha 6}$	1274.1	1509.0	0.5	0.3
K_β	1301.6	1555.6	2.0	2.0

Fig. 5.2. MgK_α XPS narrow scan spectrum from silver. Typical high resolution (0.8 eV fwhm), XPS narrow scan spectrum from a clean silver specimen, showing the detailed lineshape of the 3d photoelectron features (data courtesy of VG Scientific).

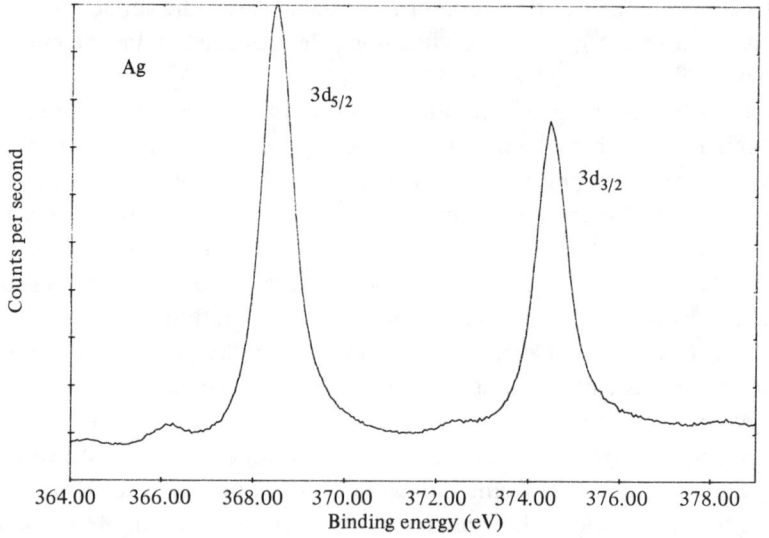

ionisation, or by direct interaction of the photoelectron with other bound electrons, resulting in a discrete (e.g. shake-up) or arbitrary (e.g. shake-off) energy loss. Final state contribution to the linewidth is often dominated by lifetime broadening. Singly ionised lifetimes are generally 10^{-14} to 10^{-15} seconds, giving rise to lifetime broadening contributions of between 0.1 and 5 eV. Final state structure in XPS may often provide useful additional qualitative or, in some cases, quantitative information about the specimen, and the various sources of final state structure have been discussed at length by previous authors (Brundle & Baker, 1981; Briggs & Seah, 1983).

Analyser contributions to the observed photoelectron lineshape and linewidth may be regarded as a negligible constant ($\Delta E < 0.1$) for most practical purposes in XPS when the spectrometer is operated in the CAE mode at high-energy resolution. Under low-energy resolution conditions, or in the CRR mode, however, due consideration must be taken of the analyser lineshape contribution, which will generally be triangular in form.

2.4 Chemical shifts

In a series of detailed and laborious studies made during the 1950s, and contrary to popular beliefs at the time, Siegbahn found that the so-called 'fixed' binding energies of the core-levels of atoms are sufficiently affected by their chemical environment to cause a detectable shift (ranging from 0.1 up to 10 eV or more in magnitude) in the measured photoelectron energy, an effect that he termed the 'chemical shift' (Siegbahn et al., 1967). So important was this discovery, that the surface specificity of XPS was largely ignored during the first decade of its development, in favour of studies concerning chemical shifts (Carlson, 1975; Brundle & Baker, 1981). It was this ability of the technique that led Siegbahn to coin the title 'electron spectroscopy for chemical analysis' (ESCA), a title that is often used, in preference to the more scientifically correct one of X-ray photoelectron spectroscopy, to this present day. In simple terms, chemical shifts arise from the variation of electrostatic screening experienced by core electrons as valence electrons are drawn towards or away from the atom of interest (Fig. 5.3).

Unfortunately, there may be many contributing factors to the observed chemical shift, and no single theory has yet been developed that has succeeded in explaining more than general trends in binding energy variations. Interpretation of observed chemical shifts is therefore best accomplished on an empirical basis, and much useful information can be acquired by comparing measured binding energies from unknowns with those of standard materials. Further reading on this subject may be found in the literature (Carlson 1975; Brundle & Baker, 1981; Briggs & Seah, 1983).

2.5 Inelastic scattering and sampling depth

Following photoionisation, the photoelectron of energy E_k (equation (1)), must travel through the solid and escape into the vacuum, *without energy loss*, before it can be energy analysed and detected as a characteristic photoelectron. If all photoelectrons were able to do this, independent of the depth at which they were created, the sampling depth of XPS would be determined by the depth to which the incident X-rays were able to penetrate, viz, several microns. In fact, the stopping power of solids for electrons is several orders of magnitude higher than it is for

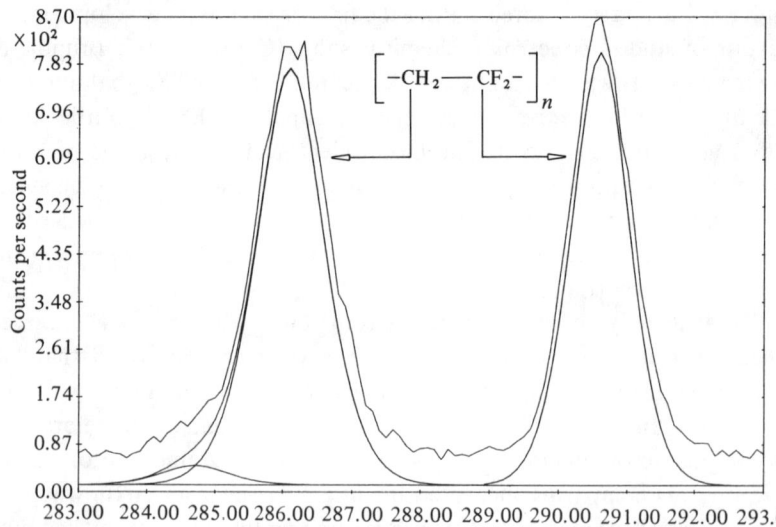

Fig. 5.3. Chemical shifts in the carbon 1s photoelectron peak from PVDF. Carbon 1s photoelectron spectrum from a specimen of polyvinylidene fluoride (PVDF), showing dominant contribution from the $-CH_2-$ (at 286.1 eV) and $-CF_2-$ (at 290.6 eV) functional groups in the polymer, and a small contamination peak (carbon or hydrocarbon) at 284.7 eV. Also shown is the computer generated peak synthesis (normalised chi squared value of 1.8) of all three components. Best fit was achieved with a Gaussian/Lorentzian lineshape mix with 53% Lorentzian character for the contamination and $-CH_2-$ peaks, and 43% for the $-CF_2-$ peak. The increase in binding energy of the $-CF_2-$ carbon 1s level is due to the reduction in electrostatic screening caused by the electronegative fluorine atoms to which it is bound. The same process may also be responsible for the reduction in linewidth (fwhm) from 1.2 eV in the $-CH_2-$ to 1.0 eV in the $-CF_2-$ peak, which may reflect an increase in core-hole lifetime.

X-rays, with the net result that electrons in the energy range 50–1000 eV will typically travel only 2–10 atomic layers before they lose energy through inelastic scattering events with other electrons, and hence cannot contribute to the characteristic photoelectron peak at energy E_k. (If they contribute to the spectrum at all, it will be to the structureless, or sometimes slightly structured, background at energies below E_k.) The probability of an inelastic scattering event occurring is determined by both the electron energy and the material through which it is travelling. Since it is a random process governed by probability, inelastic scattering is described by the standard exponential decay law:

$$I(x) = I_o \exp\left(-x/\lambda(E_k, Z)\cdot\cos\theta\right), \quad (3)$$

where I_o is the original photoelectron intensity, $I(x)$ is the intensity remaining after travelling through material of thickness x, θ is the angle of emission with respect to the surface normal, and $\lambda(E_k, Z)$ is a material and energy-dependent constant, termed the inelastic mean free path. λ represents the depth at which photoelectrons have a probability of $1/e$ of escaping without energy loss. There are numerous references to the theoretical and experimental determination of λ in the literature (Powell, 1985a, c). An empirical description of the dependence of λ on E_k and Z has been derived from an analysis of experimental data (Seah & Dench, 1979). The variation of λ with energy, for energies above 50 eV, is given by:

$$\lambda \propto E_k^{1/2}. \quad (4)$$

Some workers, however, have reported power dependencies of up to 0.79 on energy (Powell, 1985c).

Due to the exponential decay behaviour predicted by equation (3), it is not possible to arrive at a unique value for the sampling depth, d, of XPS. Since approximately 63% of the photoelectrons contributing to a particular feature must originate within a distance of $\lambda \cdot \cos\theta$ of the surface, this quantity will serve, for all practical purposes, as a measure of d. (The values for greater depths are: $2d \sim 86\%$, $3d \sim 95\%$ and $5d > 99\%$.) An important point to note is that d varies with $\cos\theta$, hence sampling depth may be controlled by varying the angle of emission (electron collection angle) from $0°$ ($d = \lambda$) to $80°$ ($d = 0.17\lambda$) (Section 4.6).

3 Historical development and basic instrumentation
3.1 Early pioneers (1887–1967)

The origins of XPS can be traced back to the first documented observation of the photoelectric effect by Hertz in 1887. From 1887 to 1967, the pedigree of XPS is dotted with such names as Rutherford, de Broglie, Einstein, Robinson, Turner and, reckoned by many to be the true pioneer of XPS, Kai Siegbahn. During the 1940s and 1950s, Siegbahn's group at Uppsala, Sweden, laid the foundations for modern-day XPS, culminating in the publication in 1967 of a text representing one of the major milestones in XPS development (Siegbahn et al., 1967). Siegbahn was awarded the Nobel prize for this work. The history of XPS has been very well documented by the workers at La Trobe University (Jenkin, 1981) and Uppsala (Siegbahn, 1985).

3.2 Commercial development

Two of the essential requirements for XPS, as we know it today, were not available until very late in the development of the technique. These were ultra-high vacuum (UHV) conditions, for the XPS experiment to take place in, and a spherical sector electrostatic electron energy analyser, for the energy analysis of the emitted photoelectrons. It was not until these facilities became available on a commercial basis, in the early 1970s, that XPS began to gain popularity as an analytical technique. During the 1970s, continuous improvements and refinements in commercial instrumentation were made, resulting in the gradual acceptance of XPS in many areas of industry as well as in academic and research institutions. More recently still, advances in multichannel detection, small-area XPS analysis, photoelectron imaging and data processing (discussed later) have all been incorporated into state-of-the-art instrumentation for XPS.

3.3 Basic requirements

All XPS spectrometers must incorporate each of the following:
(1) a UHV environment;
(2) a controlled source of X-rays;
(3) a specimen manipulation system;
(4) an electron energy analyser and detection system;
(5) a data recording, processing and output system.

These are discussed in more detail below.

3.4 UHV environment

A UHV environment is an essential requirement for almost every technique of surface analysis available today. Many of the constituent parts, such as X-ray, electron and ion sources, electron analysers and detectors will operate only under high vacuum or ultra-high vacuum conditions. In addition, the very surface sensitivity of such techniques requires that interactions between specimen and spectrometer be minimised. UHV conditions are achieved through the use of stainless steel

X-ray photoelectron spectroscopy

and glass construction materials, combined with suitable UHV pumping facilities. The details of vacuum system design and construction are beyond the scope of this book, but an excellent basic text can be found in Yarwood, 1967.

3.5 The X-ray source

The simplest type of X-ray source for XPS is one which utilises characteristic emission lines from an anode bombarded by high-energy (15 keV) electrons. The characteristic X-ray energies and linewidths of a number of suitable anode materials are listed in Table 5.3. The necessary characteristics of an X-ray source suitable for XPS are that the X-ray energy should be sufficiently high to excite core-level electrons of all elements (i.e. 1 keV or more), the X-ray spectrum should be relatively 'clean', with very few satellites or other peaks (i.e. low atomic number anode material), the characteristic X-ray linewidth should be narrow in comparison to the intrinsic core-level linewidths and chemical shifts we wish to study (i.e. < 1 eV), and the material must be suitable (conductivity,

Table 5.3. *Characteristic X-ray emission energies and linewidths (data from Briggs, 1977; Briggs & Seah 1983; Edgell, Paynter & Castle, 1985)*

Anode material	Line					
	K_α		L_α		M_ζ	
	Energy (eV)	Width (eV)	Energy (eV)	Width (eV)	Energy (eV)	Width (eV)
C	278	6				
O	524.9	4				
Ne	849	0.3				
Na	1041	0.42				
Mg	1253.6	0.7				
Al	1486.6	0.85				
Si	1739.4	1.0				
Ti	4510.9	2.0	395.3	3.0		
Cr	5417	2.1	572.8	3.0		
Cu	8048	2.5	929.7	3.8		
Zn			1011.7	2.0		
Y			1922.6	1.5	132.3	0.47
Zr			2042.4	1.6	151.4	0.77
Nb			2165.9	1.8	171.4	1.21
Mo			2293.2	1.9	192.3	1.53
Ag			2984.3	2.6	311.3	9.8

melting point) for the construction of an anode. Only two elements possess all of the above properties – magnesium and aluminium. Most commercial XPS instruments are therefore fitted with twin X-ray sources incorporating both aluminium and magnesium anodes. The advantage of having both AlK_α and MgK_α radiation available lies in the fact that photoionisation produces, not only photoelectrons, but also, via Auger decay of the electron hole formed, Auger electrons (see Chapter 4). Obviously, the Auger electron energy is independent of the X-ray energy used to create the hole, whereas the photoelectron energy is related to the X-ray energy via the Einstein relation (equation (1)). Hence, the *apparent* binding energy of Auger peaks appears to change (by 233 eV) on going from AlK_α to MgK_α radiation (or vice versa), whereas photoelectron peaks do not shift in binding energy. This feature of XPS may be used, not only to differentiate between photoelectron and Auger peaks in the spectrum, but also to resolve photoelectron and Auger peaks which may otherwise interfere with each other.

Other characteristic X-ray lines, such as silicon and titanium K_α, have been used to some advantage, since a wide variation of X-ray source energies permits non-destructive composition-depth profiling (Section 4.6) and increased flexibility of the technique (Castle & West, 1980). The X-rays generated pass through a thin ($< 1 \mu m$) aluminium window separating the X-ray source and spectrometer chambers, which serves to insulate the specimen from the high voltage on the anode, and reduce Bremsstrahlung irradiation and cross-contamination of the sample. The sample surface is thus uniformly irradiated with a low flux of X-rays over a relatively large area ($\sim 1 \, cm^2$).

3.6 Monochromated sources

The conventional twin-anode type of X-ray source described above is the most common method of excitation employed in commercial XPS instrumentation. The K_α radiation produced by such sources is sufficiently monochromatic for most applications of XPS, particularly in industrial laboratories. The twin-anode source, however, does limit the ultimate energy resolution in photoelectron spectra, due to the natural linewidth of the source (see Table 5.3), to 0.7 eV in the very best possible case. In addition, such sources give rise to satellites, ghost peaks and Bremsstrahlung radiation, which may cause assignment and interpretational problems (see Sections 2.3 and 5.4).

By employing a suitable X-ray monochromator, it is possible not only to reduce the characteristic X-ray linewidth to 0.4 eV or less, but also to remove all of the satellite, ghost and Bremsstrahlung radiation from the

X-ray photoelectron spectroscopy

source. Monochromatisation in XPS is normally achieved by first-order (AlK_α) or second-order (AgL_α) (Edgell, Paynter & Castle, 1985) back-diffraction from the (1010) face of a quartz single crystal. It has also been suggested (Siegbahn, 1985) that the third-order diffraction of TiK_α (4510.9 eV) could be used. In this case, the conditions for back-diffraction are met if

$$n \cdot \lambda = 2d \sin \theta, \qquad (5)$$

where n = diffraction order, λ = X-ray wavelength, d = crystal spacing (0.4255 nm for quartz 1010), and θ = Bragg angle.

Since the Bragg angles for Al and Ag are so similar, (Table 5.4) these anodes can be successfully combined in a twin anode monochromator, providing narrow linewidth X-rays at either 1486.6 or 2984.3 eV. The detailed theory and practice of X-ray monochromatisation are well documented in the literature (Briggs, 1977; Yates & West, 1983a). In

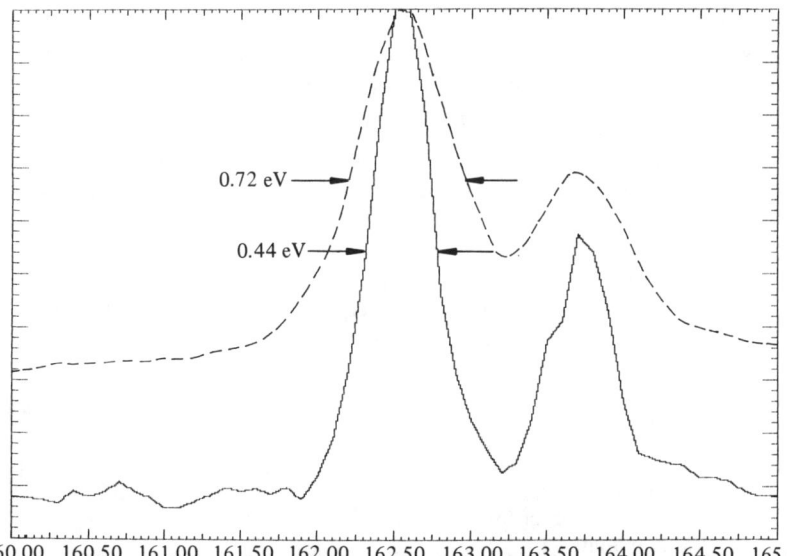

Fig. 5.4. Monochromatised AlK_α XPS narrow scan from molybdenum disulphide. Example of the energy resolution obtainable in XPS with a monochromated X-ray source. The sulphur 2 p spin–orbit split photoelectron feature from a specimen of molybdenum disulphide acquired with a monochromated AlK_α source (full curve) and a non-monochromated MgK_α source (broken curve). Energy resolution (full width at half-maximum of the $2p_{3/2}$ component) is 0.44 eV using the monochromated source, and 0.72 eV using the non-monochromated source. Note the enhanced signal to background ratio (removal of Bremsstrahlung) and lack of satellites (removal of $\alpha_{3,4}$ components) in the monochromated data (data courtesy of VG Scientific).

addition to their many advantages, however, monochromatised sources suffer from the problem of a very low X-ray flux at the specimen when compared to conventional sources.

There is some evidence to suggest that the benefits of monochromatisation (enhanced signal-to-noise ratio and improved energy resolution) may in many cases be matched by the performance of the standard twin anode sources when combined with suitable data processing (deconvolution, satellite subtraction) techniques (Koening & Grant, 1985).

For the above reasons, X-ray monochromators have, in the past, been employed almost exclusively in fundamental research applications, with very few installations in industrial analytical laboratories. The recent development of the microfocussed monochromator (see Section 6.1), however, has resulted in renewed interest in X-ray monochromatisation by photoelectron spectroscopists.

3.7 The sample

In principle, all types of sample, including gases and liquids, may be studied by XPS, but limitations on instrument design usually restrict the range to low vapour pressure ($< 10^{-6}$ Torr), solid samples. Restrictions may also be imposed on sample size, but commercial spectrometers are presently available that will accept samples with dimensions of several centimeters. Early instruments were designed to hold only one, often specially machined, sample at a time but modern spectrometers are able to handle a number of powder, fibre or bulk solid samples independently and simultaneously.

Methods of sample mounting are numerous, and are often dictated by the nature of the sample itself. Bulk solid samples may be clipped, clamped, or fixed with double-sided adhesive tape. Powders can be brushed on to adhesive tape, but better results are obtained if they are simply contained in a small metal crucible, or pressed into indium foil. Fibres may be

Table 5.4. *Monochromator parameters for aluminium, silver and titanium sources (from Yates & West, 1983a; Siegbahn, 1985)*

Anode material	Al	Ag	Ti
Characteristic line	K_α	L_α	K_α
Energy (eV)	1486.6	2984.3	4510.9
Wavelength (nm)	0.834	0.415	0.275
Diffraction order (n)	1st	2nd	3rd
Bragg angle (°)	78.5	77.5	75.7

similarly mounted or, alternatively, they may be clamped in position. In many spectrometers, the specimen is situated on a precision manipulator, such that the sample surface may be moved, relative to the X-ray source and energy analyser, with a translational and rotational precision of 10 μm and 1° respectively.

3.8 The electron energy analyser

The heart of every XPS spectrometer is the electron energy analyser. Its function is to disperse photoelectrons emitted from the specimen, according to their energies, across a detector or detector array. Electron energy analysis is discussed in Chapter 3, hence only a brief mention will be made here. Although most types of electron energy analyser have been used at one time or another for XPS applications, modern instruments employ, almost exclusively, only the spherical sector (or concentric hemispherical) electrostatic analyser (SSA). The SSA possesses a number of advantages for XPS applications. Even the smallest commercially available versions can be operated under very high energy resolution conditions (several milli-electron volts), and this ability is enhanced with increasing size of analyser. The SSA is also ideally suited to the incorporation of an input lens (increasing instrument flexibility and analyser-specimen working distance) and multidetection (effectively increasing analyser transmission) (Section 6.2). The SSA does not suffer from the specimen position dependence of energy and resolution which is a major drawback of its main rival, the cylindrical mirror analyser (CMA) (see Chapter 3).

3.9 Data recording, processing and output

Following energy dispersion in the SSA, the photoelectrons are made to enter an electron multiplier detector or detector array. Here each incident photoelectron causes a secondary electron cascade resulting in an output pulse of up to 10^8 electrons with less than 0.1 μs duration. Using conventional current amplification and pulse counting techniques it is possible to operate under count rate conditions of 10^6 s^{-1} per detector. A basic spectrometer with single detector may then be operated in analogue mode by recording the ratemeter output (count rate) versus electron energy. Data processing capability is considerably enhanced, however, by recording the energy spectrum on a dedicated microcomputer. Since the detector output is in the form of discrete pulses, XPS is ideally suited to conventional digital spectrum acquisition and processing techniques, discussed in greater detail later.

Fig. 5.5. Front elevation of a typical multitechnique XPS spectrometer. Front elevation general assembly diagram of the VG Scientific ESCALAB MkII in multitechnique configuration.

Key:

1. Fast entry specimen insertion lock (stainless steel).
2. UHV specimen preparation chamber (stainless steel).
3. UHV experimental/analysis vessel (mu-metal).
4. Viton-sealed gate valve.
5. Rotary drive to specimen transfer mechanism.
6. Titanium sublimation pump vessel (stainless steel).
7. Viewport.
8. Autocarousel motor drive.
9. High precision (X, Y, Z translation and θ tilt) specimen manipulator.
10. Twin anode X-ray source (Al/Mg).
11. UV discharge source (UPS).
12. Monochromated X-ray source (Al/Ag).
13. 2000 Å electron source (AES, SEM, SAM).
14. Scanning ion source.
15. Electron energy analyser vessel (mu-metal).
16. Detector (single or multichannel).
17. Specimen fracture stage.
18. Static, broad beam ion source.
19. High pressure gas reaction/catalysis cell.
20. 'Wobble stick' specimen transfer fork.
21. 6-specimen mobile carousel.
22. 10-specimen auto-carousel.
23. Preparation vessel specimen transfer 'railway'.
24. Binocular microscope.
25. Alternative fast-entry lock position, or optional extension chamber port.
26. Port for monochromated electron source.
27. Port for specimen heating/cooling stage.

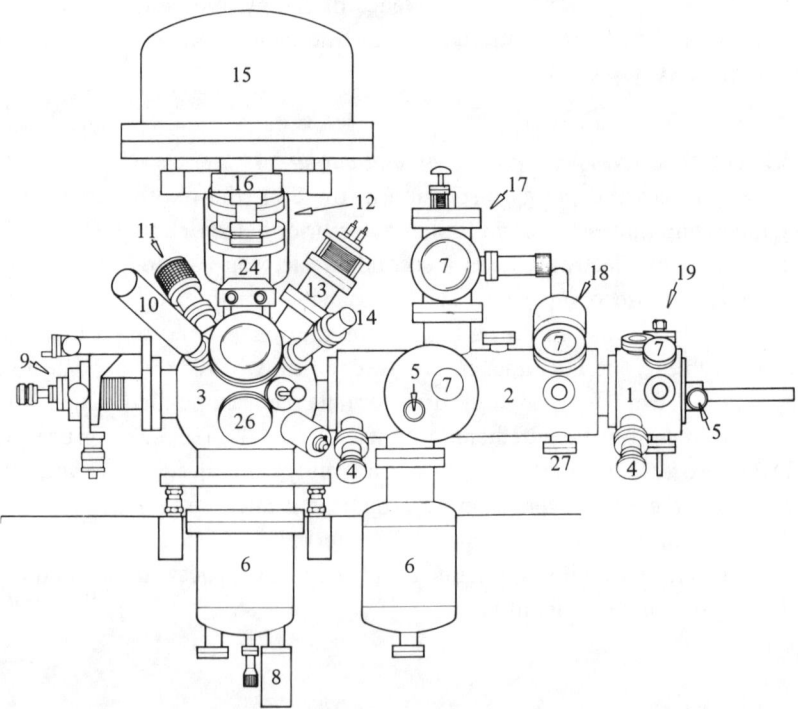

3.10 Additional facilities

In addition to the basic requirements discussed above, most modern XPS spectrometers will also be equipped with the following:

(1) a controlled inert gas ion source for ion sputtering;
(2) a specimen preparation vessel;
(3) a fast entry air lock insertion system;
(4) additional analytical techniques;
(5) an on-line microcomputer for control and processing.

The inert gas ion source is an indispensible accessory, useful both for *in-situ* sample cleaning (prior to, for example, adsorption or oxidation experiments), and for *in-situ* sputter-depth profiling. Ion erosion is common to most techniques of surface analysis and is discussed in Chapter 2.

A dedicated specimen preparation vessel, separated from the analytical chamber by a gate valve, is necessary if the spectrometer is to be used for *in-situ* experimentation (gas dosing, evaporation, fracture, etc.), as well as surface analysis. This vessel would normally be fitted in series with a fast entry air lock specimen insertion vessel, which allows sample introduction and removal within minutes without breaking vacuum in either the analytical or preparation vessels (Fig. 5.5).

4 Analytical procedures and qualitative interpretation

4.1 Specimen preparation

The degree to which any specimen preparation is carried out prior to analysis depends, obviously, on the application. In general, it is best to restrict preparation to a minimum, especially if a simple surface analysis by XPS is all that is required, in order to reduce the possibilities of surface contamination or modification by the preparation processes (Lea, 1983). In many cases, the only preparation necessary will be to cut or fracture the specimen to a size suitable for the spectrometer to handle.

In-situ preparation is also best kept to a minimum – the very first priority of every electron spectroscopist should be always to obtain an XPS spectrum of the as-received surface, before any attempt is made to modify that surface.

4.2 Operating conditions

Selection of the optimum operating conditions for XPS analyses is obviously a matter of experience – experience both of the technique and spectrometer, and of the type and nature of the specimens under study.

146   A.B. Christie

4.2.1   Vacuum

The quality of the vacuum in the experimental chamber is of vital importance in the study of fundamental properties of, for example, semiconductor and metal surfaces. XPS is used increasingly, however, as a routine analytical tool for industrial specimens, and in such cases the vacuum conditions in the spectrometer are of less importance. Indeed, many industrial-type samples (polymers, fibres, catalysts, tribological specimens) will actually limit the extent to which the vacuum conditions can be improved due to their own outgassing. In such cases, it is best to restrict the specimen dimensions to the smallest size that can easily be handled and manipulated in the spectrometer.

4.2.2   X-ray source

The normal operating conditions for non-monochromatised aluminium and magnesium X-ray sources are 15 kV accelerating voltage and 20 mA target current. Although some aluminium sources can be operated at up to 1 kW total power, it is probable that significant deterioration of the anode coating would occur during prolonged opera-

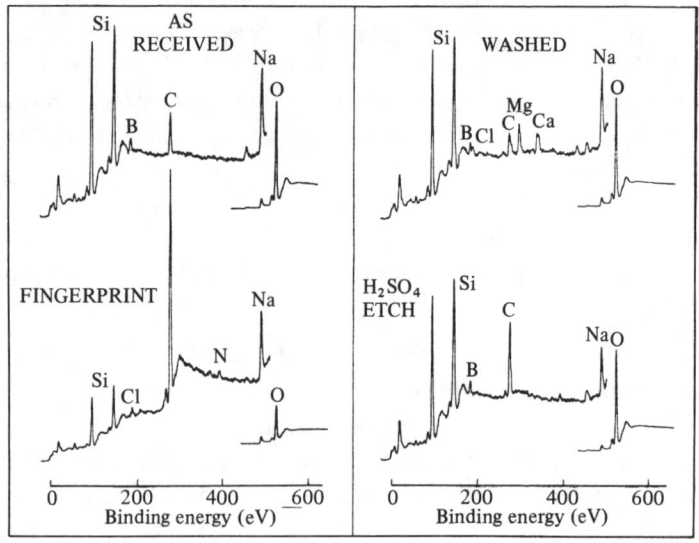

Fig. 5.6. XPS spectra of contaminated glass surfaces. Series of XPS survey spectra recorded from variously treated pyrex glass surfaces. The sequence AS RECEIVED–FINGERPRINT–WASHED–H_2SO_4 ETCH demonstrates the ability of XPS to precisely differentiate between trace contamination levels during different stages of a manufacturing or cleaning procedure. Note the magnesium and calcium residues remaining after washing, and the sodium leaching (attenuation of the sodium Auger feature) caused by the H_2SO_4 etch.

tion under these conditions. The normal operating power of standard aluminium monochromators is 600 watts (15 kV, 40 mA).

4.2.3 Electron energy analyser

The SSA is normally operated in the constant analyser energy (E_A = constant) mode for XPS, thereby yielding a constant energy resolution throughout the spectrum. For survey (wide scan) spectra, the analyser would normally be operated with a resolution of approximately 1 eV, to give fairly high transmission. For high-resolution (narrow scan) spectra, the resolution would be reduced to 0.2 or 0.4 eV, in order to resolve fine structure and lineshape detail (Figs. 5.1 and 5.2).

Operation of the SSA under optimum conditions (maximum spectrometer transmission for a given energy resolution) yields the following dependence of $E_{\frac{1}{2}}$ (the analyser full width at half-maximum, fwhm, energy resolution):

$$E_{\frac{1}{2}} = E_A \left(\frac{w}{2R_o} + \frac{\alpha_i^2}{4} \right) \tag{6}$$

where w is the input/exit analyser slit width, R_o is the analyser mean radius, α_i is the analyser acceptance half angle in the dispersion plane and E_A is the analyser pass energy (Section 6.2).

Since $E_{\frac{1}{2}}$ is proportional to E_A but not to w, it is important that experimental variation of $E_{\frac{1}{2}}$ be accomplished by varying the analyser pass energy and *not* the input/exit slit width, which should be considered an instrumental constant with α_i. Improved energy resolution may of course be gained only at the expense of sensitivity (Fig. 5.7). For the SSA, spectral intensity varies as:

$$I \propto E_A^n, \tag{7}$$

where the exponent n is determined by overall spectrometer design and mode of operation, and normally takes a value in the range:

$$1 \leqslant n \leqslant 2.$$

4.2.4 Spectrum acquisition

For analogue operation, an acquisition time of between 1 minute (narrow scan at low resolution) and 20 minutes (survey scan at high resolution) would normally be selected. For an analogue scan, it is also necessary to select the time constant (TC) of the ratemeter. The time constant determines how quickly the analogue output of the ratemeter will respond to a change in incoming pulse rate – it acts like a variable low-pass filter. Experimentally, it is found that, to avoid distorting and broadening spectral features, it is necessary to select a time constant such that

$$TC \leqslant 3 \cdot t \cdot \Delta E^{\frac{1}{2}}/W \text{ (s)}, \tag{8}$$

where t is the acquisition time in minutes, W is the scan width in eV, and ΔE is the full width at half maximum of the narrowest feature in the spectrum. The concept of a time constant is not, of course, relevant in digital operation. The important parameters in this case are the number of channels in the spectrum and the dwell time (counting time) per channel. In digital acquisition one would normally operate at one channel per eV for survey spectra and 5–10 channels per eV for high-resolution spectra, with dwell times per channel of 100–1000 ms, giving single scan times of the order of 10 s (fast, narrow scan) to 1000 s (slow survey scan). Digital acquisition has the advantage, of course, that repetitive scans can be made to increase the total number of accumulated counts and thereby (since spectral noise varies as the square root of the signal) improve the signal-to-noise ratio. In addition, a digital acquisition provides the

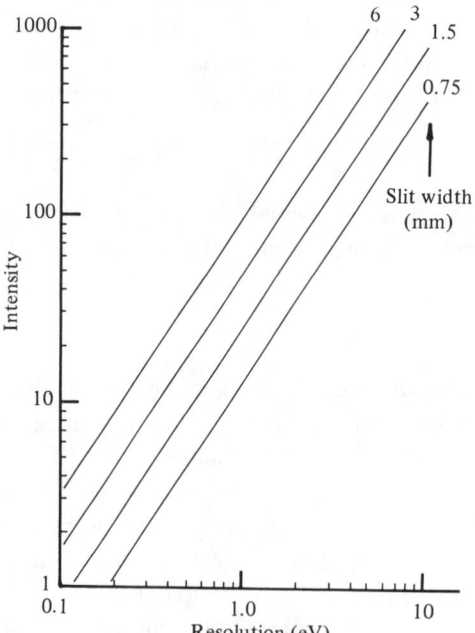

Fig. 5.7. Intensity vs energy resolution in the spherical sector analyser. Theoretical variation of intensity (peak area) with instrumental energy resolution (fwhm of the spectrometer resolution function), for various values of input/exit slit width, for an instrument similar to the VG Scientific ESCALAB MkII, assuming optimum operating conditions (see text). Spectral resolution will be determined by a convolution of the analyser resolution (above) and X-ray and specimen functions (see text).

possibilities of pre-programming a number of scan regions and acquisition conditions in advance, as well as allowing subsequent processing of the data (Section 4.5).

4.3 Binding energy measurement and calibration

One of the most important applications of XPS is in the accurate measurement of core-level binding energies. From the photoelectron spectrum, therefore, we are able to directly measure the binding energies of all core-levels, to a precision of 0.1 eV. In order to make practical use of these measurements, however, it is necessary first to carry out an accurate calibration of the spectrometer energy scale.

The detailed procedures for such a calibration are described fully in the literature (Anthony & Seah, 1984a, b), hence only a brief mention will be made here. The method involves internal calibration of the spectrometer energy scale against core-level features of known binding energy from standard materials (Table 5.5).

By convention, binding energies refer to core-level energy relative to the spectrometer Fermi level. For conducting specimens in good electrical contact with the spectrometer, the Fermi levels of both sample and spectrometer will be equivalent and absolute measurements of core-level binding energies can be made. For insulating specimens, however, it is not possible to make absolute binding energy measurement and it is necessary, instead, to rely on some internal standard for energy referencing. Although several methods of internal referencing have been suggested, the most widespread method is to use the 1s binding energy of adventitious carbon (a contaminant observed on almost every as-received surface) as a reference, at 285.0 eV (Swift, 1982).

4.4 Peak assignment

Peak assignment in XPS is generally straightforward, and very few problems arise in practice. One of the most valuable assets to every

Table 5.5. *Binding energy standards for XPS (from Anthony & Seah, 1985a)*

Material	Core-Level	Binding Energy (eV)
Copper	$2p_{3/2}$	932.67 ($\pm$ 0.02)
	$3p_{3/2}$	75.14 ($\pm$ 0.02)
Silver	$3d_{5/2}$	368.28 ($\pm$ 0.02)
Gold	$4f_{7/2}$	83.99 ($\pm$ 0.02)

150　A.B. Christie

electron spectroscopist will be a compilation of core-level binding energies for the entire periodic table, such as in Table 5.1. The same information in graphical form can be useful for quick referencing. In addition, the assignment of spectral features can be greatly aided through the use of peak library routines.

Fig. 5.8. Data processing in XPS. Examples of frequently used data processing routines in XPS, applied to the Ge2p$_{3/2}$ photoelectron peak from a 1 cm^2 germanium specimen with native oxide. (a) Raw data acquired with AlK_α radiation, showing presence of element and oxide components superimposed on a secondary electron background (broken line). (b) Spectrum A after background subtraction. (c) Spectrum B after digital smoothing with a 13-channel Savitsky–Golay cubic function. (d) Auto-peak synthesis of spectrum B, with a normalised chi squared (goodness of fit) value close to unity. Component peak parameters are:

	Element	Oxide
Binding energy (eV)	1217.9	1220.3
fwhm (eV)	1.6	2.0
Lorentz character (%)	91.5	7.6
Intensity (area) (%)	72	28

Almost identical results were obtained by auto-fitting to spectrum C (smoothed data), indicating minimal distortion due to smoothing. Attempts to fit the data with identical lineshapes for element and oxide (same % Lorentzian character) were unsuccessful (chi squared > 15), illustrating the value of including this variable in the peak-fitting parameters.

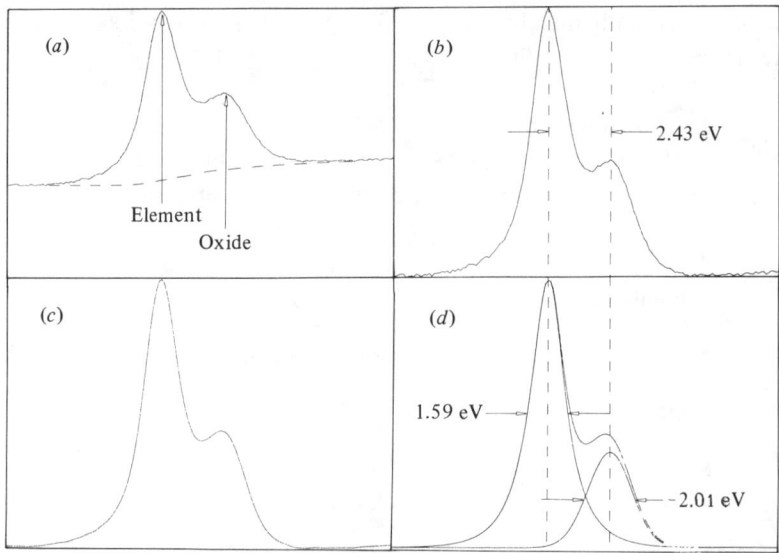

Some of the more common problems in peak assignment are due to the following:

> Auger features (often complex);
> X-ray satellites (e.g., the K_β of Al or Mg);
> energy loss features (e.g., plasmon losses, shake-up);
> 'ghost' features (due to contaminating radiation).

Auger features can be identified by the fact that they do not shift in kinetic energy (as photoelectron features do) when changing from AlK_α to MgK_α radiation or vice versa. X-ray satellites and ghost features appear at well-defined energies relative to the principal features. Most datasystems actually incorporate a satellite removal routine. By far the most useful aid in peak assignment, however, is experience.

4.5 Data processing

Data processing in XPS has been the subject of a comprehensive review by Sherwood (in Briggs & Seah, 1983), hence only a summary will be given here (see Fig. 5.8). Although it is possible to attempt some form of data processing on analogue spectra, this must be seen as a poor alternative to the processing of digital spectra. For this purpose, a dedicated on-line microcomputer is by far the best solution, although off-line processing on a mainframe can be useful in certain very specialised areas (e.g. deconvolution). Much of the early pioneering work on data processing in XPS (such as Sherwood's) was carried out on custom-built datasystems, often based on very crude and simple hardware. Despite the fact that the capital cost associated with such home-built systems is often very low, it is all too easy to underestimate the amount of labour involved in developing one's own datasystem software and interfaces (Hazell, Baker & Dearden, 1985).

Commercially available datasystems generally include most, if not all, of the following processing routines:

> spectrum display and expansion;
> spike removal;
> peak integration and area measurement;
> baseline/background removal;
> satellite removal;
> spectral comparison/addition/subtraction/ratio;
> spectrum smoothing;
> differentiation;
> deconvolution;

curve synthesis/peak fitting;
quantification.
Quantification in XPS forms the subject of Section 6.

4.6 Composition-depth profiling

Use of the electron spectrometer in the ways described above will provide information about the surface composition, averaged over about 3 nm of depth, of any particular sample. In most cases, such information is extremely valuable, and further analysis may not be necessary. However, it is also possible to use XPS to investigate the variation of composition with depth into the sample in several ways, thereby producing a composition-depth profile of the surface of interest.

4.6.1 Analysis using different core-levels of the same element

Since they appear at different energies in the spectrum, photoelectrons from different core-levels of the same element will have different mean free paths, and hence will have different sampling depths. Photoelectrons from the 3d level of germanium, for instance, will provide information to much greater depths than photoelectrons from the 2p level. When excited with AlK_α radiation, the difference in sampling depth between these two levels (from the $\lambda \propto E_k^{0.5}$ relationship, see Section 2.5) is well over a factor of two. This fact may be used to determine, from a single XPS spectrum, whether a particular species is enriched or depleted in a specimen surface, relative to the bulk.

4.6.2 Analysis using different X-ray energies

Since the photoelectron kinetic energy is related to the X-ray energy ($h\nu$) by the Einstein relation (equation (1)), then, by the same argument used above, it can be shown that the sampling depth of photoelectrons from any given core-level of any given element will increase with increasing X-ray energy. The sampling depth of photoelectrons excited from the copper 2p level, for instance, increases by a factor of 3.4 on going from MgK_α to TiK_α excitation. Thus, XPS spectra excited using suitably different X-ray sources may be used to generate crude non-destructive composition-depth profiles.

4.6.3 Analysis using different emission angles

As discussed in Section 2.5, the sampling depth in XPS is a function of the angle of emission of the photoelectrons. A spectrum recorded at normal emission, for example, would have three times the sampling depth of a spectrum recorded at 70° to the surface normal.

Acquisition at a series of different emission angles may therefore be used to provide a qualitative form of non-destructive depth profile to depths up to several nanometres, and this principle forms the basis of the auto-tilt angle-resolved XPS (ARXPS) facility available on some commercial spectrometers (Fig. 5.9).

The observed intensity is actually given by a Laplace function of the real concentration, and the real concentration vs depth functions may

Fig. 5.9. Angle-resolved XPS applied to germanium. Variation of photoelectron peak intensity (area) with emission angle (θ, with respect to the specimen normal) for the germanium specimen described in Fig. 5.8. Since the data were recorded on a VG Scientific ESCALAB MkII with $d_s/d_a = 5$ (Fig. 5.10), the absolute I vs θ data is considered reliable for values of θ up to approximately 75°.

The monotonic increase in C1s intensity from $\theta = 0$ to 75° is indicative of a carbon contamination overlayer, and this behaviour mirrors the monotonic decrease of the Ge2p (substrate) intensity over the same range. The O1s intensity is observed to rise with increasing θ up to $= 70°$, but falls markedly at greater angles, indicating an oxygen-rich subsurface layer. The Ge3d intensity is seen to exhibit a complex dependence on θ, and this may be understood in terms of the anisotropy of emission of Ge3d photoelectrons from the single crystal germanium substrate due to channelling and elastic scattering events. No structure is observed in the Ge2p ARXPS profile because these electrons (kinetic energy 270 eV) have a very low free path compared to the 3d electrons (kinetic energy 1460 eV) and hence are not significantly influenced by the substrate crystal structure. The higher mean free path of the 3d photoelectrons is reflected in the relative insensitivity of the 3d intensity (apart from the fine structure discussed above) to angle, for values of θ up to 60°.

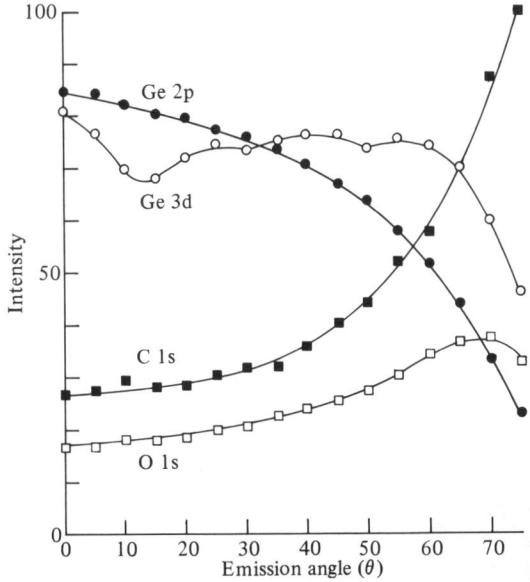

in principle be obtained from the inverse Laplace transformation, as described in the literature (Bussing & Holloway, 1985). Such numerical deconvolution of ARXPS profiles is unlikely to be of much practical value in the majority of real applications. It is probable that iterative auto-profile-fitting routines, analogous and most likely very similar in operation to the auto-peak-fitting routines for peak synthesis (Section 4.5), will be of much greater value in the quantitative interpretation of ARXPS data (Hazell, Brown & Freisinger, 1986; Watts, Castle & Ludlam, 1986).

The accuracy of ARXPS profiles will depend to a large extent on certain instrumental and operating parameters of the spectrometer, such as

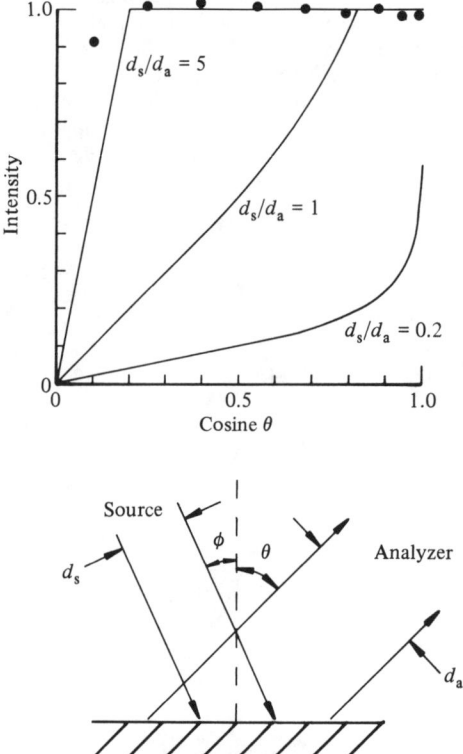

Fig. 5.10. Variation of sensitivity with emission angle in XPS. Theoretical (solid line) and experimental (filled circles) variation of sensitivity (photoelectron intensity from a clean, smooth substrate) with angle of emission (θ, measured with respect to the surface normal). Angular response calculated for various values of the ratio d_s/d_a, where d_s is the X-irradiation source dimension in the rotation plane and d_a the analysed area dimension in the rotation phase, as defined in the lower part of the diagram (experimental data taken from the VG Scientific ESCALAB MkII, for which $d_s/d_a \sim 5$).

angular resolution of the analyser input lens and spatial uniformity of the X-ray flux. For high-quality profiles a lens half-angle of 3° or less should be selected. For high sensitivity and ease of interpretation, especially towards grazing angles of emission, a high X-ray flux and large irradiated area should be selected. If the irradiated area is smaller than the analysed area (seen by the lens/analyser combination) sensitivity will fall off markedly with increasing angle of emission (Fig. 5.10).

Fig. 5.11. XPS sputter-depth profiling. XPS Concentration vs etch time profile from a multilayer chromium–nickel standard reference material specimen. The profile was produced using 2.5 keV argon as the sputtering species. Selected area analysis for XPS was achieved by input lens imaging of the photoelectrons onto a 2 mm analyser aperture (see Section 5), resulting in good depth resolution throughout the 0.5 microns of depth. The present data compare very well with the best published data using Auger electron spectroscopy with a 2 keV argon ion source, demonstrating the ability of XPS to produce high quality sputter-depth profiles under favourable conditions (data courtesy of VG Scientific).

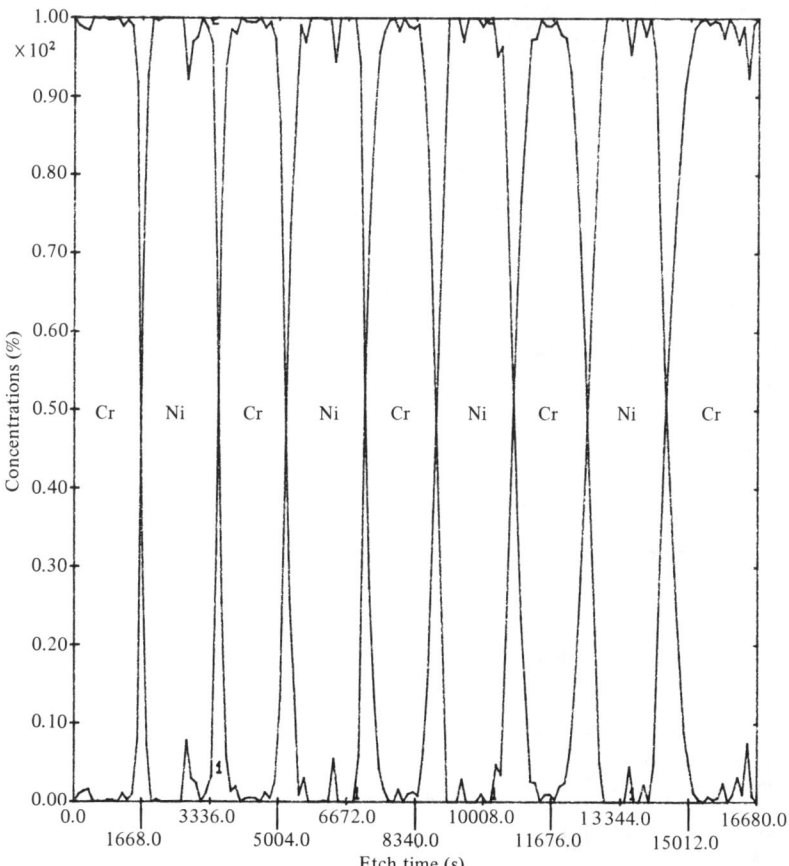

4.6.4 Sputter-depth profiling

All of the preceding methods of composition-depth profiling have the advantage of being non-destructive, and are the only practical methods where sensitive materials (polymers, organic and inorganic compounds, and most oxides) are being studied. They have a common limitation, however, in the fact that the maximum sampling depth obtainable, even under optimum conditions, is only about 10 nm. In order to probe to greater depths, using XPS, it is necessary to remove some of the sample surface, thereby exposing underlying layers for analysis.

Although techniques for removal of material external to the spectrometer have been developed (e.g. chemical or electrochemical etching), by far the most popular method is *in-situ* inert-gas ion etching of the specimen. Until recently, only two types of ion source have been widely used for sputter-depth profiling, and both have severe limitations, as far as XPS is concerned. The electron bombardment type, favoured for AES studies because of the high ion current densities and well-characterised beam that it produces, is not suitable for etching large areas of surface (essential for XPS), since the *total* ion current in the beam is inadequate. The discharge type, generally used in XPS studies, whilst it can be operated to give a high total ion current, suffers from the disadvantage of an unstable beam (such that it is virtually impossible to characterise) and a non-uniform ion current density over the sample surface. Consequently, XPS sputter-depth profiles are generally inferior, as regards depth resolution, to AES sputter-depth profiles. Recently, however, duoplasmatron ion sources, which combine the high total current of the discharge source with the uniformity and stability of the electron bombardment-type source, have been successfully applied to XPS sputter-depth profiling. A major limitation of ion etching for XPS composition-depth profiling is the inherently destructive nature of the ion beam, which leads to both chemical and topographical modification and erroneous conclusions if sufficient care is not taken in interpretation. A treatment of ion sputtering applied to surface analysis can be found in Chapter 2.

5 Quantitative surface analysis
5.1 *Basic principles*

A rigorous treatment of the theory underlying quantitative XPS is beyond the scope of this book – suitable references may be found in the literature (Seah, 1980; Wagner *et al.*, 1981; Briggs & Seah, 1983). In practice, however, the quantification of XPS spectra is quite straightforward.

Firstly, one must obtain a value for the peak intensity (peak area following background removal) of the principle spectral feature of each element detected in the XPS spectrum. A fractional atomic concentration, C_A, of element A is then given by:

$$C_A = \frac{I_A/S_A}{\sum_n (I_n/S_n)}, \tag{9}$$

where I_n is the measured peak intensity for element n and S_n is the relative atomic sensitivity factor (ASF) for that peak.

Although it involves a number of fairly crude assumptions, equation (9) does have the advantage of being almost universally applicable. Strictly, it applies only to a specimen surface which is homogeneous throughout the sampling depth of the technique, and special forms of equation (9) must be used for layered or heterogeneous specimens (Paynter, 1981). Values of S_n may be calculated from theory, or derived empirically by recording spectra from standard materials.

The former method is generally unsatisfactory, since there is considerable evidence to show that theoretical values of photoionisation cross-sections (see Section 2.2) are not necessarily reflected in experimentally measured photoelectron peak intensities because of poorly understood processes, such as shake-up and shake-off, which lead to a loss in intensity. The empirical method, although preferred, can prove extremely tedious and time-consuming unless one is working with only a small range of elements (Battistoni, Mattogno & Paparazzo, 1985). In general, XPS users tend to refer to one or more of the several published sets of experimentally derived relative atomic sensitivity factors (Nefedov *et al.*, 1975; Evans, Pritchard & Thomas, 1978; Wagner *et al.*, 1981).

5.2 *Intensity measurements*

There is a growing school of thought which believes that, of the many different methods of intensity measurement, it does not really matter which method, or combination of methods, is used, as long as one is consistent (Battistoni, Mattogno & Paparazzo, 1985). This means, also, that the method of intensity measurement for the unknown must be consistent with the method used for the standard materials. The problem associated with peak intensity measurement is centred on the spectral background. The problem has been treated in detail by Seah (in Briggs & Seah, 1983), who concludes that there are basically three different forms of background in XPS – linear, step and shifted step. For most applications, linear or step backgrounds should be used, but in certain cases of very broad or overlapping peaks there may be some advantage in using a shifted step background.

It is likely, with increased use of microcomputers and on-line data processing, that intensity measurements on state-of-the-art instruments in the foreseeable future will be carried out by direct comparison and fitting of the unknown lineshape with a library set of lineshapes from standard

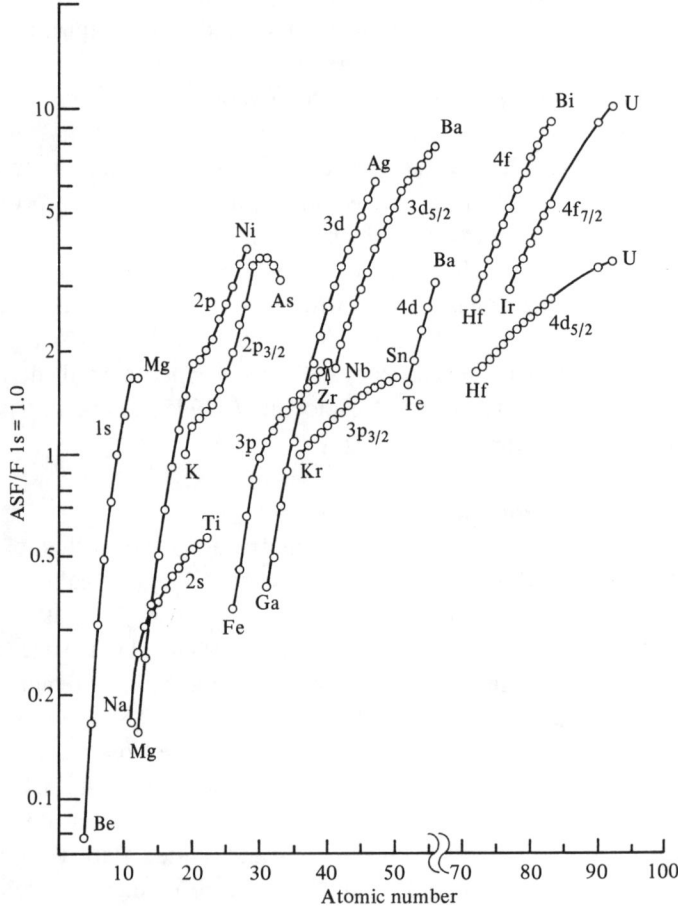

Fig. 5.12. Empirical XPS atomic sensitivity factors. Atomic sensitivity factors (peak area) relative to F1s at unity, for the most commonly encountered elements and photoelectron peaks. Atomic sensitivity factors (ASFs) valid for a spectrometer with $(E_k)^{-0.5}$ transmission (cf VG Scientific ESCALAB MkII, see Fig. 5.11), have been calculated from the data of Wagner et al., 1981. Although strictly valid only for AlK_α excitation, the probable error involved in using such literature values of ASF far exceeds the error involved in assuming their validity for MgK_α excitation.

materials, thus eliminating the need for background definition altogether. The problem of absolute intensity measurements can then be reduced to a mathematical–statistical problem of curve fitting and spectral comparison.

5.3 Instrumental considerations

In order to derive theoretical atomic sensitivity factors from photoionisation cross-sections, it is necessary to take into account a number of instrumental parameters which affect sensitivity. Foremost in importance amongst these parameters is the instrument transmission, and the way in which this varies with photoelectron kinetic energy for the spectrometer under consideration. The transmission functions of several models of commercial spectrometers have been investigated, from both theoretical and experimental points of view, by a number of workers (Richter & Peplinski, 1980; Seah, 1980; Hughes & Philips, 1982). It is unfortunate, however, that most of the present state-of-the-art commercially available XPS spectrometers have yet to be characterised.

A knowledge of the instrument transmission is not necessary if we derive experimental atomic sensitivity factors from internal standards. The instrument transmission will be identical for standard and unknown (providing the instrument is operated under the same analytical conditions), and hence its exact form is irrelevant. If we use, as most spectroscopists do, the published experimental atomic sensitivity factors derived for another instrument, then it is necessary to know the energy dependence of spectrometer transmission for both our instrument and the one on which the sensitivity factors were derived, in order to perform a transmission function correction to the data (Seah, Jones & Anthony, 1984). Atomic sensitivity factors (ASF) for use with the VG ESCALAB MkII are presented in Fig. 5.12.

6 State-of-the-art XPS
6.1 Spatial resolution

Until recently spatial (lateral) resolution in XPS has been of the order of several millimetres, and this has proven a major limitation and handicap in many areas of application. Within the last few years however, attempts have been made to improve spatial resolution both by focussing the incident X-rays and by imaging the emitted photoelectrons (Drummond, Cooper & Street, 1985). One of the earliest attempts to obtain spatially resolved XPS was made by Cazaux, but his method of transmission photoelectron spectroscopy has not been commercialised due to major problems of specimen preparation and manipulation

(Cazaux, 1977). Commercially available small-spot X-ray source photoelectron spectrometers make use of X-ray back-diffraction from a suitable quartz crystal microfocus monochromator. The conventional X-ray monochromator described in Section 3.6 is based on the Johann approximate focussing geometry. All standard monochromators employed in XPS are of this type. The Johann geometry gives excellent monochromatisation of the source, but produces a fairly large irradiated area on the specimen (approximately 0.6 mm in the dispersion plane). The perfect focussing characteristics of the Johannson geometry (where the crystal is both toroidally bent *and* mechanically polished to lie along the Rowland circle), when combined with a suitable fine-focussed X-ray anode, are capable of producing a monochromatised X-ray flux focussed into a very much smaller area. The back-diffracted X-ray spot diameter is limited by the finite energy-dispersion characteristics of the monochromator, which for Al$K\alpha$ back-diffracted from quartz yield a dispersion limited spot size (d in millimetres) of:

$$d = 5.10^{-4} Rm, \tag{10}$$

where Rm is the monochromator Rowland circle diameter in millimetres. In practice, Rm is unlikely to be reduced much below 200 mm hence the best spatial resolution that can be expected using this method is 100 μm. To achieve microfocussing of X-rays in the range 100–200 microns, however, it is necessary to reduce the total power dissipation in the anode to a very low value – approximately 10 watts total anode power for a microfocussed source compared to up to 1000 watts for a conventional monochromator. This means that in order to achieve a finely focussed source we must reduce the X-ray flux and hence also the photoelectron signal by up to two orders of magnitude.

Almost all commercial XPS instrument manufacturers now favour electron optical aperturing over the microfocus monochromator for state-of-the-art performance in spatially resolved XPS (Yates & West, 1983b). In this method, photoelectrons from a uniformly irradiated specimen are imaged through the analyser input lens onto an area-defining aperture which transmits only those electrons originating from a small selected area on the specimen. The method is often referred to as small or selected area XPS (SAX). The spatial resolution afforded by this technique is limited only by spherical aberration in the input lens, which for a well-designed lens of typical dimensions gives:

$$d \approx 180 \alpha_0^3, \tag{11}$$

where d is the aberration limited spatial resolution in millimetres and α_0 is the semi-angle of acceptance of the input lens in radians. For normal

XPS applications, where d may be in excess of 1 mm, α_0 can be increased to as much as 0.2 radians (12°). In SAX applications, α_0 may be reduced to as little as 0.05 radians (3°) and in this case the limiting spatial resolution is 22 microns. In practice, specimen photoelectron brightness will limit the useful resolution to several tens of microns, determined by the area defining aperture diameter and not by lens aberration. This has the result that the lateral resolution function in SAX has far better definition (flat topped with 22-micron edge resolution) than the expected 150-micron fwhm near gaussian resolution function of the microfocus monochromator (Fig. 5.13). Electron optical aperturing offers several further advantages

Fig. 5.13. Spatial resolution in XPS. The variation of sensitivity (transmitted intensity) across the analysed diameter under small-area XPS (SAX) analysis conditions. The thick solid line shows the measured intensity of 1000 eV kinetic energy electrons from a uniform silver specimen as a 10 keV electron beam is scanned across the surface. The fwhm of the response function is determined by the analyser input aperture, and is shown to be 150 μm in the diagram. The quality of the electron optics is illustrated by the edge resolution of the linescan, which is limited mainly by spherical aberration in the transfer lens to approximately 22 μm. The data indicate that 95% of the total transmitted intensity across the analysed diameter arises from a 150 μm region. This is in contrast to the figure of 76% intensity from the region enclosed by the fwhm of a Gaussian distribution function (indicated by the thin solid line) (data courtesy of VG Scientific).

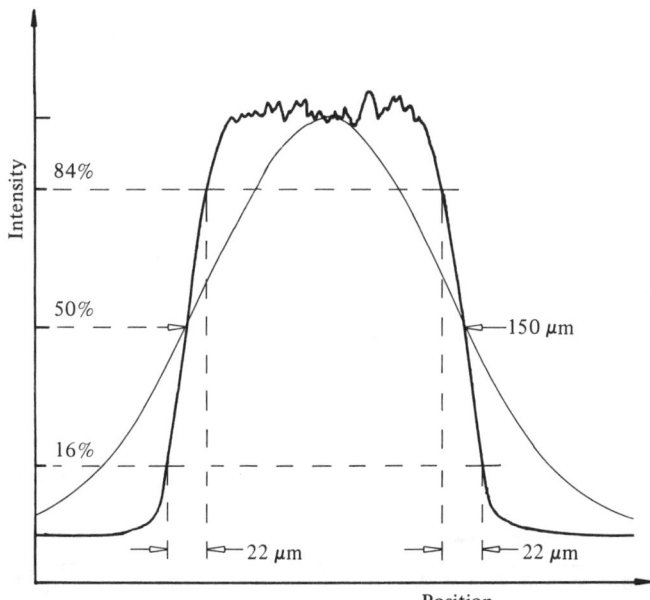

over the microfocus monochromator:

> It may be used with any source of X-radiation.
> Higher X-ray flux densities and photoelectron brightnesses are obtained with the direct radiation X-ray source.
> Ideally suited to multitechnique instrumentation.
> Relatively inexpensive addition to conventional XPS.

The provision of a high-resolution, low spherical aberration input lens for SAX operation as described above leads quite naturally into an area of photoelectron spectroscopy which has yet to reach commercial realisation – imaging XPS. One of the properties of the SSA, used in almost all commercial instrumentation, is the fact that it possesses first-order focussing in space and angle in both the dispersive and non-dispersive planes. This means that any spatial and angular distribution of monochromatic photoelectrons at the entrance plane will be reproduced to first-order at the exit plane. In combination with a spatially extended

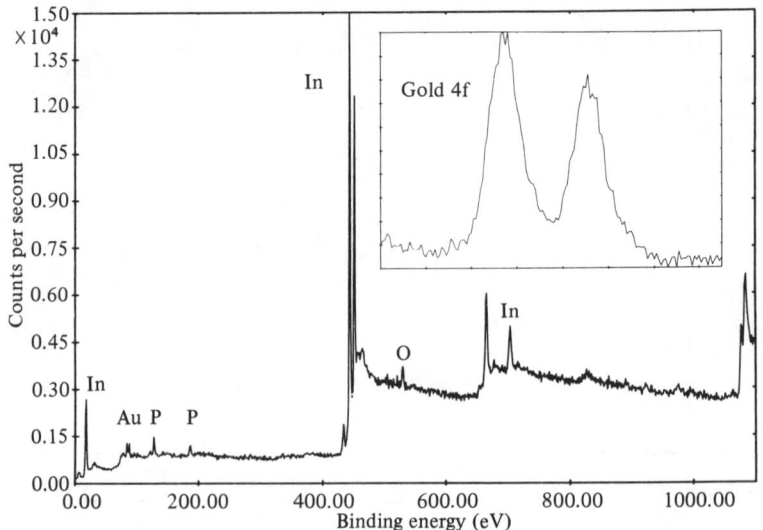

Fig. 5.14. Small-area trace analysis by XPS. XPS survey spectrum and (inset) gold 4f high resolution spectrum from an indium phosphide substrate implanted with gold. The implant was carried out in a 300 micron square area, at a dose of 10^{16} atoms cm^{-2}, representing an absolute implant level of only 10^{13} atoms. It is likely that only 10% of the implanted atoms are within the sampling depth of XPS (approx. 5 nm), hence the gold contribution to the spectra above arise from approximately 10^{12} atoms. The spectra were acquired using 300-micron selected area analysis conditions, and clearly show that the detection limit for gold in this study is of the order of 10^{11} atoms, or 30 picogrammes (data courtesy of VG Scientific).

X-ray source, a suitable input lens and two-dimensional spatially resolved detector, therefore, the SSA offers the real possibility of recording energy resolved photoelectron images under the correct conditions (Gurker, Ebel & Ebel, 1983).

Imaging XPS is likely to be one of the most fruitful and exciting areas of development over the next few years.

One other, completely different, approach exists which offers the possibility of spatially resolved XPS. The photoelectron spectromicroscope (PESM), utilises the electron optical magnification and imaging properties of a powerful, diverging magnetic field to form a photoelectron image of the specimen on a suitable channelplate multidetector (Beamson, Porter & Turner, 1981). Although the imaging properties of the instrument appear to be quite good (especially for UPS applications), the energy resolution of the technique is poor by XPS standards.

6.2 Spectrometer sensitivity and multichannel detection

Multichannel detection (MCD) energy analysers have gained widespread acceptance in XPS over the last few years. The SSA produces an energy dispersion across the exit plane of $(2R_0/E_A)\,\text{mm}\,\text{eV}^{-1}$, where R_0 is the analyser mean radius, and E_A the analyser pass energy (see Chapter 3). For any given finite exit slit width w_e we may then collect the transmitted electrons using a suitable single channel detector, with an energy resolution determined by a convolution of terms in w_e, w_i (the input slit width) and α_i (the semi-angle of acceptance of the analyser in the dispersion plane). The optimum combination (maximum transmission for a given overall resolution) is when:

$$w_e = w_i = 2R_0 \cdot \alpha_i^2, \tag{12}$$

and under these conditions:

$$E_{1/2} = \frac{0.62 E_A \cdot w}{R_0}, \tag{13}$$

where $E_{1/2}$ is the fwhm of the analyser resolution function (Briggs, 1977).

If we replace the single slit w_e with N parallel output slits, each of width w_e, across the dispersion plane then using suitable output signal processing we are able to collect the same energy spectrum as in the single channel case but at N times the collection rate. In other words, the N channel detector provides us with a factor N multiplex advantage (Fig. 5.15). This is, however, not the complete story.

For an extended source, spectrometer sensitivity at a given resolution ($E_{1/2}$) is determined by the product of etendue (G) and number of channels in the MCD (N):

$$S = G \cdot N. \tag{14}$$

G is determined by lens/analyser geometry and operating conditions as discussed in detail by Seah (Seah, 1980). For a conventional SSA/lens combination,

$$G = \frac{w_i \cdot L \cdot \Omega}{M^2}, \tag{15}$$

Fig. 5.15. Schematic cross-section of lens, analyser and detector in XPS. Input lens limits acceptance from the specimen to a cone of semi-angle α_o, and projects these electrons onto the input slit/aperture, width w_i, of the energy analyser. Here (or in the lens if a retarding lens is used) electron refraction occurs due to retardation, and the acceptance angle in the dispersion phase of the analyser (α_i) may be very different to α_o. In any case, α_i is generally limited by the lens output (retarding lens system) or by the analyser input (non-retarding lens). Transmitted electrons are dispersed in energy across the analyser exit plane, but energy resolution is limited by α_i and w_i. The shortfall in electron trajectory caused by α_i is given by $2R_o\alpha_i^2$, where R_o is the analyser mean radius. Optimum transmission vs resolution characteristic are when $w_e = w_i = 2R_o\alpha_i^2$. The configuration for MCD operation, with increase in transmission equivalent to the number of channels, N, is shown inset.

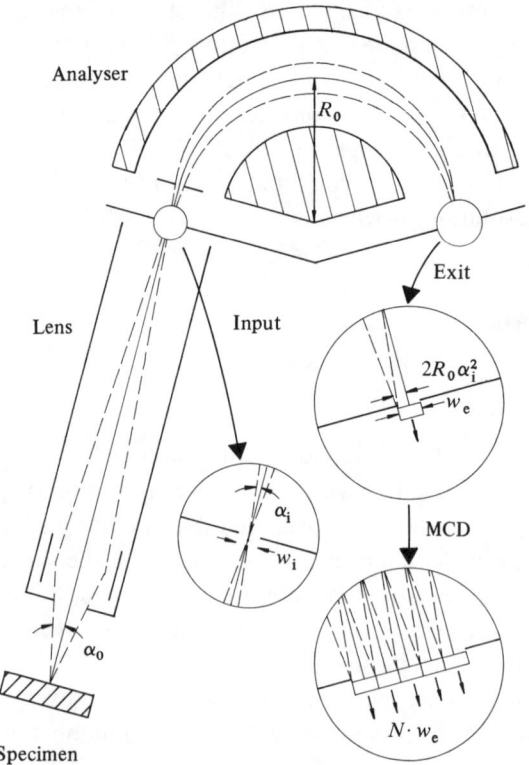

where L is the input/exit slit length, M the input lens magnification and Ω the solid angle of collection of the lens/analyser combination. If α_0 is the half-angle of acceptance of the lens and β is the half-angle acceptance of the analyser in the non-dispersive plane, then there are three regimes of operation, and Ω is given by:

Regime (1)

$$\Omega_1 = \pi\alpha_0^2 \quad \text{for} \quad \alpha_0 < M \cdot \alpha_i \left(\frac{E_A}{E_K}\right)^{1/2}, \tag{16}$$

Regime (2)

$$\Omega_2 = \pi\alpha_0 \cdot M \cdot \alpha_i \left(\frac{E_A}{E_K}\right)^{1/2}$$

$$\text{for} \quad M \cdot \alpha_i \left(\frac{E_A}{E_K}\right)^{1/2} < \alpha_0 < M\beta \left(\frac{E_A}{E_K}\right)^{1/2}, \tag{17}$$

Regime (3)

$$\Omega_3 = \pi M^2 \alpha_i \beta \left(\frac{E_A}{E_K}\right) \quad \text{for} \quad \alpha_0 > M\beta \left(\frac{E_A}{E_K}\right)^{1/2} \tag{18}$$

(see Seah, 1980, for a complete discussion of the above arguments).

Combining equations (12)–(18), we may express S in terms of the input lens parameters and total MCD area, A (where $A = N \cdot w \cdot L$):

$$S_x = \Omega_x \frac{w \cdot L \cdot N}{M^2}, \tag{19}$$

Regime (1)

$$S_1 = \frac{\pi \cdot \alpha_0^2}{M^2} \cdot A, \tag{20}$$

Regime (2)

$$S_2 = \frac{0.9\pi \cdot \alpha_0}{M} \cdot \left(\frac{E_{1/2}}{E_K}\right)^{1/2} \cdot A, \tag{21}$$

Regime (3)

$$S_3 = 1.1\pi \cdot \beta \cdot (N \cdot L \cdot R_0)^{1/2} \left(\frac{E_{1/2}}{E_K}\right) \cdot A^{1/2}. \tag{22}$$

We see that for regimes 1 and 2 (corresponding to β unfilled) S is *independent* of N, i.e., sensitivity is determined *only* by total detector area (A) and not by individual channel width (w) or number of channels (N).

Most modern commercial spectrometers (Anthony & Seah, 1984b) are expected to operate only in regimes 1 and 2 (β never filled), hence any improvement in sensitivity due to MCD operation is strictly due to the effective increase in detector area only. Provided the input slit w_i is always filled by the source, the gain in sensitivity is independent of whether the MCD is split into three or 300 separate channels.

In general, the MCD active area will be limited by the physical size of the analyser (a useful rule of thumb is $N \cdot w < R_0/4$) hence sensitivity will increase with analyser size for all operating regimes. MCD operation may cause unexpected problems or compromises in performance. The most popular form of MCD, the position sensitive detector (PSD), may typically operate with a dead time of 2–5 ms. This means that the maximum total random counting rate, if we wish to avoid pulse pile up and loss of linearity in the detection system, is of the order of 20–50 kcps. This compares with typical counting rates up to 1 Mcps for a single channeltron detector. In addition, the modes of operation of certain other types of MCD may result in increased background counting rates and spectral noise, reducing any real advantage over the single channel performance.

6.3 *Automated sample handling and analysis*

As the XPS spectrometer evolved from a fundamental research tool into an analytical instrument for routine applications, increasing demands for automated multi-specimen handling, analysis and data reduction were placed upon it. With the majority of new spectrometers now entering industrial rather than academic environments these capabilities have become even more important. In quality control (QC) applications in particular automated operation is a prime requirement in order to accommodate a high specimen throughput with the minimum of operator skill and supervision. The obvious areas of automation of the XPS spectrometer may be summarised as follows:

6.3.1 *Specimen introduction and manipulation*

This requires the incorporation of highly sophisticated, high reliability mechanical transfer devices into the entry, preparation and experimental chambers. Such automation is generally desirable in QC environments where large numbers of physically similar specimens require unattended analysis. Severe constraints and restrictions on overall system flexibility and adaptability will generally be incurred as a result.

6.3.2 *Analytical procedures*

On single-specimen instruments this allows complete specimen analysis following a pre-set experimental menu which may include survey spectrum, high resolution spectra and depth profiling. Incorporation of computer-controlled motor drives on the specimen stage/manipulator allows automation of XPS linescan data acquisition (an area that will increase in popularity with the development of SAX) and angle resolved (auto-tilt) XPS data acquisition.

On multi-specimen instruments automated control of a specimen carousel allows sequential sample analysis (spectral acquisition, profiling) during unattended overnight operation. The ability to specify different analytical conditions and procedures for each specimen increases the flexibility of this facility.

6.3.3 Data reduction

Essentially a software capability, automated data reduction or batch processing allows pre-selected sequences of data manipulation (e.g. smoothing, curve fitting, quantification) routines to be applied in turn to batches of similar experimental data with the minimum of operator interaction. Once again, extremely valuable in QC applications of XPS.

References

Anthony, M.T. & Seah, M.P. (1984a). *Surface and Interface Analysis*, **6**, 95.
Anthony, M.T. & Seah, M.P. (1984b). *Surface and Interface Analysis*, **6**, 107.
Battistoni, C., Mattogno, G. & Paparazzo, E. (1985). *Surface and Interface Analysis*, **7**, 117.
Beamson, G., Porter, H.Q. & Turner, D.W. (1981). *Nature*, **290**, 556.
Bowling, R.A. & Larrabee, G.B. (1983). *Analytical Chemistry*, **55**, 133R.
Briggs, D. (1977). *Handbook of X-ray and Ultraviolet Photoelectron Spectroscopy*, London: Heyden.
Briggs, D.& Seah, M.P. (1983). *Practical Surface Analysis*, New York: Wiley.
Brundle, C.R. & Baker, A.D. (1981). *Electron Spectroscopy: Theory, Techniques and Applications*, Vols. 1–4 London: Academic Press.
Bussing, T.D. & Holloway, P.H. (1985). *J. Vac. Sci. Tech.*, **105**, 114.
Carlson, T.A. (1975). *Photoelectron and Auger Spectroscopy*, London: Plenum.
Castle, J.E. (1986). *Surface and Interface Analysis*. **8**, 137.
Castle, J.E. & West, R.H. (1980). *Journal of Electron Spectroscopy and Related Phenomena*, **19**, 409.
Cazaux, J. (1977). *Journal de Physique*, **38**, L473.
Christie, A.B., Sutherland, I. & Walls, J.M. (1983). *Surface Science*, **135**, 225.
Drummond, I.W., Cooper, T.A. & Street, F.J. (1985). *Spectrochimica Acta*, **40B**, 801.
Edgell, M.J., Paynter, R.W. & Castle, J.E. (1985). *Journal of Electron Spectroscopy and Related Phenomena*, **37**, 241.
Einstein, A. (1905). *Annalen der Physik*, **17**, 132.
Evans, S., Pritchard, R.G. & Thomas, J.M. (1978). *Journal of Electron Spectroscopy and Related Phenomena*, **14**, 341.
Gurker, N., Ebel, M.F. & Ebel, H. (1983). *Surface and Interface Analysis*, **5**, 13.
Hazell, L.B., Baker, C. & Dearden, D.P. (1985). *Surface and Interface Analysis*, **7**, 150.
Hazell, L., Brown, I.S. & Freisinger, F. (1986). *Surface and Interface Analysis*, **8**, 25.
Hofmann, S. (1986). *Surface and Interface Analysis*, **9**, 3.

Hughes, A.E. & Philips, C.C. (1982). *Surface and Interface Analysis*, 4, 220.
Jenkin, J.G. (1981). *Journal of Electron Spectroscopy and Related Phenomena*, 23, 187.
Koening, M.F. & Grant, J.T. (1985). *Surface and Interface Analysis*, 7, 217.
Lea, C. (1983). *Trends in Analytical Chemistry*, 2(5), 118.
Matienzo, L.J. & Shah, T.K. (1986). *Surface and Interface Analysis*, 8, 53.
Nefedov, V.I., Sergushin, N.P., Salyn, Y.V., Band, I.M. & Trzhakovskaya, M.B. (1975). *Journal of Electron Spectroscopy and Related Phenomena*, 7, 175.
Paynter, R.W. (1981). *Surface and Interface Analysis*, 3, 186.
Paynter, R.W., Castle, J.E. & Gilding, D.K. (1985). *Surface and Interface Analysis*, 7, 63.
Powell, C.J. (1985a). *Surface and Interface Analysis*, 7, 256.
Powell, C.J. (1985b). *Surface and Interface Analysis*, 7, 263.
Powell, C.J. (1985c). *J. Vac. Sci. Tech.*, A3(3), 1338.
Reilman, R.F., Msezane, A. & Manson, S.T. (1976). *Journal of Electron Spectroscopy and Related Phenomena*, 8, 389.
Richter, K. & Peplinski, B. (1980). *Surface and Interface Analysis*, 2, 161.
Scofield, J.H. (1976). *Journal of Electron Spectroscopy and Related Phenomena*, 8, 129.
Seah, M.P. (1980). *Surface and Interface Analysis*, 2, 222.
Seah, M.P. & Anthony, M.T. (1984). *Surface and Interface Analysis*, 6, 230.
Seah, M.P. & Dench, W.A. (1979). *Surface and Interface Analysis*, 1, 2.
Seah, M.P., Jones, M.E. & Anthony, M.T. (1984). *Surface and Interface Analysis*, 6, 242.
Siegbahn, K. (1985). *Photoelectron Spectroscopy: Retrospects and Prospects*, introductory lecture given at the Royal Society Meeting on *Studies of the Surfaces of Solids by Electron Spectroscopy: Recent Trends*, 28 February 1985.
Siegbahn, K., Nordling, C., Fahlman, A., Nordberg, R., Hamrin, K., Hedman, J., Johannson, G., Bergmark, T., Karlsson, S.E., Lindgren, I. & Lindberg, B. (1967). *ESCA: Atomic Molecular and Solid State Structure Studied by means of Electron Spectroscopy*. Uppsala: Almquist & Wiksells.
Swift, P. (1982). *Surface and Interface Analysis*, 4, 47.
Thompson, M., Baker, M.D., Christie, A. & Tyson, J.F. (1985). *Auger Electron Spectroscopy*, New York: Wiley.
van Oostrom, A. (1984). *Vacuum*, 34, 881.
Wagner, C.D., Davis, L.E., Zeller, M.V., Taylor, J.A., Raymond, R. H. & Gale, L.H. (1981). *Surface and Interface Analysis*, 3, 211.
Wagner, C.D. & Joshi, A. (1984). *Surface and Interface Analysis*, 6, 215.
Walls, J.M. & Christie, A.B. (1982). In: *Surface Analysis and Pretreatment of Plastics and Metals* (ed, D. Brewis), London: Applied Science.
Watts, J.F., Castle, J.E. & Ludlam, S.J. (1986). *Journal of Materials Science*, 21, 2965.
Webb, R.T. & Brinen, J.S. (1983). *Surface and Interface Analysis*, 5, 89.
Yarwood, J. (1967). *High Vacuum Technique*, London: Champman & Hall.
Yates, K. & West, R.H. (1983a). *Surface and Interface Analysis*, 5, 133.
Yates, K. & West, R.H. (1983b). *Surface and Interface Analysis*, 5, 217.

6

Static secondary ion mass spectrometry

J.C. VICKERMAN

1 Introduction

Secondary ion mass spectrometry is well known as a technique for high-sensitivity element analysis. The technique is basically an application of the sputtering phenomenon known to physicists since the 1890s. It consists of bombarding a surface with a beam of high-energy primary ions which, after transferring some of their energy to the lattice atoms, produce a cascade of atomic collisions within the solid, which results in the emission of secondary ions that can be analysed mass spectrometrically. These secondary ions are characteristic of the composition of the surface. In earlier days the technique was known as ion probe. Today its high sensitivity has made it a vitally important technique in the semiconductor industry. However, to be able to detect elements in the parts per billion region and below, high primary beam fluences are required which remove a number of monolayers per second. Clearly such analytical conditions, usually referred to as dynamic SIMS, are not suitable for surface analysis. In the late 1960s Benninghoven showed that it was possible to adapt the technique to maintain the surface integrity for periods well in excess of the analysis time. In this mode of operation the primary beam currents used should not exceed $5\ \text{nA cm}^{-2}$ and should preferably be 1 nA or below. Such conditions yield monolayer lifetimes of several hours to several days! As a consequence very high sensitivities are required of the detection system. Modern high-transmission mass filters and pulse counting equipment fulfil these requirements, resulting in a surface sensitive mass spectrometry, usually known as static SIMS or SSIMS.

The importance of static SIMS for surface analysis lies in the possibility it offers for studying not only the elemental composition but also the *chemical structure* of surfaces. This is because cluster ions are emitted as well as elemental ions. As a result a surface fragmentation pattern is generated. Thus all the chemical characterisation facilities associated with organic mass spectrometry should, in principle, be available to surface analysis.

There are a number of variants on the basic technique depending on the type of samples to be studied and the information required.

Many materials which are of research interest or are of technological and commercial importance are electrical insulators. In common with many other surface analytical techniques which utilise charged particles as the probe species, ion bombardment results in sample charging, which in the SSIMS experiment can cause spectral loss or at least severe instability. Low-energy electron neutralisation is frequently used in an attempt to eliminate the problem. Such a procedure is not always successful and, in addition, electron bombardment may cause decomposition or desorption. In the mid 1970s it was shown by Vickerman and co-workers that the charging problem could be virtually eliminated by the substitution of a beam of neutral particles for the primary ion beam and, in consequence, insulator materials could be routinely analysed (Surman et al., 1982). This modification of SSIMS is sometimes referred to as fast atom bombardment mass spectrometry, FABMS, or FABSIMS. Although it was originally developed for surface analysis its first spectacular application was for the analysis of involatile organic and bio-organic molecules (Barber et al., 1981 and Surman & Vickerman, 1981). However, it is becoming an increasingly important tool in applied surface analysis.

Spatial resolution is a requirement which is often of considerable significance in many surface problems. Although some degree of lateral resolution (c. 300 μm) was possible in dynamic SIMS, the ability to chemically characterise very small areas or features or to map the chemistry of a sample under SSIMS conditions only became possible in the early 1980s with the development of liquid-metal ion sources. These sources can be very finely focussed to yield submicron (down to 500 Å) spots on the surface, thus giving rise to a *chemical microscopy*.

2   The sputtering phenomenon

The basic process by which a high-energy particle impacts on a solid surface giving rise to the emission of secondary ions whose kinetic energy distribution maximises at very low energies (2–10 eV) has been the subject of much experimental and theoretical investigation. The mechanism or *mechanisms* involved are, however, far from understood. The field is plagued by partial experimentation and theoretical models whose predictions have been elevated to fact with little or no experimental support. There are a number of excellent reviews (see Chapter 2 and the

Static secondary ion mass spectrometry 171

reviews by Williams 1979, 1982). Here the basic experimental parameters underlying the SSIMS process will be outlined, and *some* of the mechanistic models which attempt to describe the sputtering process will be briefly described.

2.1 *The experimental parameters*

In static SIMS we are concerned with bombardment by low-energy (100 eV–10 keV), low-flux density (< 5 nA) primary beams. The most significant variables involved firstly are the sputter yield, S, of secondary particles emitted per primary particle impact; and secondly the ionisation probability, R^+, which is the probability that a given particle will be emitted as an ion. Clearly, as the following equation indicates, the sensitivity of the technique for a particular ion depends on the sputter yield and the ionisation probability:

$$i_S^M = i_p S R^+ \theta_M \eta,$$

where i_S^M is the secondary ion current of an element or species M, i_p is the primary particle flux, θ_M is the fractional coverage of m and η is the transmission of the analysis system. The sputter rate for the emission of atoms is very much greater than that for ions (c. 10^3 greater) so it is this parameter which primarily determines the rate at which a surface is damaged during analysis. Typically, S will lie between 1 and 10 so if it is assumed that the density of atoms in a surface is 10^{15} cm^{-2}, the life-time of the topmost monolayer, t_m, is given by

$$t_m = \frac{10^{15}}{S x i_p},$$

where i_p is expressed as the number of primary particles incident cm^{-2}s^{-1}. Thus when $i_p = 10^{-9}$ A cm^{-2} ($\sim 10^{10}$ particles cm^{-2}s^{-1}) and $s = 1$, then $t_m \sim 10^5$s. These are good conditions for static SIMS.

If we are to understand the SSIMS phenomenon we need to be able to describe and explain the factors which control the sputter yield and ionisation probability. Experimentally it has been shown that secondary ion yield;

(a) increases with beam energy;
(b) increases with beam flux;
(c) increases with primary particle mass;
(d) maximises at an impact angle of about 70° to surface normal.

Of course the rate of surface damage also follows these parameters. It is generally assumed that ion yields and damage rates are independent of the charge state of the primary particles. This, however, may not be the case since it has been shown in the author's laboratory that ion bombardment of insulating materials results in significantly higher surface damage than by atom bombardment.

In common with dynamic SIMS it has also been shown that

(i) positive ion yields of different elements sputtered from a common matrix exhibit an inverse exponential dependence on the ionisation potential of the sputtered atom (there is some evidence for a similar dependence of negative-ion yield on electron affinity);
(ii) positive-ion yields are greatly enhanced in the presence of oxygen or other electronegative species at the surface;
(iii) ion energy distributions generally peak at rather low energies (5–10 eV) and exhibit low-level tails up to several hundred eV.

These observations have been mainly recorded for elemental ions but they are also generally true for cluster ions.

Little detailed investigation has gone into the characteristics of cluster ion emission; however Benninghoven has shown in his so-called valence model that the yields of positive and negative oxidic secondary ions (MO_n^+, $n = 0, 1, 2$ and MO_n^-, $n = 1, 2, 3, 4$) can be described by Gaussian functions (see Fig. 6.21):

$$S_{MeO_n^+} = S_{max}^+ \exp -\frac{(G^+ - K)^2}{2\gamma^2},$$

$$S_{MeO_n^-} = S_{max}^- \exp -\frac{(G^- - K)^2}{2\gamma^2}.$$

K is the 'fragment valency' ($= 2n + q$) or the formal valency of the metal of the emitted molecule, MO_n^q. The 'lattice valencies' G^+ and G^- vary with the elements and with the metal valency state at the surface. γ is a constant which is approximately constant for all metal oxides, so the shape of the Gaussian function hardly changes with the metal. The maximum yield of M^+ decreases sharply with increasing metal mass, i.e.,

$$S_{max}^+(m_{metal}) = \text{const.} \, m_{metal}^{-2.4}$$

and the maximum yield of negative ions S_{max}^- does not vary with mass.

2.2 The mechanism of secondary ion emission

The general outline of the process is widely accepted. The primary particle impacts with the surface atoms, and symmetry considerations suggest that at least three surface atoms will be involved. Energy is transferred and collision sequences between atoms in the near-surface region are initiated. Some energy will be dissipated into the bulk of the solid, whilst some collision sequences or cascades return to the surface causing the emission of secondary ions or atoms. It will be clear that in general these emitted particles will be released at a point remote from the point of initial impact, and that the collision resulting in emission will be of far lower energy than that of the primary particle. The extension of the emission model beyond this point is subject to considerable disagreement. Most models have been proposed to account for small subsets of data; for example, that ion yields apparently rise with surface work function, or ion yields vary with the host matrix, and attempts to develop generalised mechanisms often become lost in their own complexity or in oversimplification.

The major question is whether the two processes of emission and ionisation occur simultaneously or consecutively. Clearly, a complex mixture of collisional dynamics and quantum mechanics is involved.

Although surface annealing is thought to occur within 10^{-10} s, in the region of the primary particles impact zone the crystalline structure of the surface will be considerably disrupted. It would not seem to be appropriate, therefore, to consider the electronic structure to be that of the perfect solid. In other words models which rely on a normal solid band structure are not likely to be valid. It is probably more satisfactory to consider that the area of emission is amorphous with a continuum of energy states.

Many treatments consider that the ionisation probability of a departing atom is determined by an atom–surface interaction within a few tenths of nanometres from the surface. The mechanism of the interaction has been variously described. The *perturbation model*, due to Norskov and Lundquist (see Williams, 1979), suggests that the ionisation probability is a maximum closest to the surface where atom–surface coupling varies most strongly. Of course this coupling also produces efficient neutralisation. At some distance from the surface, the coupling is weaker and the ionisation probability is less but the *ion* escape probability begins to be finite. Thus the probability of creating and observing an ion maximises in a region some few Ångstroms from the surface. Thus the ionisation probability is given by

$$\gamma^+ \alpha \exp[-(I-\phi)/\varepsilon_0],$$

where I is the ionisation potential, ϕ is the work function and $\varepsilon_0 = hav/C\pi$ where h is Planck's constant, a is a range factor $\sim 2\,\text{Å}^{-1}$, v is the velocity of the departing atom. For metals this model rationalises the well-known dependence of yield on ionisation potential and on adsorption or matrix effects which give rise to work function changes. However, for non-metals where the valence levels of the departing atom correspond to an energy in the band-gap this model does not predict surface atom-coupling.

A development of the above model is the *surface excitation model*. Resonant electron transfer between the surface and the departing atom occurs up to a distance of 5–10 Å. For ionisation to occur and the ion to escape it is necessary that a valence level in the departing atom is isoenergetic with a *vacant* level in the surface *below* the photothreshold (Fermi level in a metal or valence band edge in a semiconductor or insulator). The probability (P_e) of this occurring is greatly enhanced by the surface exicitation resulting from the primary ion impact. This will, as we have suggested, result in a transient continuum of energy states. Thus the ionisation probability will be given by

$$\gamma^+ = P_e/(1 - P_e).$$

This model unifies discussion of metal and insulator surface by focussing attention on the local transient disorder at the sputtering site and thus relating ionisation probabilities to the local photothreshold. Local changes in the surface matrix as a consequence of, for example, oxygen adsorption and incorporation, which have been reported to enhance *both* positive and negative ion emission, can be explained by this model.

Ion emission from ionic solids has been addressed specifically by the *bond-breaking model*. The key assumption is that the ion state is the ground state, and the probability of observing an ion depends simply on whether the ion state remains energetically favoured over the neutral state as the

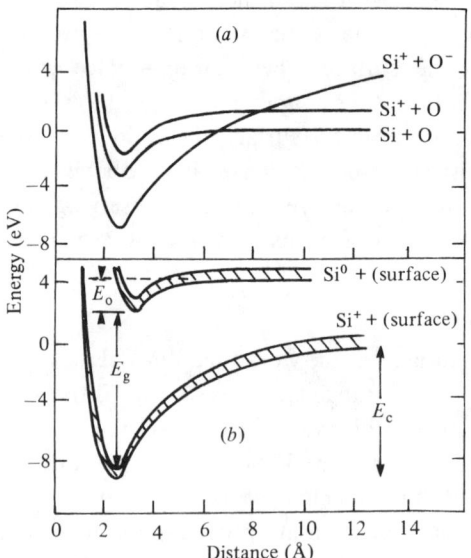

Fig. 6.1. Potential energy curves for the ejection of excited and ionised Si from (a) a gas-phase SiO molecule and (b) solid SiO$_2$.

ion moves away from the surface. This depends on the distance at which the ion and neutral potential energy surfaces cross. If this happens at an atom–surface separation of less than 10 Å, the ion and neutral states will be strongly mixed. The features which influence the crossing distance are (a) the ground state energy of the associated species, (b) the difference between the ionisation potential and electron affinity of the positive and negative ions, and (c) the range of the electrostatic interaction. Sometimes gas phase ion-pair interactions have been used to understand the effects. These are really not appropriate. In Fig. 6.1 it can be seen that level crossing occurs in the gas phase for the SiO system whereas it does not from solid SiO_2. The effect of oxidation on ion yield can then be thought of in terms of moving the curve-crossing further from the surface.

These three models view the mechanism of ion emission from different angles but the underlying concepts are similar, differing mainly in the sophistication of the atom–surface interaction.

A rather different approach is the *molecular model* due to Thomas (see Williams, 1979, 1982) and an extension of it the *nascent ion molecule model* due to Plog. Here it is suggested that the rapid electronic transitions which occur in the surface region will neutralise any ions before they can escape. This underlying assumption is based on a theoretical argument with as yet no experimental support. Secondary ions are thought to result as a

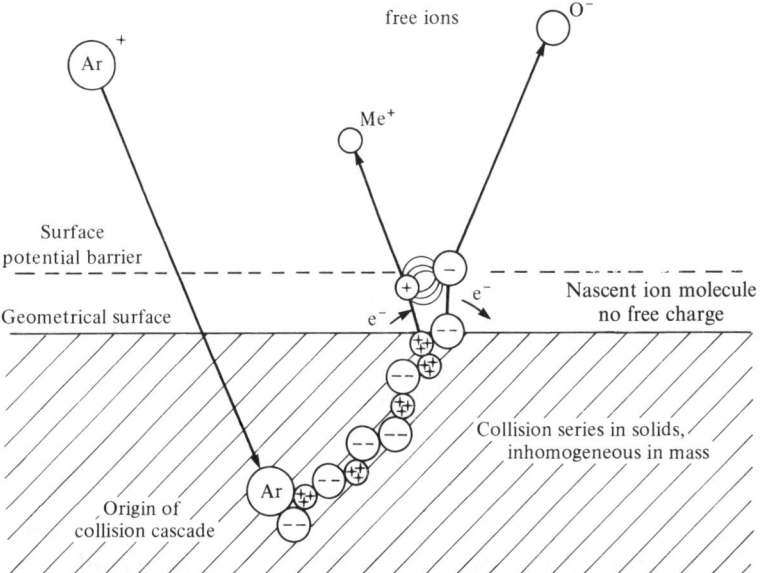

Fig. 6.2. Secondary ion emission by dissociation of 'nascent ion molecule' (C. Plog & W.Gerhard, Z. Phys. B **54**, 71, 1983).

consequence of dissociation of sputtered neutral molecular species some distance from the surface (Fig. 6.2). There are three steps in the process:

(a) energy transfer in the collision cascade in the solid;
(b) collision energy transfer between the atoms of the molecule leaving the surface. This reservoir of energy will give rise to dissociative ion formation;
(c) charge exchange during dissociation of the molecule.

The treatment leads to a mass dependence of Me^+ yield from oxides and chlorides which mirrors that derived from the S^+_{max} values derived from Benninghoven's valence model.

The concept of the molecular model is a helpful one, especially in the understanding of the emission of cluster ions. The quantitative derivation is not possible without serious assumptions and speculative constants. It illustrates the weakness of all the theoretical approaches. They have each been derived to fit a particular set of data and their general validity is difficult to accept in other than qualitative terms.

3 Instrumentation

The heart of any SIMS apparatus is a combination of a primary beam source, and a mass spectrometer to analyse the sputtered ions. Static SIMS is in the main concerned with surface studies and it is therefore wise to analyse in UHV conditions to eliminate the possibility of surface contamination. A standard arrangement is shown in Fig. 6.3. This system consists of an analysis chamber; a preparation chamber where samples can be etched, cleaved or chemically treated; and a fast entry lock. The latter enables samples to be rapidly loaded into the system without degrading the vacuum conditions in the analysis chamber. Sample transfer (in the illustrated system) is by a 'railway system' although many other possibilities exist.

3.1 Primary beam

The first requirement for static SIMS is a controllable primary beam source having a beam density variable from 10 pA to 5 nA cm^{-2}. Until quite recently the primary beam was always provided by a fairly low energy (less than 10 keV) ion gun, but the need to analyse insulators and attain high spatial resolution has given rise to other types of beam source:

3.1.1 Ion beams

Since low beam currents (10^{-11} to 10^{-8} A cm^{-2}) are required for static SIMS, most systems use electron bombardment to produce the ion beam. The beam is then accelerated to energies between 0.5 and 4 keV. To maximise the secondary ion yield the impact area is usually large (c. 0.1 cm^2) compared to the field of view of the analyser. To reduce the possibility of sample contamination by impurity ions from the source filament or residual gases the beam should be mass filtered by a Wien filter or a small radio-frequency (RF) quadrupole. In SSIMS inert gas ion beams are usually used to preserve the chemistry of the surface being studied. An impact angle of about 70° to the surface normal has been found to give maximum yield.

3.1.2 Atom beam

A neutral primary beam is usually produced by passing a high-energy ion beam through a chamber containing a high pressure (10^{-4} mbar) of inert gas. A proportion of the fast ions (up to 10% depending on the chamber length and the gas pressure) lose their charge by capturing an electron from the atoms or molecules randomly moving in the chamber. Although these charge-exchanged particles have lost their charge their kinetic energy and direction of motion are essentially unchanged; consequently a neutral high-energy beam emerges from the charge exchange

Fig. 6.3. An ultra-high vacuum (UHV) static SIMS system (courtesy VG Ionex).

chamber. The ion component of the beam is removed using electrostatic deflection plates.

It is frequently advantageous to have both an ion and atom beam source available. Van den Berg and Vickerman have described a combined ion–atom source which can be rapidly switched from ion beam to atom beam operation and is mass filtered in both modes, Fig. 6.4. It is ideally suited for SSIMS. Beam energy can be varied between 500 eV and 2000 eV. The beam flux densities are controllable from 10 nA cm^{-2} in ion mode and 1 nA cm^{-2} in atom mode down to 1 pA cm^{-2}, and the beam diameter at the target surface is 3 mm in both modes.

3.1.3 Liquid metal ion beam

These sources typically are metals which are liquid near room temperature, for example gallium, indium and caesium. There are two

Fig. 6.4. Mass filtered ion-atom primary beam source F – source filaments; S – source plate; Ex – ion extracter plate; EL1 – einzel lens 1; WF – Wien filter and charge exchange chamber; EL2 – einzel lens 2; CM – current monitor; A – apperture plate; X and Y – deflector plates. (from A. Brown, J.A. van den Berg & J.C. Vickerman, *Spectrochimica Acta*, (1984).

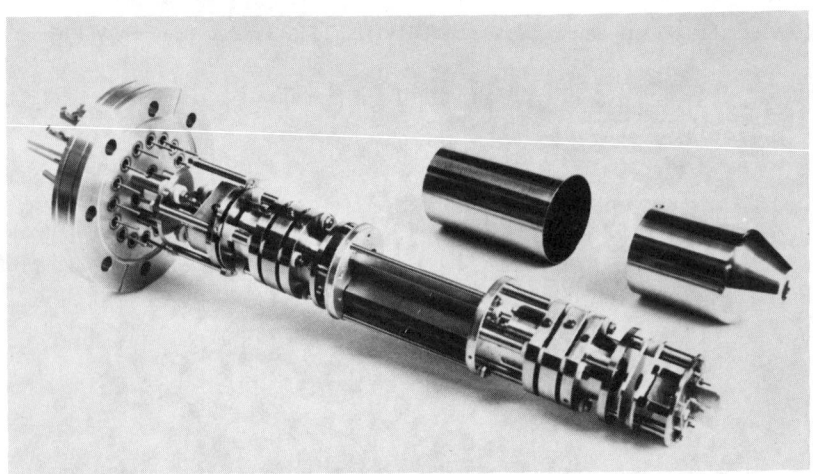

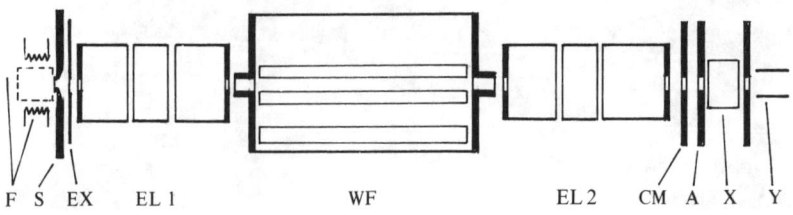

general methods of operation. In the first the liquid metal flows through a fine aperture and, under the influence of a high positive electrostatic field relative to its surroundings, forms a cone from the apex of which intense positive-ion emission occurs. Alternatively, the liquid metal wets and flows over a fine needle. A high positive field is established between the needle tip and an extractor and again intense positive-ion emission of liquid metal ions occurs. The result is a very high brightness, highly collimated ion source (Prewett & Jeffries, 1980) (see Chapter 2). Such a beam can be steered and focussed by conventional ion optics to produce a submicron beam diameter (down to 500 Å) at the target. Currents of about 1 nA can be obtained into less than a 1 μm spot.

3.2 Mass analyser

In the original development of the technique by Fogel magnetic mass spectrometers were used; however today the majority of SSIMS systems utilise quadrupole mass spectrometers. The requirement of good vacua because surface studies are involved is probably the main reason for this. It is much easier to enclose the compact quadrupole filter in a UHV system than to provide good pumping for a magnetic spectrometer.

3.2.1 Quadrupole analyser

Mass analysis is performed by establishing a radio frequency (RF) electric field between the rods and superimposing a varying DC field. It is this latter component which provides the mass filtering effect. Despite its popularity the quadrupole filter has a number of problems which are peculiar to it and which have to be carefully borne in mind when using it as a SSIMS analyser:

(a) Although mass spectra can be obtained rapidly and easily the relative intensities of detected ions depend rather critically on the operational parameters of the quadrupole. This is because the transmission and mass resolution of the analyser are very sensitive to the axial and transverse velocities of the incoming ions. There are no generally accepted criteria for defining the operating conditions as between different quadrupoles and hence it is difficult to obtain absolute ion abundances.

(b) Large fringing fields exist at the entrance and exit of the filter. These arise as a result of the non-ideal termination of the mass-analysing fields produced at the quadrupole rods. Their presence can severely impair the performance of the analyser as a consequence of ion trapping and the reflection of slow-moving ions. Brubaker has successfully treated this problem by the addition of a separate set of short quadrupole rods at the entrance to the main analyser. Only the RF component is applied to these rods. As a consequence increases in transmission of up to three orders of magnitude for slow-moving high-mass ions were observed. Many modern SSIMS quadrupoles are now equipped with these segmented rods and their performance can often match magnetic sector instruments.

(c) A further complication is the large spread of kinetic energy of the secondary ions produced in the sputtering process. Ions having high kinetic energies cannot be effectively mass filtered by the analyser and, consequently, the mass resolution is degraded. This problem can be alleviated by using an energy filter to select a small energy 'window' of ions. However, this procedure introduces a further problem since secondary ions are characterised not only by mass and charge state but also by the *shape* of their kinetic energy distribution. The more complex the secondary ion cluster the sharper this distribution has been found to be. Thus as Fig. 6.5 shows the observed spectrum can be drastically influenced by the width and position of the energy window selected.

3.2.2 Energy analyser

As indicated above when a quadrupole mass analyser is used it is necessary to include some form of energy analysis of the ions before

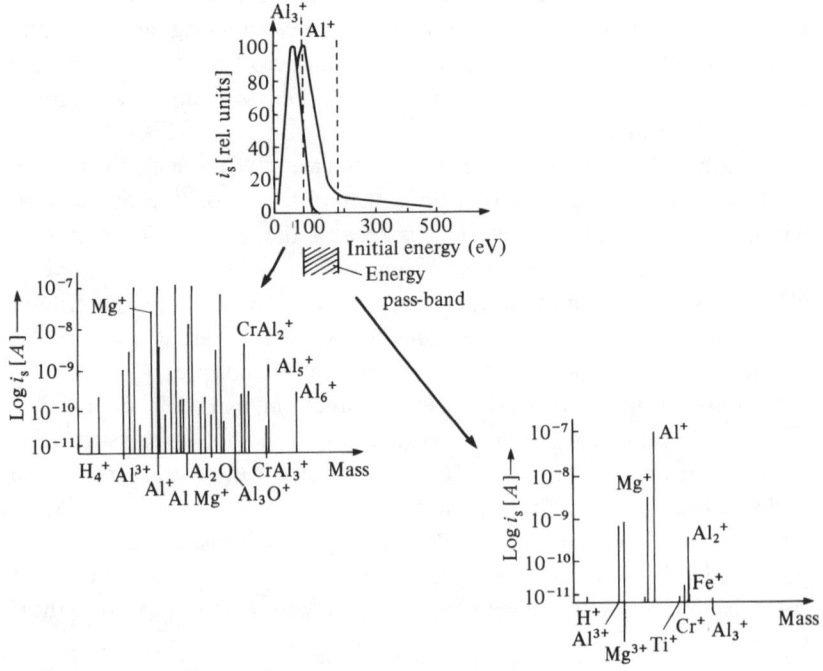

Fig. 6.5. The effect of the position and band-pass of an energy filter on the detected secondary ion spectra. This is a consequence of the differing energy distributions of the secondary ions.

they enter the filter. This has two benefits: firstly, direct line of sight into the analyser is removed, thus reducing background intensity due to sputtered neutrals, etc; secondly, high-energy secondary ions are rejected, thus improving the resolution performance of the analyser. Although the latter is extremely important, it is also necessary to guard against severe energy discrimination since, as indicated in Fig. 6.5, this can distort the spectrum obtained. Ideally an energy window of about 5–12 eV should be accepted. Thus quite simple energy analyser devices have usually been proposed. In the early development of the static SIMS methods by Benninghoven, crude energy filtering was carried out by operating the mass filter at some potential above that of the target, and retarding the ions at entry. This procedure is possible with most modern quadrupoles since they have the facility, known as pole bias, for varying the potential of the rod assembly. However, there is the danger that the acceptance, and hence the sensitivity, of the mass filter will vary widely as the time the ions spend in the fringing fields is varied.

Although energy analysis is necessary in SSIMS it is desirable to have some arrangement which will also collect ions from a *large area* and from a wide range of emission angles in order that adequate sensitivity can be obtained with a small primary beam current. This is a very difficult problem and most energy analysers currently in use are extremely crude and owe their design parameters mainly to practical considerations and experience rather than any theoretical ion trajectory calculations. These include cylindrical electrostatic sectors, cylindrical mirror, parallel plate analysers, and retarding–accelerating lens systems with central stop. The most popular have probably been variants on the latter two, illustrated in Fig. 6.6.

It can be seen that the quadrupole SSIMS analyser system is complex with many variables. The transmission of secondary ions by this spectrometer system may be as high as 1% but it is important that in the static

Fig. 6.6. A variant on a parallel plate analyser and ion collector. Note the filament for residual gas analysis.

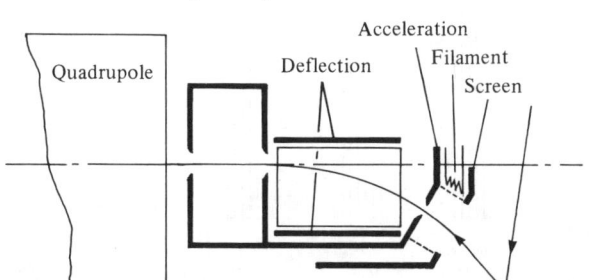

mode this is maximised. However, sensitivity and mass resolution are interrelated and can be affected by any of the variables mentioned. It is important to establish a procedure which ensures maximum sensitivity, optimum resolution and good reproducibility. The problem has been overcome to some extent by setting up the spectrometer to give an agreed spectrum using a standard compound. The procedure proposed by the UK SIMS Users' Forum is outlined below.

3.2.4 Standard quadrupole operating conditions

The method is simply to record the intensities of selected peaks from a reference material in the positive and negative ion mode. The materials chosen for static SIMS were recommended by the UK SIMS Users' Forum to be a disc of KBr or a thin layer of carbon black deposited on a metal surface. The relative intensities of the peaks, Fig. 6.7, are dependent on the mass resolution and transmission of the analyser and the energy of the secondary ions.

There are three variables on the mass analyser itself: *mass resolution* which sets the ratio between the DC and RF voltages (V_{DC}/V_{RF}) applied to the quadrupole rods; ΔM which allows accurate zeroing of the DC voltage amplifiers at zero mass (by applying a V_{offset}) to reduce unwanted signal that may be transmitted through the filter when RF fields collapse to zero; *pole bias* which sets the energy of ions entering the quadrupole, and clearly affects the input position of ions into the filter. Each of these

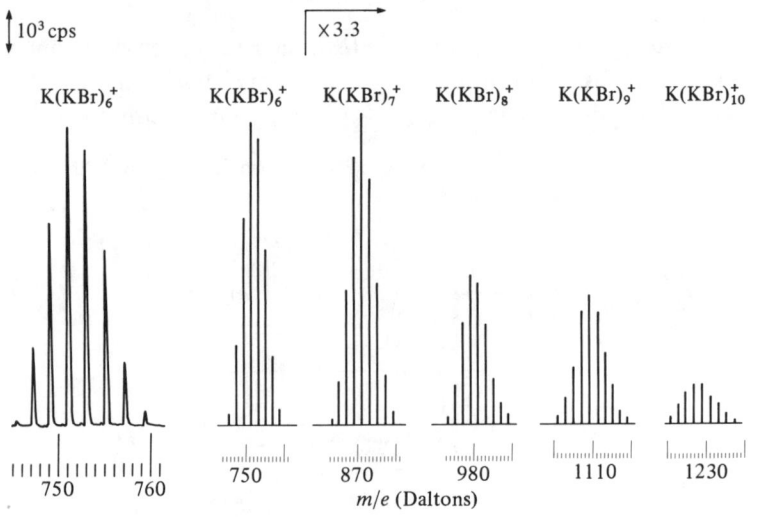

Fig. 6.7. A partial spectrum of KBr using fast atom bombardment and analysis by a VG MM12–12 quadrupole mass spectrometer.

parameters affects resolution: the first two via the ratio $(V_{DC} + V_{offset})/V_{RF}$; pole bias because maximum resolution is proportional to (time of flight)^2. It is recommended, therefore, that *mass resolution* and **M** should be adjusted to give mass widths of 1 dalton measured at 5% peak height on the major peaks in the KBr spectrum. *Pole bias* interacts with the settings of the energy filter and target bias, but it should be set such that it gives the best composite between peak shapes consistent with the 5% definition of resolution, and maximum peak heights.

The *energy filter* and *target potential* setting control energy acceptance of the analyser system. Single-element peaks from solids exhibit a broad energy spectrum, whereas multiatom ions have a narrower energy distribution. Thus it is necessary to set an energy acceptance which is sufficiently wide to ensure a minimum energy discrimination but narrow enough to give the 5% resolution. Thus it is recommended that the filter is first set to maximise the $K(KBr)^+$ signal and then energy resolution is coarsened consistent with the required mass resolution. If the target bias is adjustable a mix of ion focussing and ion energy variation is possible. Thus having set the analyser and energy filter the target bias may be adjusted for maximum intensity of $K(KBr)^+$.

Clearly, the other components of the SSIMS instrument will influence the absolute ion yields. The primary beam source parameters will be crucial: whether ions or neutrals, the beam energy, the beam angle of incidence, the beam fluence and the beam impact area.

3.2.5 *Other types of mass analyser*

Although the quadrupole is most commonly used for SSIMS it will be apparent that it is not ideal. The *double-focussing magnetic sector instrument* offers many advantages as far as the mass spectrometry is concerned. Since ions are accelerated into the spectrometer with energies of several keV the energy acceptance of the analyser is much greater, and mass discrimination is much less of a problem; consequently ion transmission can be better than 1%. Ultimate mass resolution is greater, an important consideration for the complex mass spectra which can arise in surface analysis. However, the size of double-focussing spectrometers has in the past, ruled out their application to SSIMS in all but a few cases; however this is now seen as a promising area of development.

It is evident, however, that SSIMS is limited by the proportion of the secondary ions which are actually detected by the analyser, $< 1\%$ by quadrupoles and $< 10\%$ by magnetic analysers. In the early 1980s the possibility of utilising the *time of flight* (TOF) *mass spectrometer* began to be explored. This type of analyser had been used by Togerson and

Macfarlane for plasma desorption mass spectrometry in which involatile organics were sputtered by the 50–100 MeV particles resulting from the fission decay of ^{252}Cf. This was a linear TOF analyser and resolution was degraded by the large kinetic-energy spread of the sputtered ions. However, the energy compensated TOF mass spectrometer proposed by Poschenrieder (1972) which incorporates an electrostatic section in the flight tube, is ideal for SSIMS. Ions of different energy may now take trajectories of different lengths and, with a suitable geometry, the slower speed of the low-energy ions will be exactly compensated by the shorter flight path taken. Thus the flight time is then only dependent on the mass of the ions and is independent of their exact energy. The great advantage of this analyser is its enormously greater transmission $< 30\%$, and its almost simultaneous detection of all ions. This yields a potential increase in sensitivity over the quadrupole analyser of 100–1000 and possibly more, at the expense of a rather more complex primary beam system, since a pulsed beam arrangement is required, Fig. 6.8.

3.3 Data aquisition

All modern SSIMS instruments require the use of computerised data systems both for control purposes, the primary particle beam and the mass analyser, and for data acquisition. Programs are required for

Fig. 6.8. An imaging time-of-flight SIMS instrument with energy compensating analyser and a pulsed primary ion beam (courtesy VG Ionex).

Static secondary ion mass spectrometry

standard mass spectral analysis, peak switching, depth profiling of particular elemental or cluster ions, and for ion mapping to give chemical images of surfaces. The latter can be recorded photographically or by computer imaging.

4 The application of SSIMS in surface studies

SSIMS is extremely versatile in its range of applications. Unlike many surface analytical techniques it is as easily applied to surface problems in practical materials technology as to fundamental studies of surface reactivity. This section will describe some of the problems which have been investigated to illustrate the potential of the technique.

Before these are presented, however, it is appropriate to describe the methods of application and to consider a number of the criticisms which are levelled at the technique.

4.1 Methods of application

There are two broad methods of SSIMS operation distinguished by their spatial resolution. The first is what may be termed macroanalysis, the second is SSIMS imaging or SSIMS microscopy.

4.1.1 SSIMS macro-analysis

Here the chemistry of a large area (~ 0.1–$0.3\,\text{cm}^2$) of the surface of the sample requires to be analysed. With due recognition of the limitations of the mass analyser being used, positive- and negative-ion spectra will be obtained which give a qualitative picture of the elemental composition and chemical structure of the surface. Elemental ions are frequently easy to pick out from their characteristic isotope patterns. The more complex cluster ions which contain chemical information as to surface bonding and structure are more difficult to interpret. A start has been made on building up a library of standard surface mass spectra of well-characterised materials which will be of immense importance for the full characterisation of surfaces.

Organic contamination of surfaces is frequently evident when they are first introduced into the spectrometer. This can be recognised by clusters of peaks separated by 12–14 amu of decreasing intensity up the mass scale. Such contamination, or indeed other surface layers, may be removed by gentle ion or atom beam etching. This also provides the possibility of depth profiling to probe the chemistry of the subsurface layers.

A further type of experiment is possible with most SSIMS systems, namely surface analysis followed by a surface treatment in a preparation chamber followed by further analysis. This is particularly useful in fundamental studies or when seeking to simulate some process which may affect the surface chemistry of a material.

4.1.2 SSIMS microscopy

Here analysis is required with high spatial resolution usually using a liquid-metal ion source. When such a source is coupled to a secondary ion mass spectrometer, SIMS analysis of submicron surface features is possible; or if the mass analyser is tuned to a particular mass and the beam scanned across the surface, elemental or molecular imaging becomes possible at very high spatial resolutions to yield SEM type images *but with detailed chemical content*, Fig. 6.9.

A useful by-product of the technique is the generation of good quality secondary electron images whose contrast and depth of field is superior to SEM (Fig. 6.10).

Although in principle it is possible to use static conditions with these sources to analyse or image without causing significant surface damage, care has to be exercised, as will be indicated in some of the following sections, because the low transmission of the mass analyser may require high primary beam densities into a small area to generate sufficient ions for detection. This can lead to an operating regime where damage in the analysed area is very great.

Fig. 6.9. A chemical image, $^{12}C^-$ of a fly's eye. Magnification × 100 obtained using a liquid gallium source operating at 10 keV and 1 nA.

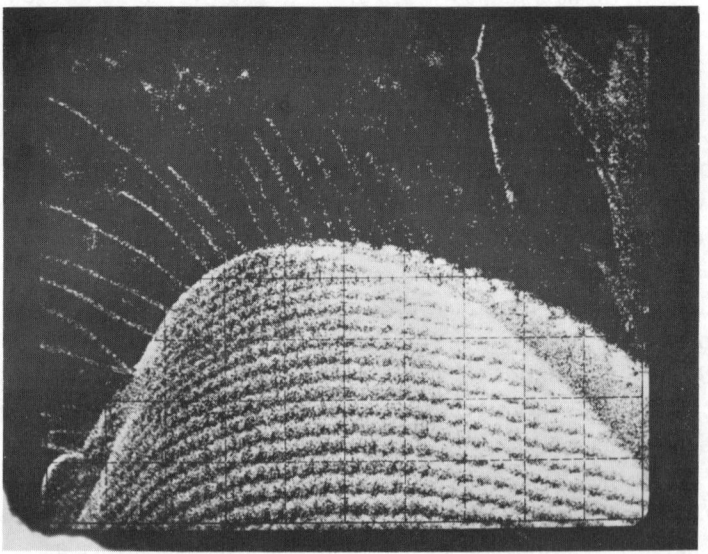

4.2 Criticisms of SSIMS

The first criticism is the problem of quantisation. Secondary ion emission is very sensitive to the electronic state of the atom or molecule to be ionised and to the substrate from which they are emitted. Thus

Table 6.1. *Monolayers removed as function of ion current and magnification*

Assume:	Sputter yield = 2						
	Frame time = 100 s						
	Nos. frames = 1 (1028 × 1028 points)						

I_p	1 pA	0.05 nA	0.1 nA	0.5 nA	1 nA	5 nA	10 nA
MAG							
100	10^{-4}	5×10^{-3}	10^{-2}	5×10^{-2}	10^{-1}	5×10^{-1}	1
200	4×10^{-4}	2×10^{-2}	4×10^{-2}	2×10^{-1}	4×10^{-1}	2	4
500	3×10^{-3}	1.5×10^{-1}	3×10^{-1}	1.5	3	15	30
1000	0.01	0.5	1	5	10	50	100
2000	0.04	2	4	20	40	200	400
5000	0.25	12.5	25	125	250	1250	2500
10 000	1	50	100	500	1000	5000	10 000
20 000	4	200	400	2000	4000	20 000	40 000

Fig. 6.10(a) A weld edge in stainless steel illustrating the crystallographic contrast available from ion-induced secondary electron images. Micrograph obtained using a gallium microprobe operating at 10 kV. (b) A secondary electron micrograph obtained using a gallium gun of an integrated circuit and its wire bonding illustrating the depth of field obtainable. The image allows the probe to be focussed onto a specific micro-area for high-sensitivity analysis.

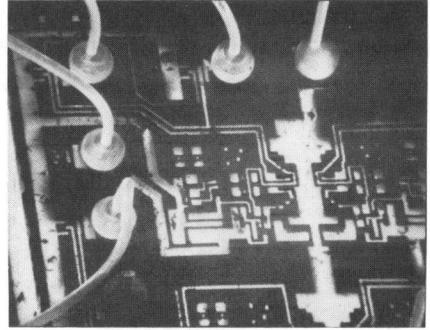

(a) (b)

careful calibration procedures have to be carried out if accurate quantitative data is required. This is usually possible and is certainly worthwhile because SSIMS can be orders of magnitude more sensitive than other techniques.

The second criticism is more general; it concerns the 'violence' of the SIMS process. It is interesting to note that in fact the power dissipated in the surface during a typical static SIMS experiment is several orders less than in XPS or Auger. However, at the point of impact it is certainly true that considerable lattice disruption will occur. It is important to remember however that (a) under static conditions no spot on the surface is hit more than once, (b) surface annealing occurs in fempto-seconds, and (c) the secondary ions *detected* are of low translational energy and in the main are emitted from points remote from the point of initial impact as a result of subsurface cascades, so these ions will be unaffected by the primary collision with the surface. Basic work on adsorbate systems, which will be described in the next sections, shows that even such delicate surface structures can be studied and the mass spectra yield unequivocal data on the adsorbate chemical structure (Brown & Vickerman, 1983).

The third criticism concerns the validity of the assumption that the atoms in a cluster ion arise from nearest neighbour atoms in the surface. Some workers have suggested, mainly on the basis of computer simulation studies with their inevitable simplifying assumptions, that clusters are formed as a result of recombination of atoms as they leave the surface (Winograd, 1981). It seems rather unlikely that the density of atoms leaving the surface in static SIMS would be sufficient to make this a probable process. If such a process were to occur to a significant extent, the relationship between the mass spectra and surface chemical structure would be at best rather indirect. The mass of data to date on a wide range of inorganic and organic materials is that there is a clear and direct relationship between the spectra and chemical structure.

4.3 Basic studies of surface reactivity

Studies of adsorption and molecular reactions at solid surfaces utilising the modern surface physics techniques have dominated much of surface science over the last two or three decades. The rationale is clear. An understanding of the chemistry and physics of the adsorbate state should unlock the mysteries of catalysis and all the economic benefits that would entail. When static SIMS became feasible the prospect of mass spectrometric information on the surface state to add to all the other surface spectroscopic data was very attractive. However, it was quite

4.3.1 CO adsorption on metals

Studies of CO adsorption on metals important as catalysts have shown clearly that SSIMS is sensitive to adsorbate state and is able to monitor CO coverage, the adsorbate bond energies and the adsorbate chemical structure (Bordoli et al., 1979).

The earliest studies investigated the adsorption of CO on a series of polycrystalline metals. It was known from other studies using electron spectroscopy that the adsorption of CO at 300 K would be molecular or dissociative depending on the adsorption energy. Where the enthalpy of adsorption is less than $250\,kJ\,mol^{-1}$ as it is on Cu, Pd and Ni, the adsorption is molecular; where it is greater, as on W, it is dissociative, whilst adsorption on Fe lies between and both forms are observed. Fig. 6.11 shows that SSIMS can distinguish clearly between the two. Dissociative adsorption is characterised by MC^+, MO^+, M_2O^+ and M_2C^+ secondary ions, whilst the molecular adsorbate state is identified by MCO^+ and M_2CO^+ ions.

The next stage was to relate the MCO^+ and M_2CO^+ intensities to the surface coverage of CO adsorbate species. This was not an easy task. The secondary ion intensities, as has been shown earlier, are not sensitive *only* to the quantity of surface CO; they are also influenced strongly by the

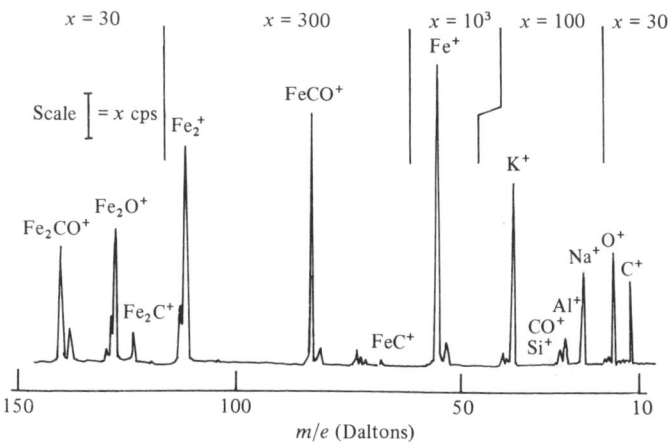

Fig. 6.11. SIMS spectrum recorded at equilibrium when iron is exposed to 10^{-8} Torr of carbon monoxide. Primary ion current density $= 5 \times 10^{-10}$ A cm^{-2}, energy $= 3\,keV$.

work function change which accompanies adsorption. Thus as CO coverage increased on a surface the secondary ion intensities increased non-linearly as a consequence of *both* of these parameters. There is no satisfactory theoretical model which would allow the two parameters to be unravelled. An empirical approach was found to be very successful. It was argued that whilst the M_xCO^+ species were sensitive to both coverage and work function the M_x^+ ions also increased but only as a consequence of the work function change. If the ratio M_xCO^+/M_x^+ was taken perhaps it would only be sensitive to coverage. Since some CO is removed as MCO^+ and some as M_2CO^+, total coverage should be represented by $\Sigma_x(M_xCO^+/M_x^+)$. The linear plots of this function against coverage demonstrated that SSIMS data was sensitive to adsorbate coverage, (Fig. 6.12). There is a break at a relative coverage of about 0.5 in the plots for most of the surfaces. It is at this point that the structure of the adlayer usually changes and becomes 'compressed'. The gradient change is correlated with this change. Indeed this was one of the first indications that SSIMS might also be sensitive to the adsorbate *structure*.

It can be seen that the gradient of the linear relationship is surface dependent. It seems to be a function of the enthalpy of adsorption. The ability to monitor coverage by SSIMS automatically led to the possibility of measuring the enthalpy of adsorption by the isosteric heat

Fig. 6.12. The variation of the sum of intensity ratios, $(\Sigma_n M_n CO^+/M_n^+)$, as a fraction of relative coverage for CO adsorption at 300 K on Ru(001). Ni(111) and Pd(111).

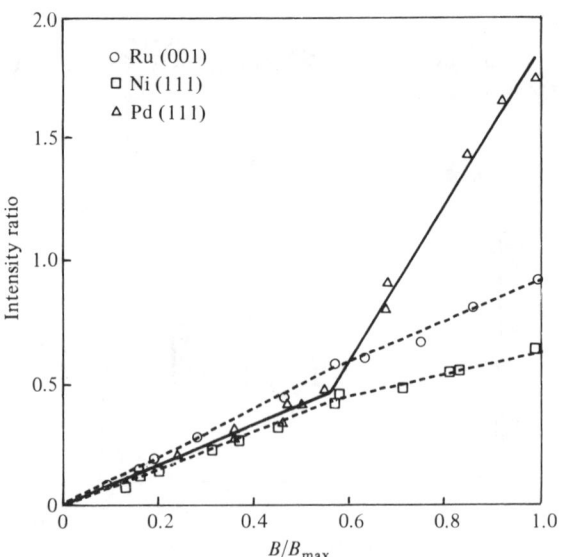

method. Thus the sum of ion ratios was monitored as a function of temperature for particular ambient pressures of CO.

From these plots isosteric heats were derived which were in very good agreement with those derived by other methods, Fig. 6.13.

Thus SSIMS had been able to provide information on the mode, the coverage and the energetics of adsorption. Mass spectrometry is, however, best known for its ability to provide data related to the molecular chemical structure. CO can adsorb into linear, bridged or triply bridged surface states. By carefully selecting single crystal faces which were known from vibrational spectroscopy to adsorb CO in specific states it was possible to show empirically, by comparative studies utilising vibrational spectroscopy and SSIMS on these metal surfaces, that the relative populations of the MCO^+, the M_2CO^+ and the M_3CO^+ ions can be used to identify the state of the CO on the surface. In other words a SSIMS 'fragmentation pattern' of each state can be defined as in Table 6.2, which allows the distribution of each type of adsorbed CO on a surface to be identified and quantified. This latter facility is quite unique, even vibrational spectroscopy cannot with any certainty quantify the relative coverages of adsorbed species.

These observations on 'ideal model' systems have important implications for catalyst analysis. First, they show that the technique does not destroy

Fig. 6.13. The coverage dependence of (1) the isosteric heat of adsorption, ΔH_{ads}, from the sum of the SIMS intensity ratios (Δ), (2) the energies of desorption, E_d, ($\square$) and (3) the literature plus error bars are shown. (from A. Brown & J.C. Vickerman, *Vacuum*, **31**, 429, 1981).

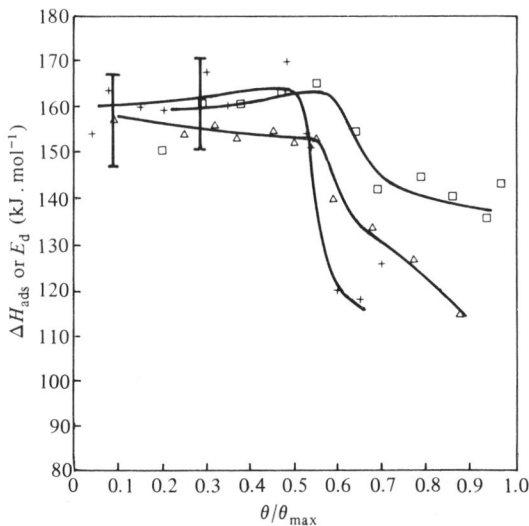

information on very delicate structures; if adsorbate structures are stable then there is a good prospect for all other surfaces. Second, we have very strong evidence that SSIMS is sensitive to the adsorbate chemical structure. Third, although it is necessary to analyse the data with care, the various contributions to secondary ion intensities can be dealt with to yield

Table 6.2. *SSIMS 'fragmentation pattern'*

Surface	MCO^+	SSIMS M_2CO^+	M_3CO^+	CO bonding IR/HREELS/LEED assignment
Cu(100)	0.9	0.1	—	Linear only
Ru(001)	0.9	0.1	—	Linear only
Ni(100)	0.8	0.2	—	Linear + bridged
Ni(111)	0.6	0.4	—	Linear + bridged
Pd(100)	0.3	0.6	0.1	Bridged
Pd(111)	0.3	0.4	0.3	Triply-bridged

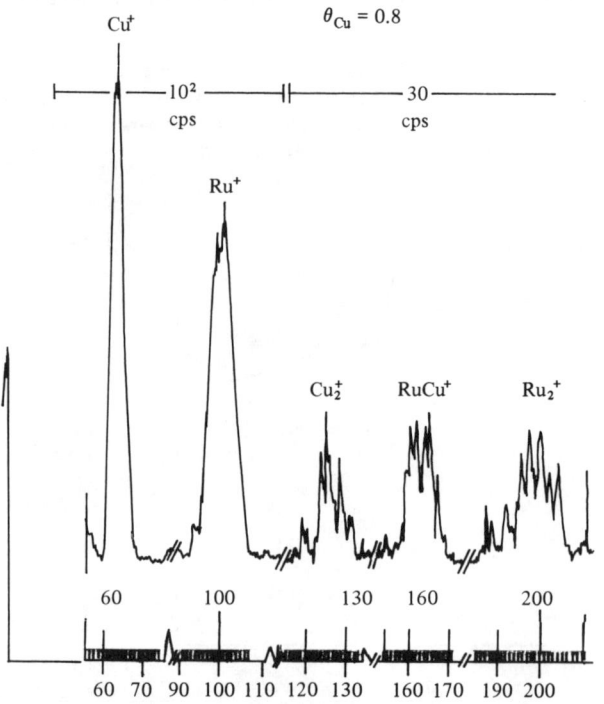

Fig. 6.14. A SSIMS spectrum of a Cu/Ru bi-metallic surface $\theta_{Cu} = 0.8$ at low mass resolution to increase sensitivity.

valuable quantitative or semiquantitative information on the coverage and state of the surface layer.

4.3.2 Structure of thin metal films

Similarly detailed descriptions of the surface are accessible from SSIMS studies of metal adsorption. The adsorption of Cu on Ru(001) is a good example of a basic study (Brown & Vickerman, 1984a, b). Since alloy formation does not occur the Cu stays on the surface of the Ru. A typical SSIMS spectrum of the bimetallic surface is shown in Fig. 6.14. It can be seen that in addition to the Cu^+ and Ru^+ secondary ions a number of cluster ions are evident: Ru_2^+, Cu_2^+, $CuRu^+$, $RuCu_2^+$, $CuRu_2^+$. A study of these ions as a function of Cu surface coverage and preparation temperature has furnished a detailed picture of the surface state. It is not possible to go into a discussion of all the features. Cu is essentially an adsorbed species. It was deposited on the Ru surface held at either 540 K or 1080 K. For both conditions as a function of Cu coverage the Ru^+ secondary ion intensity increased and then fell. There are two effects at work: first a work function increase occurs when Cu adsorbs on Ru and this increases the secondary ion yields; second, as Cu coverage increases the Ru surface is *covered* reducing the number of Ru atoms which can be removed. Fig. 6.15 shows that on the 1080 K surfaces the Ru^+ signal

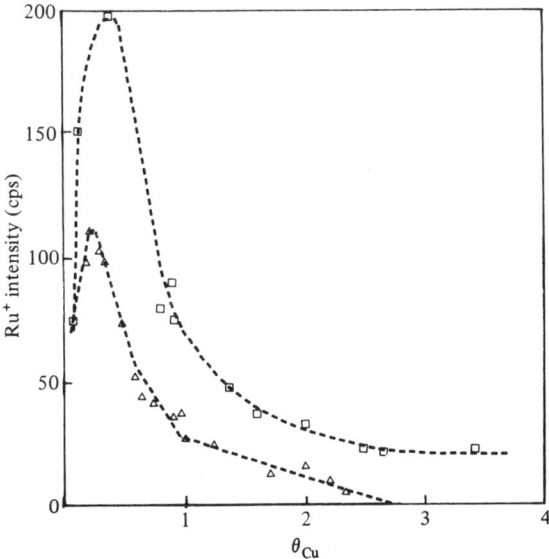

Fig. 6.15. The variation of the intensity of Ru^+ with Cu coverage for Cu/Ru bi-metallic surfaces prepared at 540 K (□) and 1080 K (△).

reached zero after the equivalent of about two monolayers had been deposited.

The emission of Ru^+ ions never reached zero on the 540 K surfaces. Thus it was concluded that clustering of Cu tends to occur on the low-temperature surfaces, whereas a more atomic distribution occurred at the higher temperatures resulting in a smooth overlayer. The cluster ions Ru_2^+ and $RuCu^+$ confirmed this and provided further information. It has been shown for CO adsorbate systems that the ratio MCO^+/M^+ is proportional to the coverage of CO; Fig. 6.16 shows that this also applied for Cu on Ru up to one monolayer. Beyond one monolayer the ratio became constant, and $RuCu^+$ disappeared after two monolayers on the 1080 K surfaces, as would be expected for smooth statistical growth. The lower gradient for the $RuCu^+/Ru^+$ for the 540 K surfaces suggested that some of the Cu does not contact with Ru and thus Cu clustering is implied, indeed it also indicated that some 25% of the Ru surface was not covered with Cu. These conclusions were fully supported by the Ru_2^+/Ru^+ ratio which fell much more rapidly with coverage on the 1080 K compared to the 540 K surfaces.

Thus the picture given of the surfaces by SSIMS is of a random dispersal of Cu after the high-temperature preparation resulting ultimately at high Cu coverage in smooth layer-by-layer growth; after low-temperature

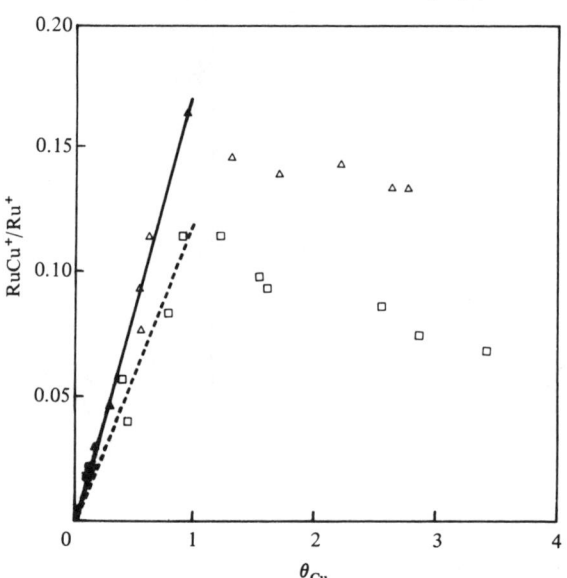

Fig. 6.16. A plot of the secondary ion ratio $RuCu^+/Ru^+$ as a function of Cu coverage for Cu/Ru bi-metallic surfaces prepared at 540 K (□) and 1080 K (Δ).

Static secondary ion mass spectrometry

preparation Cu clusters are formed even at low Cu coverage and this results in three-dimensional islands as surface Cu increases.

It is interesting and significant that these different modes of Cu adsorption give rise to significant changes in the mode of CO adsorption on Ru. Here again SSIMS provided significant and unique information. Temperature-programmed desorption and SSIMS can indicate that Cu reduces the amount of CO adsorbed; work function measurements can show that Cu markedly reduces the extent of charge movement towards the adsorbing CO but SSIMS showed, via the 'fragmentation pattern' ($MCO^+/\Sigma M_x CO^+$), that Cu tends to push CO away from the usual linear sites found on Ru into bridge sites, Fig. 6.17.

4.3.3 Surface reactions

Surface reactivity is not only concerned with static adsorption studies, there is also an obvious interest in reactions at surfaces. SSIMS has shown itself to be extremely useful in following the transformation of adsorbed species. An especially interesting example is a kinetic study of the hydrogen–deuterium exchange in the adsorbed ethylidene species (Creighton, 1984). Ethylene is known to dehydrogenate on Pt(111) near room temperature to form the ethylidene species, but its reactivity is difficult to study unless a mass spectral technique can be applied. By monitoring the four adsorbed ethylidene isotopes, C_2H_3, C_2H_2D, C_2HD_2 and C_2D_3, which is extremely easy with SSIMS using the CH_3^+, CH_2D^+, CHD_2^+ and CD_3^+ secondary ions, it is possible to determine their time dependence as adsorbed ethylidene reacts with gas phase D_2. Using static conditions and taking care that there is no contribution from ions formed

Fig. 6.17. The variation of $RuCO^+/RuCo^+ + Ru_2CO^+$ at θ/θ_{max}^{Ru}, -0.2 with Cu coverage for the 540 K (O) and 1080 K (Δ) series (from A. Brown, J.A. van den Berg & J.C. Vickerman, *Proc. 8th Int. Congr. Catal.*, Berlin, 1984).

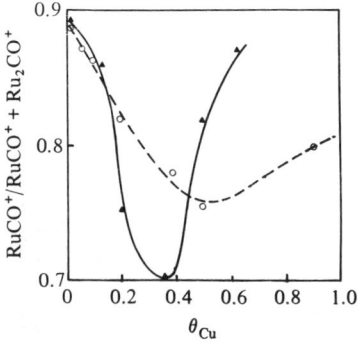

in the primary ion source, the variation of the ions with time is shown in Fig. 6.18.

A simple kinetic model can be set up to account for the observations, which assuming that adsorbed D quickly reaches a steady-state concentration, can be easily solved as shown in the lower part of Fig. 6.18. Of course it is of fundamental interest to know whether or not the dehydrogenation step precedes or follows the deuteration step. SSIMS again sheds light on

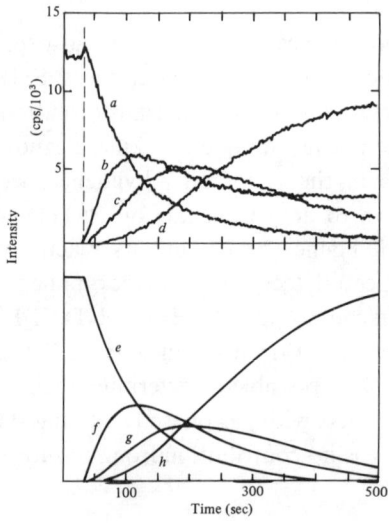

Fig. 6.18. Upper panel: A plot of the secondary ion intensities during the isotope exchange reaction at 383 K; curves a–d are for $m/e = 15$, 16, 17 and 18, respectively. Incident Ar^+ current density is 7 nA/cm^2. Lower panel: Calculated results for the model discussed in the text using $(k\theta_D)^{-1} = 87$ s. Curves e–h are for θ_{X0}, θ_{X1}, θ_{X2} and θ_{X3}, respectively (from Creighton, 1984).

Fig. 6.19. Schematic paths for isotope exchange in ethylidene (from Creighton, 1984).

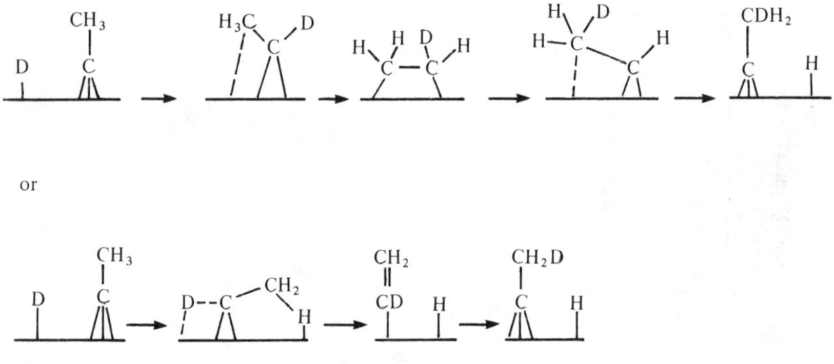

this matter. If some of the ethylidene is dehydrogenated by heating to about 450 K and then H_2 is admitted, the temperature lowered and the CH_3^+ ion monitored, there is no evidence that the resulting surface species can be rehydrogenated. Thus it is concluded that deuterium is added to the ethylidene before loss of hydrogen or in a concerted process as in Fig. 6.19.

4.4 Oxidation of metals

This is an area which has received considerable attention and SSIMS has made a significant and unique contribution. Much of the work was done by Benninghoven and his group. SSIMS has been particularly helpful in unravelling the early stages of oxidation. The secondary ion spectrum was observed as a function of the admission of oxygen to a clean metal surface. By way of example, the oxidation of chromium illustrates the main features of the analysis (Benninghoven, 1975).

Fig. 6.20 shows the variation of intensity for a number of significant secondary ions with oxygen dose. The Cr_2^+ and CrO_2^+ ions pass through a maximum, whereas the other ions reach a plateau. To clarify the significance of this behaviour after a steady state had been achieved the oxidised layers were progressively sputtered away and again the secondary ion yields were monitored. CrO^+ decayed to zero very rapidly whilst CrO_2^+ fell more slowly, never reaching zero. CrO_2^- did not begin to fall

Fig. 6.20. Increase of secondary ion intensities during exposure of a clean chromium surface (from A. Müller & A. Benninghoven, Surface Science, 39, 427, 1973).

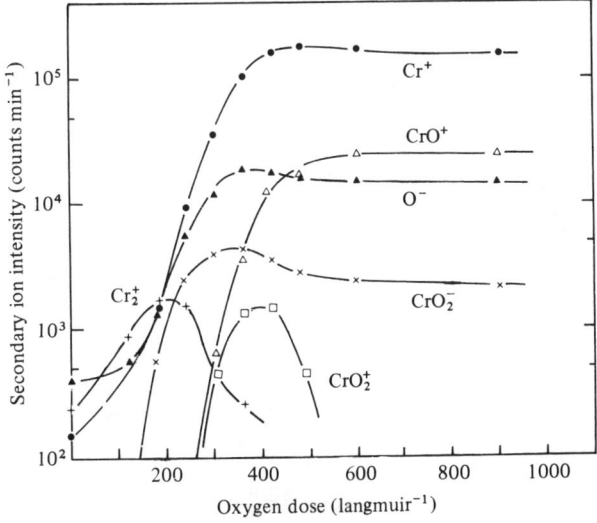

for some time and even then it decayed slowly. Cr_3^+ began to increase when CrO^+ was lost. Similar observations were recorded for a number of other metals and led to a general picture of the early stages of surface oxidation. It was concluded that an initial dose of about 20–50 L is required before the oxide layer begins to form. In the 50–100 L range, an oxide layer emitting CrO_2^- is formed at the surface. In this layer the metal atoms have a low valence state. It was found that after fairly low exposures to oxygen, the decay curve of the CrO_2^- species during sputtering followed an exponential law:

$$\theta(t) = \theta(0)\exp(-t/t_m)$$

where $t_m = \theta_0/SV$; $\theta(t)$ is the relative coverage at time t; $\theta(0)$ is the relative coverage at time zero; t_m is the average lifetime of a monolayer; S is the sputtering rate for a monolayer and V is the primary particle flux. This dependence implies that the oxide was not more than one monolayer thick. As the oxide layer developed there was an initial increase in the emission of Cr^+, Cr_2^+, and Cr_3^+ owing to the increase in ionisation

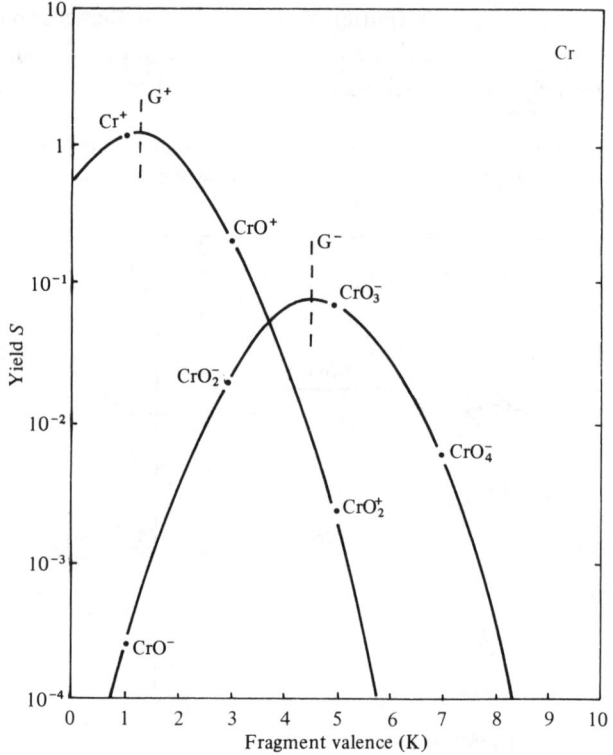

Fig. 6.21. Valence model: application to $CrO_n^{\pm}$ emission.

potential which is correlated with the increasing surface work function as the oxidation process proceeds. This initial rise was followed by a fall in Cr_2^+ and Cr_3^+, indicating the decrease in the number of adjacent Cr atoms in the surface as the oxide layer developed.

At oxygen exposures in excess of 200 L the emission of CrO^+ and CrO_3^- developed and the CrO_2^- decreased, suggesting the formation of a new oxide phase of which CrO^+ is characteristic. When this phase was removed by sputtering, the CrO_2^- phase reappeared and the Cr_2^+ and Cr_3^+ emissions increased. Thus it was concluded that there are two steps in the initial oxidation of clean metals, generating two distinct phases having characteristic spectra.

The appearance of different characteristic oxygen containing ions $M_mO_n \pm$ as surface oxidation proceeds can be described by Benninghoven's Valence Model mentioned earlier, Fig. 6.21 shows that after an exposure to 100 L oxygen the relative intensities of the positive and negative ions can be represented by two Gaussian distributions. Clearly as oxidation continues G^0, the average valence of the metal atoms in the oxide lattice will tend to increase thus influencing the relative yields of the different secondary ion clusters.

Surface oxidation can occur as a consequence of the exposure of metals to other gases. For example, a SSIMS study of the interaction of methanol with clean aluminium at 273 K (Tindall & Vickerman, 1985) showed that the first effect was an increase in the emission of O^- and $AlO_x^\pm$ and loss of

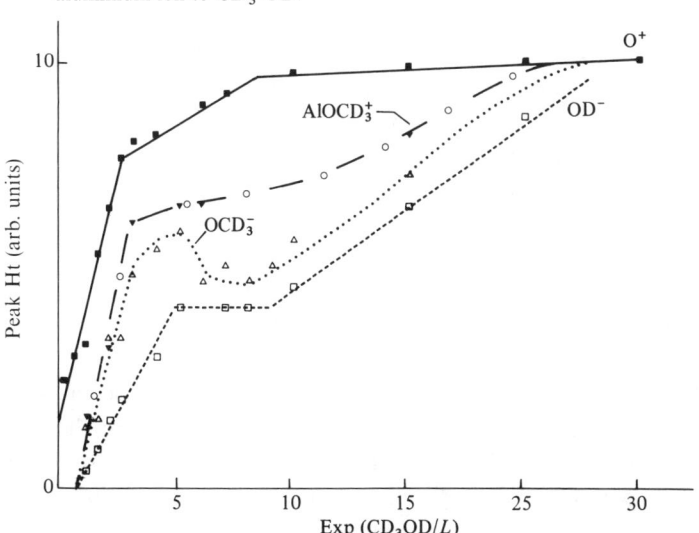

Fig. 6.22. SSIMS secondary ion yields as a function of exposure of a clean aluminium foil to CD_3OD.

Al_2^+. This was accompanied by the appearance of methane in the gas phase. Thus it was concluded that methanol decomposed to oxidise the surface. After this initial oxidation phase other secondary ions began to appear, in the case of CD_3OD OD^- and OCD_3^-, Fig. 6.22.

There was no evidence for the presence of the molecular ion which is detected when methanol is physisorbed. It was concluded that these ions reflect the formation of surface hydroxide as methanol dissociatively adsorbed on the oxide surface initially by abstraction of the hydroxyl proton. This was confirmed by the fact that CD_3OH produced OH^- and CD_3OD yielded OD^- ions. The CH_3O fragment was then thought to adsorb on a surface oxidised Al site since both CH_3O^- and $AlOCH_3^+$ were observed. The study demonstrated that an oxide layer is necessary before methoxide can be formed. The stability of the surface methoxide species so formed was monitored by raising the temperature of the aluminium and monitoring the secondary ion intensities. By 373 K all the methoxy species and most of the surface hydroxide had decomposed.

The considerable body of work utilising SSIMS for basic studies of surface reactivity provides clear support for the conclusion that the surface mass spectra are sensitive to the surface chemical structure. Clearly, this has significant implications for the use of SSIMS in surface studies of materials of practical technological importance. Some illustrative applications will now be considered.

4.5 Catalyst characterisation

The surface activity of catalysts is dependent on the surface composition and chemical state. The ability to analyse these parameters is crucial to the understanding of catalyst behaviour, the monitoring of catalyst reproducibility during preparation and in catalyst trouble shooting. The sensitivity of SSIMS/FABMS to elemental composition and to surface chemical structure suggest that it would be an ideal technique for catalyst characterisation.

4.5.1 Alloy catalysts

An interesting series of papers which clearly illustrates the application of SSIMS to the understanding of metal alloy catalyst is due to Fleisch, Delgass, Winograd and their co-workers, 1979. They have utilised SSIMS and XPS to characterise the surface state of FeRu Fischer–Tropsch catalysts before and after use. The catalysts were studied in the main in the unsupported state in the form of powders. Various compositions from pure Ru to pure Fe were prepared and investigated after reduction in H_2 at 573 K. At this stage XPS suggested that there was surface enrichment of Fe and that some of that Fe was still in the 2+ and 3+ state. Ru, however, was completely reduced. SSIMS is sensitive only to the top two layers at most and it confirmed that surface segregation of Fe was very significant at low Fe contents, suggesting that a very high proportion of the Fe was located at the surface. However, a complete skin of Fe does not cover the surface because Ru^+ ions are still detectable, the implication was that Fe islands form at the surface. This application illustrates the complication in obtaining quantitative data in SSIMS. The sensitivity of SSIMS varies from element to element, e.g. the yield ratio between Ru^+ and Fe^+ is 2.3. That is easy to account for. However, when Fe is combined with Ru there is also a work function change which affects the secondary ion yield. In this case it appeared to reduce it rather severely. For qualitative purposes the authors assumed that the Fe and Ru ion yields were affected to equal extents – a reasonable approach but still an assumption. Other secondary ions were noted, particularly $FeRu^+$, which seemed to confirm alloy formation or at least bonding between Fe and Ru because the ion was not emitted from a physical mixture of Fe and Ru powders. Thus SSIMS gives quite a detailed picture of the alloy surfaces.

Of equal interest are the surface effects consequent on exposing the surfaces to a CO/H_2 mixture. The Ru and very low Fe-containing alloy (3%) showed stable long-term Fischer–Tropsch activity with no evidence from XPS or SSIMS of carbon deposition. However, the Fe alloy, which as we have seen does have a high surface Fe content, did show definite selectivity changes from C_1 to C_2 hydrocarbons. Higher Fe compositions (25–67%) resulted in very high carbon build-up. XPS estimates suggested up to 40 Å; SSIMS certainly showed that metal secondary ions became undetectable and C^+, CH_x^+, C_2^+, $C_yH_x^+$ ions were the predominant ion species. Catalytic activity fell by a factor 2–3, and selectivity moved to lower molecular weight products. The carbon overlayer was rather stable and required a high temperature, 673 K, treatment in H_2 to remove it. CO hydrogenation over Fe was a little different; SSIMS showed that the Fe ions did not disappear, carbon was deposited but a fair amount was incorporated to form carbide.

4.5.2 Zeolite catalysts

FABMS has been utilised to monitor surface segregation of silicon or aluminium after preparation and treatment of zeolite catalysts (Dwyer et al., 1982). These were semiquantitative studies using depth profiling to probe the subsurface region to obtain an estimate of the extent of segregation. First it was necessary to calibrate for the relative sensitivities to Si and Al. A range of standard zeolite materials from faujasite to ZSM-5 provided a concentration range of Si/Al = 1–18. A series of three depth measurements on each zeolite yielded a linear relationship between the secondary ion ratio Si/Al and the bulk concentration ratio. This relationship has also been demonstrated for the negative cluster ion ratio SiO_2^-/AlO_2^-, Fig. 6.23.

The effect of a number of treatments on the surface Si/Al has been studied. Analysis of steamed zeolites revealed surface layers enriched in aluminium due to migration during steaming at elevated temperature, Fig. 6.24.

Treatment with $SiCl_4$ can lead initially to surface enrichment with aluminium, whereas dealumination using EDTA leads to silicon rich surfaces. These surface modifications have a significant effect on, for example, the cracking of n-hexane. Phosphorus, boron and magnesium are effective in enhancing the shape-selectivity of pentasils. FABMS has been used to investigate the surface concentration of such additives.

4.5.3 Supported oxide catalysts

The use of secondary ion cluster to characterise the chemical structure of a catalyst surface has been illustrated rather clearly by a study of a rhenium oxide – alumina metathesis catalyst (Coverdale et al., 1983).

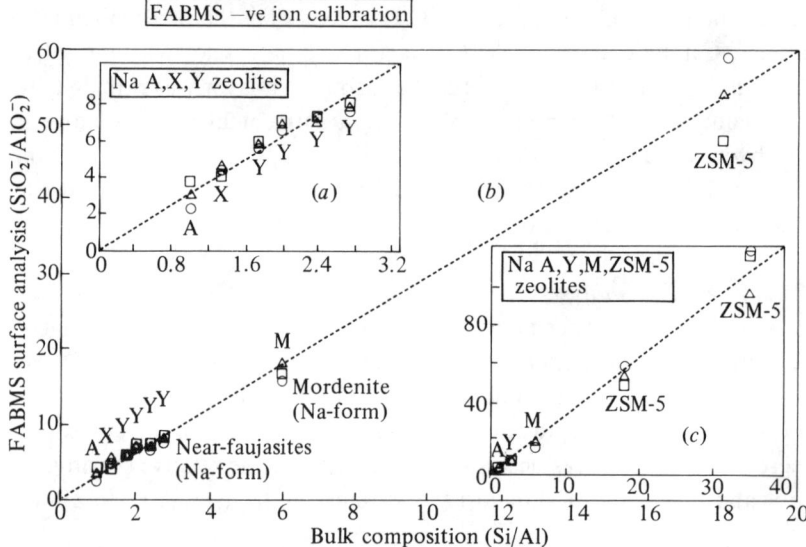

Fig. 6.23. Correlation between secondary ion ratio Si/Al^+ and bulk chemical analysis (Si/Al) for synthetic zeolites; Δ = first layer; O = after first bombardment (fluence = 75 μA min); □ = after second bombardment (fluence = 150 μA min) steam at 600 °C for 14 h (from A. Brown & J.C. Vickerman, *Surface and Interface Analysis*, **6**, 1, 1984a).

Two pairs of catalyst samples were prepared by impregnating gamma alumina with aqueous ammonium rhenate (VII), one pair were only dried at 110 °C and consisted of 5 and 17.5% w/w NH_4ReO_4 on alumina, whilst the other pair were also heated at 520 °C to produce Re_2O_7 at the surface, again in 5 and 17.5% w/w loadings. The negative-ion FABMS spectrum of the supported rhenium species were compared with spectra of unsupported bulk compounds to identify the chemical nature of the surface species. Table 6.3 presents the intensities of a selection of the most significant ions relative to the ReO_2^- ion observed from the supported samples and from Re_2O_7 and NH_4ReO_4.

Table 6.3. *Rhenium oxide-alumina metathesis catalysts*

Fragment	NH_4ReO_4	ION Fragment intensities relative to ReO_2^-				
		5% NH_4ReO_4 /Al_2O_3	17.5% NH_4ReO_4 /Al_2O_3	Re_2O_7	5% Re_2O_7 /Al_2O_3	17.5% Re_2O_7 /Al_2O_3
ReO^-		0.096	0.04	0.089	0.063	0.07
ReO_2^-	1	1	1	1	1	1
ReO_3^-	4.8	3.48	5.21	1.70	12.3	16.2
ReO_4^-	1.06	1.5	2.64	24.5	8.48	9.9
$Re_2O_4^-$		0.022	0.020	0.055	0.066	0.021
$Re_2O_5^-$		0.082	0.08	0.23	0.16	0.046
$Re_2O_7^-$		0.024	0.02	0.013	trace	trace
$Re_2O_{10}^-$			0.01	0.016	0.014	0.014
$Re_3O_7^-$			0.01	0.046	0.015	0.011

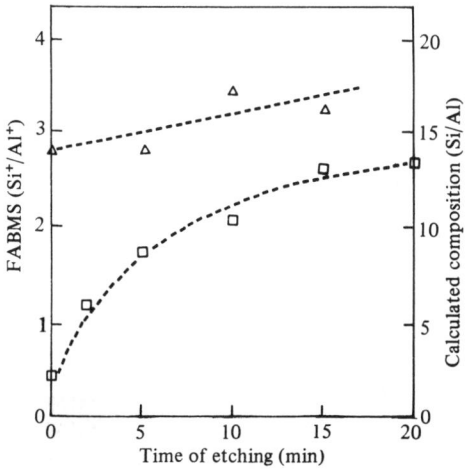

Fig. 6.24. Depth profiles of H-ZSM-5 △ = untreated, □ = heated in steam at 600° for 14 h.

There are quantitative similarities between the ion pattern observed from supported and unsupported NH_4ReO_4 and between supported and unsupported Re_2O_7. However, there is clear evidence for the presence of Re_2O_7 on the supported NH_4ReO_4 samples. A lattice of discrete ReO_4 tetrahedra would not produce significant secondary ion clusters containing more than one rhenium atom. In contrast, and this demonstrates vividly the power of SSIMS/FABMS in this area, the molecular chain structure of Re_2O_7 should and does produce characteristic $Re_xO_y^-$ ion fragments with $Re_2O_5^-$ and $Re_3O_7^-$ predominating. The authors concluded that decomposition of NH_4ReO_4 on alumina produces aggregates of Re_2O_7 lying on ReO_4 clusters which are in intimate contact with the alumina. A further interesting feature was concluded from a comparison of the data from the two pairs of catalysts. Table 6.4 shows the positive and negative intensity ratios ReO_2/AlO for the four catalysts.

For the NH_4ReO_4 pair there is marked increase with loading. It is concluded that production of Re_2O_7 is accompanied by aggregation such that the surface area of exposed rhenium and alumina remains the same as loading increases.

The above studies show the tremendous potential these techniques possess for elucidating the composition and surface structure of catalyst materials. It can be expected that their use will grow considerably in the next few years.

4.6 Polymers

Organic materials are of considerable importance in a wide range of surface technologies – surface coatings, packaging, lubrication, etc. There are special problems associated with their analysis utilising SSIMS; nevertheless the mass spectral nature of the SSIMS spectrum makes it an extremely attractive technique for this type of analysis. However, polymer materials also highlight the difficulties associated with the technique.

The major technique used over the past decade for surface analysis of polymers has been X-ray photoelectron spectroscopy. Although XPS

Table 6.4. *Intensity ratios for fragments ReO_2 and AlO from Rhenium oxide-alumina catalysts*

Catalyst	$ReO_2^+ AlO^+$ 187_{Re}	$ReO_2^- : AlO^-$ 187_{Re}
5% NH_4ReO_4/Al_2O_3	0.0043	0.041
17.5% NH_4ReO_4/Al_2O_3	0.0191	0.248
5% Re_2O_7/Al_2O_3	0.0092	0.152
17.5% Re_2O_7/Al_2O_3	0.110	0.149

provides multielement and chemical state information, the technique is not adequate for molecular structure studies of polymers, since the chemical states of the typical constituent elements, i.e. carbon, oxygen, silicon, do not yield easily resolvable chemical shifts. This lack of molecular specificity, the inability to detect hydrogen and photoelectron escape depths averaging over several atomic layers, led a number of workers to explore SSIMS as an alternative technique. In principle SSIMS has a high potential for surface analysis of polymers, since it offers a high degree of molecular specificity with high signal-to-noise ratio and a sampling depth of only 1–2 atomic layers. These features of the technique have been demonstrated for non-volatile organic compounds (Surman & Vickerman, 1981).

A good example of the ability of SSIMS to provide precise molecular information compared to XPS was provided by a comparison of the spectra obtained from the n-butyl, sec-butyl, isobutyl and tert-butyl polymers (Gardella & Hercules, 1981). The XPS results did not provide a clear method for distinguishing between such structural isomers since the C_{1s}/O_{1s} peak areas yield identical values and an examination of the spectra for C_{1s} and O_{1s} revealed no differences between the isomers. SSIMS, however, showed very clear differences and two features were used to differentiate between the isomers. Firstly the relative intensities of the 57 dalton peak (butyl cation-$C_4H_9^+$) ranged from 0.12 for n-butyl, 0.41 for sec-butyl, 0.22 for isobutyl to 1.0 for the tert-butyl polymer. This follows the expected stability series for primary (n), secondary (sec, iso) and tertiary carbonium ions. Secondly an examination of the relative intensities of peaks due to bond breaking events in the butyl group were used to differentiate between sec-butyl and isobutyl polymers.

The processes involved were illustrated as follows:

Scheme 1

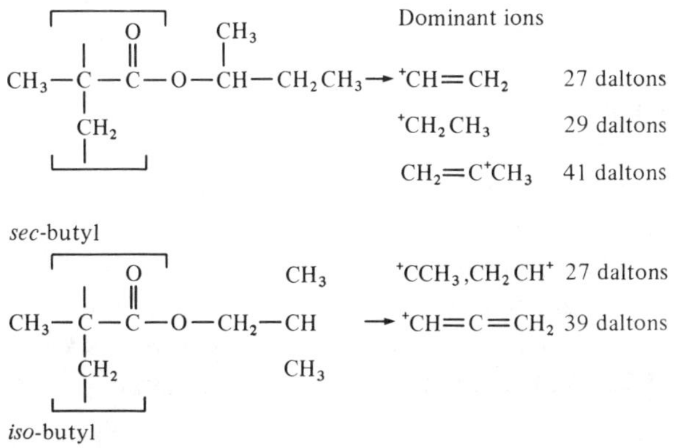

$$\begin{array}{c} \overset{O}{\underset{CH_2}{CH_3-C-C}}-O-\overset{CH_3}{\underset{}{CH}}-CH_2CH_3 \end{array} \rightarrow \begin{array}{l} {}^+CH=CH_2 \quad 27 \text{ daltons} \\ {}^+CH_2CH_3 \quad 29 \text{ daltons} \\ CH_2=C^+CH_3 \quad 41 \text{ daltons} \end{array}$$

sec-butyl

$$\begin{array}{c} \overset{O}{\underset{CH_2}{CH_3-C-C}}-O-CH_2-\overset{CH_3}{\underset{CH_3}{CH}} \end{array} \rightarrow \begin{array}{l} {}^+CCH_3, CH_2CH^+ \quad 27 \text{ daltons} \\ {}^+CH=C=CH_2 \quad 39 \text{ daltons} \end{array}$$

iso-butyl

The observed SSIMS spectra confirmed the above analysis with the 27, 29 and 41 dalton ions having the highest relative intensities from poly(sec-butyl methacrylate) and 27 and 39 ions the highest relative intensity from poly(isobutyl methacrylate).

Coincident with the benefits of SSIMS there are a number of experimental problems which may limit its applicability to polymer surfaces. These are (a) the expected high ratio of ion beam damage, and (b) the need for charge neutralisation leading to (c) uncertainty as to surface potential. The common method of charge neutralisation, namely flooding of the sample with low energy electrons, leads to a possibility of electron stimulated desorption (ESD) of secondary ions. Clearly any systematic assessment of polymers by SSIMS should address these problems carefully. Briggs and co-workers (1982, 1983) have sought to do this in their very careful studies of polymer surfaces. Initial experiments on polystyrene (PS) and polytetrafluoroethylene (PTFE) revealed the complex problems involved in obtaining SSIMS spectra from polymers using an ion beam charge neutralised at the surface by a defocussed electron beam (SIMS/EN). These have been listed earlier. Ion emission produced by the electron beam itself was found to be negligible under the current densities employed (1–4 nA cm^{-2}, 700 eV) for PS, but for PTFE significant emission of characteristic molecular ions such as CF_3^+ occurred. A far more serious problem was the time-dependent variation in relative intensities of molecular species from PS. For example, the 91 dalton ($C_7H_7^+$) ion was shown to decay in intensity far more rapidly than, say, the 51 dalton ($C_4H_3^+$) ion. It was recognised that there were two possible reasons for this effect. Firstly, ion beam damage producing changes in the chemical state of the surface was considered. In an XPS study the C_{1s} shake-up satellite, a measure of the aromaticity of the surface, was monitored as a function of ion fluence during a sputtering experiment. Ion-beam-induced damage was indicated by the loss of the C_{1s} satellite after a flux of 1.6×10^{14} ion cm^{-2} at 4 keV. Lower ion beam doses would therefore be expected to produce significant changes in the SSIMS spectrum since the information from XPS is averaged over several layers.

The second problem arose from changes in surface potential due to insufficient charge neutralisation. This caused in the relative peak intensities changes in the SSIMS spectrum. The reason lies in the differing energy distribution curves for ions of different mass (and hence molecular composition) ejected from the polymer surface. A series of ions ranging from 15 daltons to 83 daltons, originating from paraffin were monitored and shown to have significantly different intensity versus kinetic energy profiles (Fig. 6.25). As a consequence any change in surface potential would move the position of the peaks in the energy distribution curves with respect to the energy acceptance window of the quadrupole mass spectrometer, and would give rise to a change in relative intensity of the ions.

The relative importance of surface potential drift and ion beam damage could not be resolved until FABMS was applied to the problem. This SSIMS technique does not suffer as much from sample charging and hence surface potential drift can be ruled out as an important factor. Time dependent studies on polystyrene using argon ion and argon atom bombardment in separate experiments showed that the initial decay rates for the 91 dalton ion were rather different, Fig. 6.26.

Clearly beam damage processes can account for changes in relative intensity of ions emitted from these polymer surfaces. The decay rates

Fig. 6.25. Normalised energy distribution curves for positive ions of different mass ejected from this (conducting) film of paraffin wax; 4 keV Me^+ primary ions (0.2 nA into $\sim 0.3\,cm^2$).

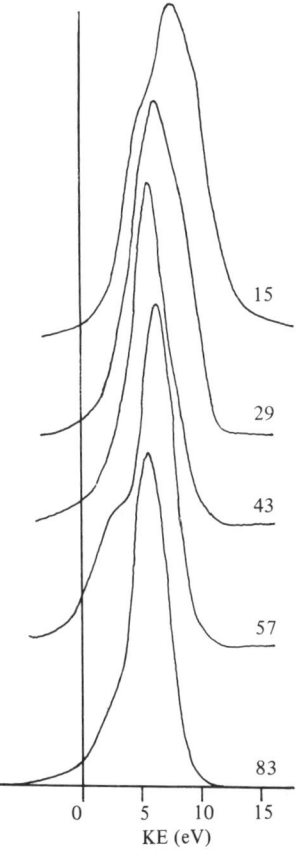

themselves were some 100–1000 times faster than would be expected for inorganic materials. Organic molecular solids will be very sensitive to high energy particle impact. They do not have large high-energy lattices to dissipate the transferred energy. However, under ion bombardment the intensity decayed four to five times *more rapidly* than under atom bombardment. It would appear, therefore, that in addition to the collisional effects there is a charge contribution to the damage mechanism.

Almost identical spectra were, however, obtained from polyethylene oxide by ion and atom bombardment, Fig. 6.27(a),(b). This polymer is distinguished by the ability to rapidly anneal out surface damage. Although spectral decay is observed as with polystyrene, if the ion or atom beam is switched off for a while, when analysis recommences the secondary ion intensities have returned to their initial values. This behaviour is probably

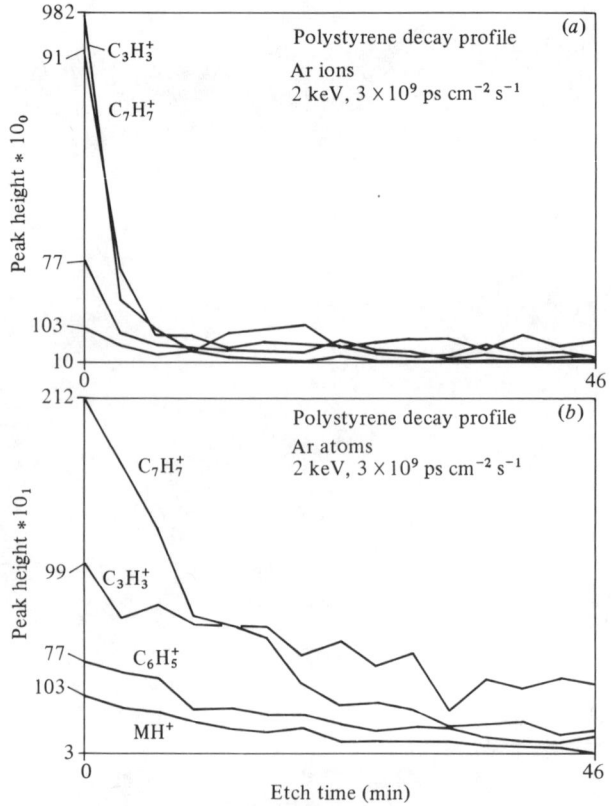

Fig. 6.26. The decay of secondary ion yield from polystyrene as a function of time for argon ion (with electron neutralisation) and argon atom bombardment.

due to its high monomer mobility. Thus although this is apparently good evidence for the character of the secondary ion spectra being independent of primary particle charge state, it is important that SSIMS and FABMS spectra are compared on an equivalent basis, namely using very low beam fluences when damage is small.

The molecular specificity of SSIMS/FABMS techniques has allowed 'fingerprint spectra' to be identified for a series of polymers with very

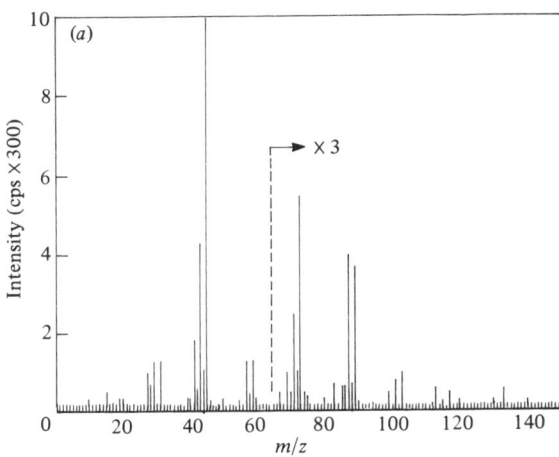

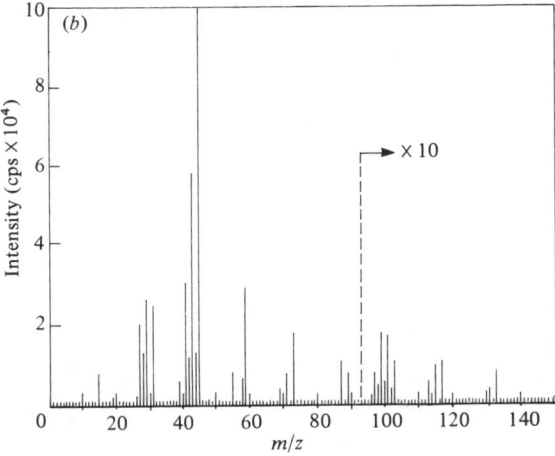

Fig. 6.27. (a) SIMS/EN spectrum of poly(ethylene oxide) (PEO) using 4 keV Ar$^+$ (1 nA cm^{-2}); FSD = 3×10^3 cps, (b) FABMS spectrum of PEO using 2 keV Ar atoms (5×10 particles cm^{-2} s^{-1}); FSD = 10^4 cps (from A. Brown & J.C. Vickerman, *Surface and Interface Analysis*, **6**, 1, 1984).

different chemical structures ranging from low-density polyethylene to Nylon 6 and polyethyleneterephthalate (PET). The most important feature is that positive-ion spectra are readily interpretable using conventional electron ionisation mass spectrometry rules. A good example of this is the spectrum of PET, Fig. 6.28, where the major ions can be rationalised by the following fragmentation sequence:

Scheme 2

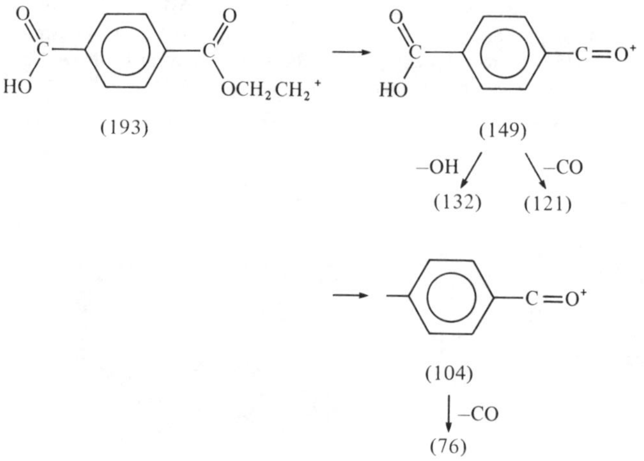

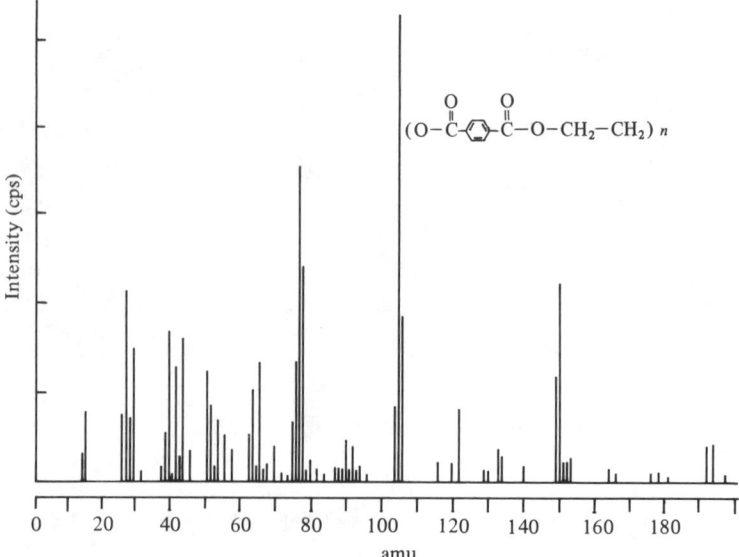

Fig. 6.28. SIMS positive-ion spectrum of polyethylene-terephthalate (PET).

Static secondary ion mass spectrometry 211

Fig. 6.29. Secondary electron (*a*) and secondary ion images $^{56}\text{Fe}^+$ (*b*) and $^{91}\text{C}_7\text{H}_7^+$ (*c*) from polymer coated iron powder. Conditions: 6 keV Ga$^+$, beam current 0.4 nA.

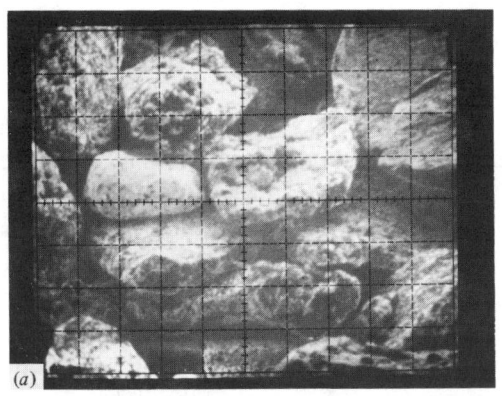

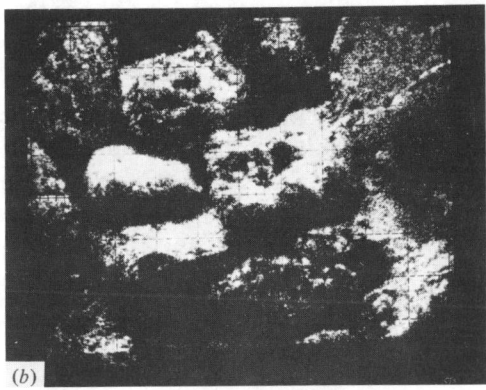

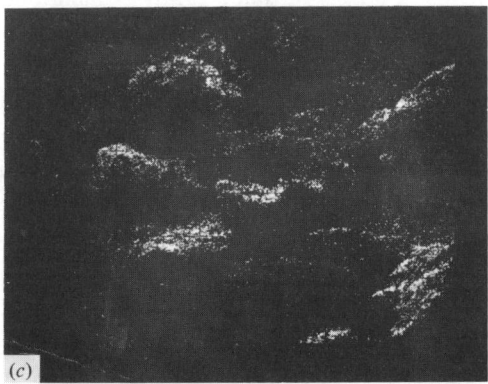

The high spatial resolution capability of SSIMS has been used to monitor small heterogeneous surface features ($< 100\,\mu$m) on polymer fibres (Briggs, 1983). Characteristic mass spectral features allowed chemical mapping of the distribution of silicone oils on a low density polyethylene substrate using a rastered beam of argon ions ($< 30\,\mu$m spot size). The implications of this work are considerable since the ability to map the *molecular* chemistry of surface layers is of paramount importance in surface analysis and beyond the power of other surface techniques. The advent of the liquid metal ion sources with spot sizes typically $< 0.5\,\mu$m have potentially opened up a whole new field of surface microanalysis of polymers by static SIMS imaging. An example is shown of a study of a protective polymer coating on an iron powder. Fig. 6.29(*a*) shows the ion induced secondary electron image of the iron particles at a magnification of 300. The particles size is about $100\,\mu$m. Fig. 6.29(*b*) shows the ^{56}Fe$^+$ SSIMS image; whilst Fig. 6.29(*c*) is an image of the ^{91}C$_7$H$_7{}^+$ cluster from the surface coating. Comparing (*b*) and (*c*) it is clear the coating is not uniform, although care must be taken in interpretation since the analyser to sample geometry can affect the collection efficiency from rough samples.

The advance of the technique will be limited by the beam damage effects together with the low secondary ion transmission of quadrupole analysers mentioned in section 4.1.2. Table 6.1 showed that as the magnification required increases, and thus sufficient ion yield is demanded from smaller and smaller areas at the surface, the erosion rate in the analysed area will increase rapidly. Thus static SIMS conditions are not possible at high magnification because of the very rapid damage rate of polymers. It is here that the potential of the time of flight analyser should be realised with its much higher transmission.

4.7 Oxide materials

In addition to the analysis of oxide catalysts mentioned in the earlier section, SSIMS is of great value in the surface characterisation of other oxide based materials such as glasses, refractories, semiconductors, insulators, etc. As with polymers the major problem with such insulating materials is, of course, the build-up of charge produced by the primary ion beam. Mobile ions, such as Na$^+$, which frequently occur in oxide glasses, are known to diffuse under the surface electric fields which build up under ion bombardment even when electron neutralisation is used. A solution to the above problems is provided by the use of a neutral primary beam. Studies of SiO$_2$ have shown that neutral particle bombardment prevents migration of sodium ions.

Most oxide materials are insulators or semiconductors. However it is still important to be aware of the electronic state of the materials when using SSIMS to quantitatively monitor the surface composition. A basic study of $Sn_{1-x}V_xO_2$ solid solutions illustrates the point (Pomonis & Vickerman, 1981). Using atom bombardment the surface composition was monitored via the V^+/Si^+ ion ratio as the composition was varied from $x = 0$ to $x = 0.1$. Up to $x = 0.03$ there was a linear rise of V^+/Si^+ with x. Above $x = 0.03$ there was a similar relationship, but between these two regions there was a factor 5 fall in the relative sensitivity to vanadium. It is known from electrical measurements that electron delocalisation occurs above $x = 0.03$. The fall in sensitivity to V^+ can be attributed to a concurrent fall in the ionisation probability. Similar observations have been reported for alloys. Alloying can increase the ionisation probability by a significant factor (see Williams, 1982).

Whilst the electronic factor must be borne in mind glass surfaces can be examined using fast atom beams for both quantitative elemental and molecular information under truly *static* conditions. The mass spectrum of the surface contains unique chemical structural information and this has particular importance in the fields of surface weakening of glasses and coating technology. The positive-ion FABMS spectrum of a three-

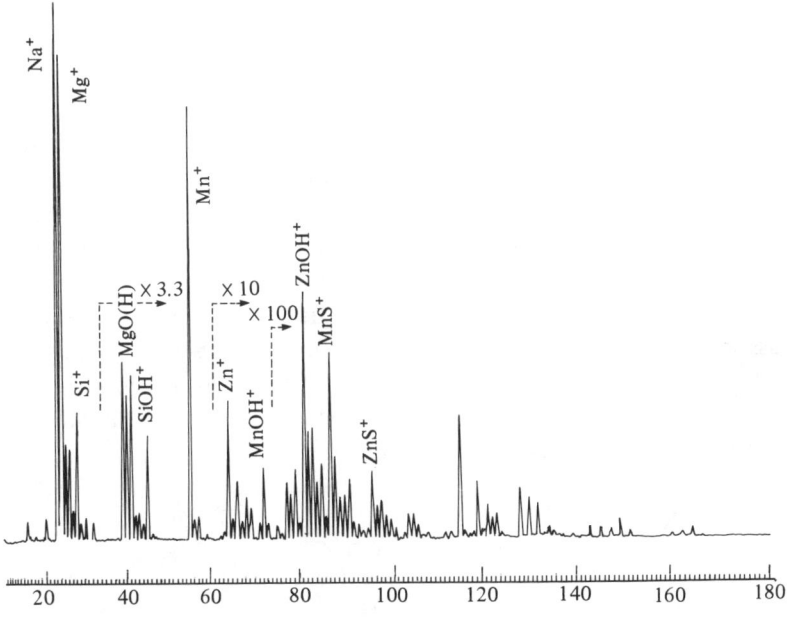

Fig. 6.30. FABMS positive-ion spectrum of a Mn doped ZnS thin film on tin oxide-coated soda lime glass: Ar atoms, 2 keV, 5×10^{10} particles $cm^{-2} s^{-1}$.

component glass consist not only of the elemental ions, Si^+, Ca^+ and Al^+ but of cluster ions which reflect the surface chemistry.

The hydroxylated nature of the surface is demonstrated by the presence of intense $SiOH^+$ (45 daltons) and $CaOH^+$ (57 daltons) ions. The state of surface hydroxylation could be increased by exposure to water vapour or decreased by outgassing in vacuum. The dehydroxylation process can be monitored using the ratio of $SiOH^+/SiO^+$ ion intensities and heating the glass surface over the range 280–520 K. A further feature of the spectrum is the occurrence of ternary cluster ions such as $SiAlO^+$ (71 daltons), $CaAlO^+$ (83 daltons) and $CaSiO_2^+$ (100 daltons). These seem to reflect the state of surface vitrification, since upon prolonged heating to 750 K all these spectral features are lost. Subsequently only binary ions, e.g. CaO^+, SiO^+, AlO^+, Si_2O^+ were observed indicating separation of the glass into its component oxides.

An example of the use of FABMS in analysing a coating on an oxide glass is given in Fig. 6.30. This shows the positive-ion spectrum of a tin oxide coated soda lime glass with a surface thin film of manganese doped zinc sulphide (approx 1% Mn) used in electroluminescent displays. As well as the expected intense ions Mg^+, Si^+, K^+, Mn^+ and Zn^+ a number of molecular ions such as $MnOH^+$ (72 daltons), $ZnOH^+$ (81, 83, 85 daltons) MnS^+ (87 daltons) and ZnS^+ (96, 98, 100 daltons) characterise the spectrum. The presence of OH containing ions again demonstrates a degree of surface hydroxylation, while MnS^+ indicates chemical interaction of the dopant with surrounding matrix.

5 Conclusion

This chapter has illustrated the power and potential of static SIMS for surface characterisation in the fields of pure and applied research. The SSIMS phenomenon is complex, and it is far from being understood, but experimental scientists have shown that it can be applied with great benefit without waiting for a full description of the mechanistic subtleties. Empirical experience shows that it has much in common with analytical mass spectrometry. This is its power. Yet its future is not only as a surface mass spectrometry. The development of new mass analysers which promise very considerable increases in sensitivity, together with the introduction of highly microfocussed primary beam systems means that SSIMS also has an exciting future as a *chemical microscopy*.

References

Barber, M., Bordoli, R.S., Sedgwick, R.D. & Taylor, A.N. (1981). *Nature* (London), **293**, 270.

Benninghoven, A. (1975). *Surface Science*, **53**, 596.

Bordoli, R.S., Vickerman, J.C. & Wolstenholme, J. (1979). *Surface Science*, **85**, 244.
Briggs, D. (1983). *Surface and Interface Analysis*, **5**, 113.
Briggs, D. & Wootton, A.B. (1982). *Surface and Interface Analysis*, **4**, 109.
Brown, A. & Vickerman, J.C. (1981). *Vacuum*, **31**, 429.
Brown, A. & Vickerman, J.C. (1983). *Surface Science*, **124**, 267.
Brown, A. & Vickerman, J.C. (1984a). *Surface and Interface Analysis*, **6**, 1.
Brown, A. & Vickerman, J.C. (1984b). *Surface Science*, **140**, 261.
Brown, A., van den Berg, J.A. & Vickerman, J.C. (1984). *Spectrochimica Acta*.
Coverdale, A.K., Dearing, P.F. & Ellison, A. (1983). *J. Chem. Soc. Chem. Commun.*, **567**.
Creighton, J.R. (1984). *Surface Science*, **138**, L137.
Dwyer, J., Fitch, F.R., Qin, G. & Vickerman, J.C. (1982). *J. Phys. Chem.*, **86**, 4574.
Fleisch, T., Delgass, W.N. & Winograd, N. (1979). *J. Catal.*, **56**, 174.
Gardella, J.A.Jr & Hercules, D.M. (1981). *Anal. Chem.*, **53**, 1879.
Müller, A. & Benninghoven, A. (1973). *Surface Science*, **39**, 427.
Ott, G.L., Fleisch, T. & Delgass, W.N. (1979). *J. Catal.*, **60**, 364.
Plog, C. & Gerhard, W. (1983). *Z. Phys.*, **B54**, 71.
Pomonis, P. & Vickerman, J.C. (1981). *J. Chem. Soc. Faraday Disc.*, **72**, 247.
Poschenrieder, W.P. (1972). *Int. J. Mass Spectrum and Ion Phys.*, **9**, 357.
Prewett, P.D. & Jeffries, D.K. (1980). *Institute of Physics Conference Series*, **54**, 316.
Surman, D.J. & Vickerman, J.C. (1981). *J.C.S. Chem. Comm.*, p. 324.
Surman, D.J., van den Berg, J.A. & Vickerman, J.C. (1982). *Surface and Interface Analysis*, **4**, 160.
Tindall, I.F. & Vickerman, J.C. (1985). *Surface Science*, **149**, 557.
Williams, P. (1979). *Surface Science*, **90**, 588.
Williams, P. (1982). *Applications of Surface Science*, 13, 241.
Winograd, N. (1981). *Prog. Solid State Chem.*, **13**, 285.

7

Dynamic secondary ion mass spectrometry

D.E. SYKES

1 Introduction

Secondary ion mass spectrometry (SIMS) is arguably the most powerful and versatile of all the surface analytical techniques. As it has developed into today's dynamic SIMS instrumentation (sometimes known as the ion-microprobe or ion-microscope) it offers extremely high-sensitivity quantitative elemental analysis with detection limits in the ppm–ppb range. Not only is the technique element specific, covering all elements from H to U, but it is also isotope specific. Lateral resolution of less than 1 μm can easily be achieved and in model systems depth resolution of less than 2 nm has been demonstrated. However, as with all analytical techniques, SIMS cannot provide the answer to every problem but in some areas it excels and can give information that is not accessible by any other means. It is the object of this chapter to introduce the basic principles behind the SIMS technique, briefly to review the quantitative aspects, to consider the type of instrumentation currently available, and to review the practical considerations of SIMS analysis, illustrating these with practical examples and finally presenting a range of typical applications carried out. The range of applications is by no means exhaustive and for a fuller range of topics the reader is referred to the bibliographies by Yin (1980, 1981) and to the conference proceedings of the biennial SIMS conferences SIMS I, II, III, IV etc.

2 Basic principles

2.1 Secondary ion emission

The basis of any SIMS analysis is the sputtering process (see also Chapters 2 and 6). A beam of primary ions impinges onto the solid surface of the sample and the atoms in the surface of the sample are sputtered off into the surrounding vacuum. When a primary ion strikes the surface of the sample it can be back-scattered or, more probably, will penetrate into the surface of the sample, losing its initial energy through a series of elastic and inelastic collisions with the atoms of the sample, until it finally comes to rest at some depth below the surface. This collision cascade process is illustrated schematically in Fig. 7.1. Sputtering takes place when atoms near the surface of the sample receive sufficient energy from the cascade to escape from the surface of the sample. These sputtered particles can be ejected as atoms or molecules in a neutral, excited or ionised state, but it is only the ionised species that can be used for SIMS analysis.

The escape depth of the sputtered species is generally of the order of a few angstroms for the primary ion energies used for SIMS analyses, with the use of higher primary beam energies leading to greater escape depths. However, a small but finite probability exists that material from a greater depth can escape from the sample. Although the emission of the secondary

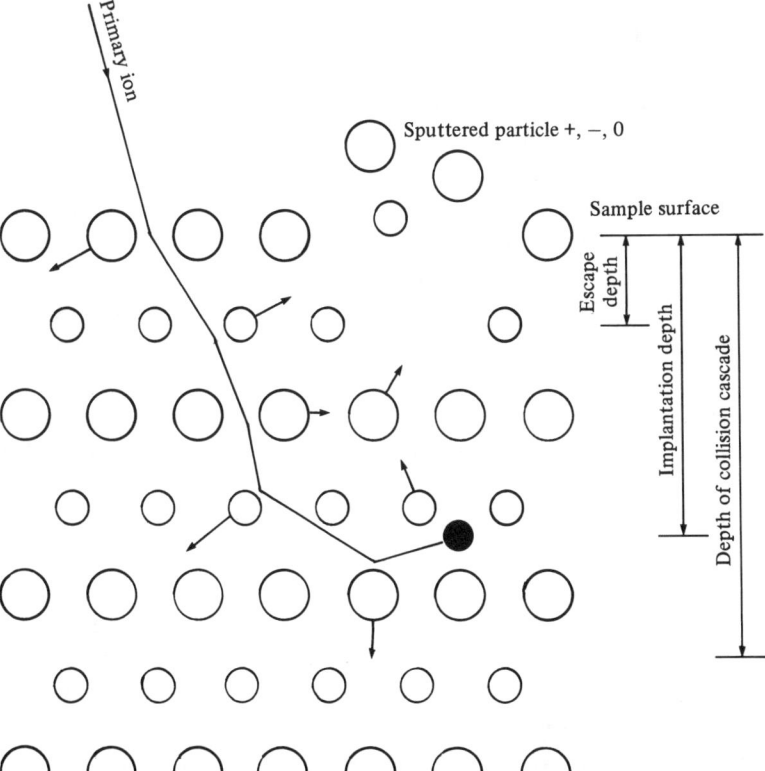

Fig. 7.1. Schematic diagram of the sputtering process.

species takes place predominantly from the surface layer the depth resolution of the SIMS technique is influenced by the collision cascade, as the cascade produces atomic mixing of the species within the subsurface region. Consequently, when species are sputtered from the surface they may not be occupying the sites that they originally occupied in the unsputtered sample. In addition, the primary ion becomes implanted into the sample surface with the possibility of chemically modifying the substrate of the sample. As dynamic SIMS is a destructive technique and the surface of the sample is continuously being eroded, it is important to recognise that once an equilibrium situation is reached the implanted primary ions and collision cascade region are always present and move through the sample ahead of the analysed surface. The implantation depth of the primary ions and the depth of mixing produced by the collision cascade will increase with increasing primary ion beam energy.

2.2 The secondary ion current

The number of ions sputtered from the surface of the sample provide a means for measuring the composition of that sample. For an ideal analytical technique the number of ions of any species sputtered from a sample should be linearly related to concentration, should not change from element to element (no elemental sensitivity variation), should be independent of the sample being analysed (no matrix effect), and should be independent of the primary bombarding species.

In such a situation measurement of secondary ion currents for each species would be all that was required for an absolute quantification of the mass spectrum. Unfortunately for the analyst, none of these conditions are fulfilled for the SIMS technique, although a linear dependence of signal with concentration is observed for dilute systems. That is, for situations where the concentration of a particular element does not exceed a few atom per cent the secondary ion signal is linearly proportional to the concentration of that element. For example for P in iron and nickel matrices non-linear dependence of signal with concentration is found above 10% (Takadoum et al., 1984).

2.3 Secondary ion yields

The secondary ion yield of a surface is strongly influenced by the chemical and electronic properties of that surface; for example, the secondary ion yields produced by argon ion bombardment of some clean metal surfaces have been shown to be up to 1000 times lower than those of the same elements when the surface is fully oxidised (Benninghoven, 1975).

In the case of argon ion bombardment of silicon, Wittmaak (1980) has demonstrated an enhancement of the Si^+ ion intensity by three decades as the partial pressure of oxygen in the sample chamber is increased from 2×10^{-6} Pa to 2×10^{-4} Pa. This effect, the enhancement of secondary ion yields as a function of oxygen incorporation into the sample surface, is deliberately exploited in dynamic SIMS analysis by using a primary ion beam of oxygen ions which implants into the sample during the sputtering process. Oxygen acts as a yield enhancing species for the electropositive elements which produce predominantly positive secondary ions. Storms *et al.* (1977) have shown that under oxygen bombardment the positive ion yields of electropositive elements such as Mg, Ca, V, Ti, Cr, etc., are at least three orders of magnitude greater than the positive-ion yields of the electronegative elements C, S, O, Se, As, for elemental systems. However, if caesium primary ions are used in place of oxygen ions and negative secondary ions are detetcted then the situation is reversed and the electronegative elements have the greater ion yields. There are some elements, notably Zn and Cd, which have relatively poor secondary ion yields both as $+ve$ and $-ve$ ions. However, these elements have their highest ion yields as the CsM^+ molecular species. The relative secondary ion yields of a range of elements in GaAs and InP matrices are presented in Fig. 7.2 for $+ve$ and $-ve$ ion detection under Cs^+ and O_2^+ bombardment.

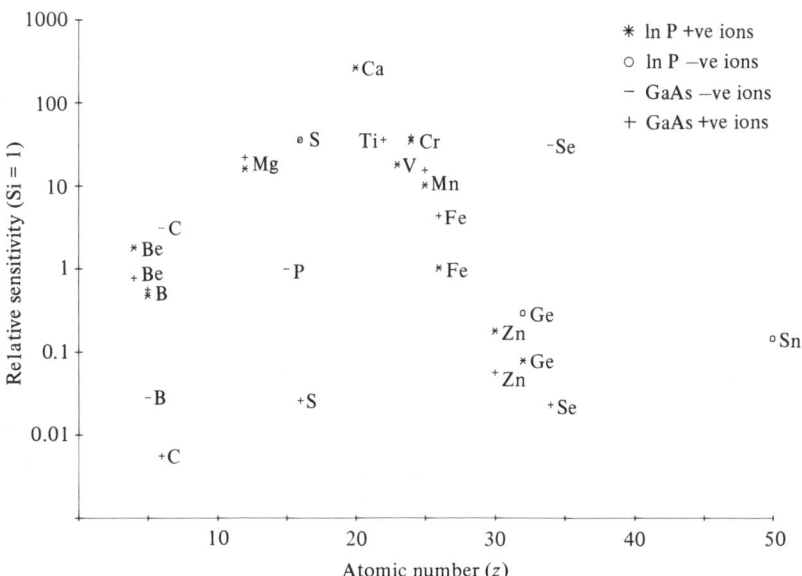

Fig. 7.2. Relative secondary ion yields of some elements in GaAs and InP matrices (normalised to silicon).

Despite this wide variation in secondary ion yields, by judicious choice of the analysis conditions, that is, primary ion species and polarity of the secondary ions, the relative sensitivities to the various elements can be reduced to a level that is no worse than the other surface analytical techniques.

2.4 The matrix effect

SIMS, as well as having a wide variation in secondary ion yields between different elements, also has strong variations in the secondary ion yield from the same element in different matrices. The matrix effect is graphically illustrated in Fig. 7.3(a) and (b) which shows depth profiles of Be and Mg ion-implanted through a thin Si_3N_4 layer on GaAs. The energies of the implanted species have been chosen so that the peaks of the implant distributions correspond to the Si_3N_4/GaAs interface. Ideally the atomic distributions of the two species should be continuous functions with the characteristic ion implant shape, as shown in Fig. 7.3(c), Be implanted directly into GaAs. As can be seen from the diagram Mg shows very little change in ion yield between the two matrices whilst Be has an order of magnitude lower ion yield in GaAs than in Si_3N_4. Another

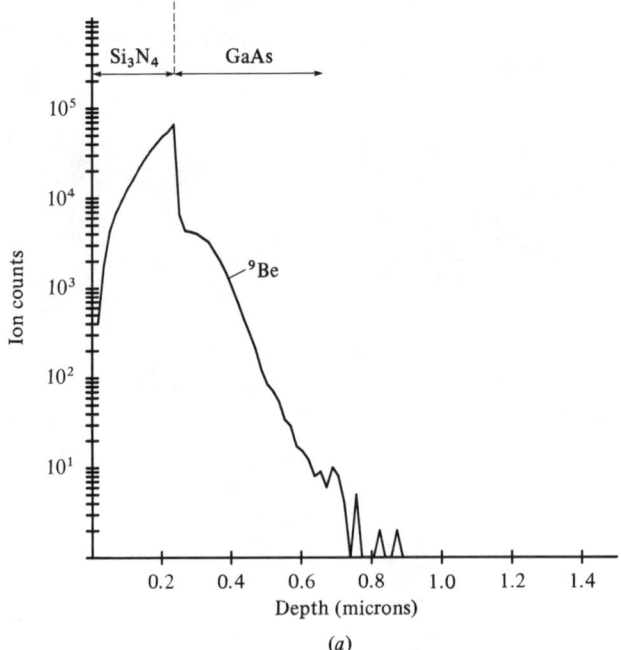

Fig. 7.3. Be (a) and Mg (b) ion implanted into Si_3N_4/GaAs structure showing the strong matrix effect. (c) Be implanted into GaAs.

(a)

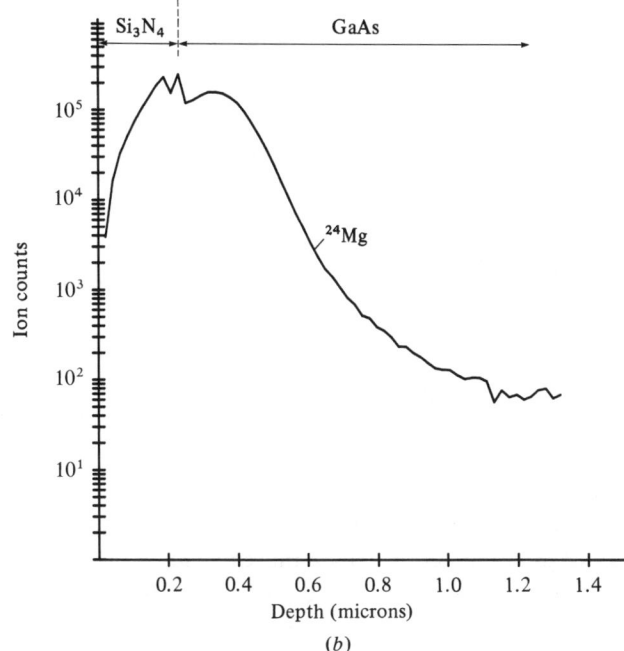

(b)

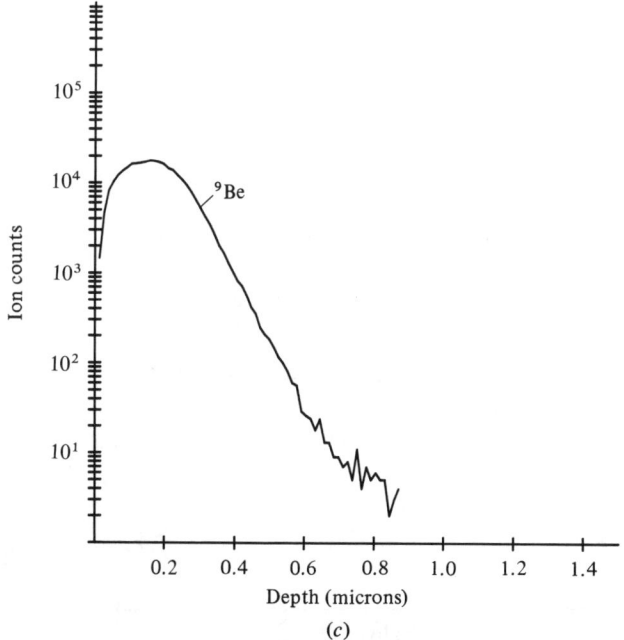

(c)

example of the matrix effect is shown in Fig. 7.4 which shows a matrix element profile through a GaAs/GaAs/Ga$_{1-x}$Al$_x$As interface. In both of these systems the arsenic content is 50 atom% of the material; however, the SIMS profile shows an increase in the As$^+$ signal on entering the GaAlAs layer despite the fact that the sputtering rate of GaAlAs is lower than GaAs, so that less arsenic is actually being removed from the surface per unit time.

The existence of the strong matrix effect can be a disadvantage when dealing with a multiphase system if attempting to provide quantitative information. However, for homogeneous systems, such as glasses and semiconductors quantitative analyses are straightforward provided suitable calibration standards are accessible.

2.5 Preferential sputtering

The phenomenon of preferential sputtering (whereby, in a two-element system in which one element has a higher sputtering rate than the other, the surface becomes enriched in the element with the low sputtering rate) requires corrections in the interpretation of surface concentrations as measured by the other surface analytical techniques Auger electron spectroscopy (AES), and X-ray photoelectron spectroscopy (XPS). In SIMS, once an equilibrium situation is reached the sputtered material leaving the surface has the same composition as the bulk material – preferential sputtering is not a problem in SIMS analysis. Consider a two-component system containing elements A and B and that the sputtering rate of element A, RA is x times greater than that of element B, RB. In the bulk material the concentrations of A and B are C_A and

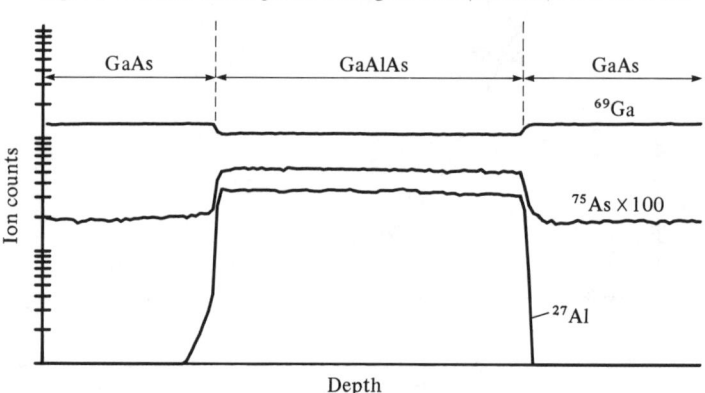

Fig. 7.4. Matrix element profile through a GaAs/GaAlAs/GaAs interface.

C_B thus

$$C_A + C_B = 1,$$

but once the surface is sputtered the atoms of element A will be removed at a greater rate than element B, leading to surface concentrations of A and B, S_A and S_B given by

$$\frac{S_A}{S_B} = \frac{C_A}{xC_B},$$

and $S_A + S_B = 1$, so

$$S_A = \frac{C_A}{xC_B + C_A} \text{ and } S_B = \frac{xC_B}{C_A + xC_B}.$$

The number of particles leaving the surface will be given by

$$\sum n = S_A \cdot R_A + S_B \cdot R_B$$

per unit time per bombarding particle as $R_A = xR_B$,

$$\sum n = \frac{C_A}{C_A + xC_B} xR_B + \frac{xC_B}{C_A + xC_B} R_B = \frac{xR_B}{C_A + xC_B}(C_A + C_B).$$

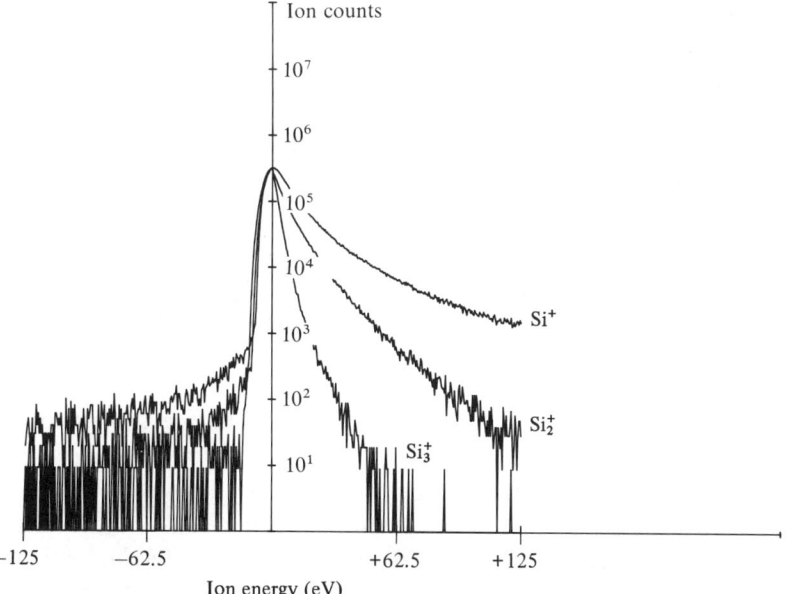

Fig. 7.5. Energy distribution of Si^+, Si_2^+ and Si_3^+ ions sputtered from pure silicon.

Therefore the proportion of particles of element B leaving the surface

$$\frac{nB}{\sum n} = \frac{S_B R_B}{\sum n}$$

$$= \frac{xC_B}{C_A + xC_B} R_B \bigg/ \frac{xR_B}{C_A + xC_B}(C_A + C_B)$$

$$= C_B,$$

and similarly for element A.

2.6 The ion energy distribution

The energy distribution of the sputtered ions is an important factor which has a significant influence upon practical analysis. The general form of the energy distribution is a low-energy peak with a significant high-energy tail extending to several hundreds of eV in energy. The shape of the ion energy distributions obtained from a pure silicon sample for the Si^+, Si_2^+, Si_3^+ ions are shown in Fig. 7.5. (The signal intensities have been normalised to allow comparisons to be made, normally the molecular ions would have progressively lower intensities.) The energy distribution obtained for the atomic ion (Si^+) is considerably broader and flatter than those for the molecular ions; the energy dependence of the molecular ion becoming more marked as the complexity of the molecule increases. In some special cases the ion energy distributions can appear to have a significant component of ions with a negative energy. This is an experimental artifact produced by a combination of the way that the ion energy distribution is measured and the fact that these ions are created after leaving the sample surface (Holland & Blackmore, 1983). As a general rule atomic species have slowly varying energy distributions whilst molecular species show a sharp energy distribution.

The implications of the ion energy distribution to practical analysis and system configuration will be discussed in subsequent sections; however, it should be noted that by careful selection of the energies of the ions analysed the mass spectrum produced can contain a mixture of elemental and molecular information or primarily elemental information.

3   Quantification

The information contained in a SIMS mass spectrum recorded from a sample can give quantitative information about that sample, provided that it is uniform in composition within the analysed volume. If a second phase is present then the interpretation must be considered qualitative because of the strong matrix effects characteristic of SIMS analyses. Quantitative SIMS analyses can be performed on glasses (generally a good system for SIMS analysis as they are amorphous, homogeneous single-phase systems saturated with oxygen), minerals and metals, provided that no second phases are present and, perhaps most importantly, semiconductor systems.

Two separate approaches have been developed to quantify SIMS data, one based upon a physical model of the secondary ion emission process and providing semiquantitative information and a wide range of samples; the other using a calibration factor approach requiring standard samples of similar compositions to the unknown thus requiring a large suite of standard samples.

3.1 Quantification using physical theories

Various physical models have been proposed to explain the secondary ion yield effects observed in practice; these models may then be employed to relate measured secondary ion currents to the atomic concentration of the elements present in the sample. For example, see Rudenauer (1977), Werner (1980) for more comprehensive discussions. Of these models the most successful have been the thermodynamic model of Anderson & Hinthorne (1973) and a simplification of it by Morgan & Werner (1976).

The model of Anderson and Hinthorne, known as the LTE (local thermal equilibrium) model, proposes that when the primary ion interacts with the sample surface a dense plasma in thermal equilibrium exists from which the secondary ions are emitted. Within this plasma, positive ions, negative ions, electrons and neutral atoms are considered to be in equilibrium. The generation of positive ions (and similarly negative ions) can then be described by the equation

$$M^0 \rightleftharpoons M^+ + e^-,$$

with a reaction constant Kn^+ given by

$$Kn^+ = \frac{n_M^+ n_{e^-}}{n_{Mo}}$$

where n_M^+, n_{Mo} and n_{e^-} are the number of ions, neutrals and electrons respectively. Kn^+ can be calculated using the Saha–Eggert ionisation equation. Thus

$$\frac{n_M^+ n_{e^-}}{n_{Mo}} = \frac{2\pi m k T^{3/2}}{h^2} \frac{2Z}{Z^0} \exp|-(E_i - \Delta E_i)/kT|, \tag{1}$$

where k is Boltzmann's constant, h is Planck's constant, m is the mass of the electron, Z^+, Z^0 are the partition functions of ions and atoms respectively, E_i is the ionisation energy, ΔE_i is the depression in ionisation energy due to the plasma, and T is temperature. This expression relates the number of ions, $n_M^+ = I_s$ (the measured ion current), to the number of neutrals, $n_{Mo} \alpha C_M$ (the concentration) of a particular element in the plasma to two parameters describing the plasma: the electron density n_e^- and the temperature T. The remaining parameters in the expression can be determined from other sources. So if the concentrations of at least two elements in the sample are known, the parameters n_e^- and T can be calculated and hence by substitution into (1) the concentrations of others elements can be determined from their measured secondary ion currents.

The local thermal equilibrium model is considered to be accurate to within a factor of two of the true concentrations of almost any element in a solid sample. This is despite the fact that there is no physical meaning to the parameters T and n_e^- as used in the model. Even at the highest primary ion beam currents employed in dynamic SIMS, only one primary ion/surface interaction occurs in any one event and the time between primary ion events is much greater than the interaction time. Thus only a few atoms in the sample surface are excited at any time and a dense plasma in thermal equilibrium is difficult to justify.

Morgan & Werner (1978), have further simplified the LTE model. By assuming that the sample constituents are sputtered as neutral atoms they reduce the number of fitting parameters to one, the temperature T. They observed that, provided there was sufficient oxygen present in the system to saturate the ion yields, the parameter T became a constant for a given matrix with no dependence upon primary ion mass, energy or current density.

3.2 Quantification using standards

When suitable standards are available quantitative SIMS analyses can be achieved using a calibration factor approach. The technique relies upon the availability of a range of uniform standards with a composition similar to that of the unknown sample. For a given matrix the concentration of an element X can be expressed as

$$C_X = K_X \cdot I_X / f_X,$$

where I_X is the measured secondary ion current, f_X is the isotopic abundance of the isotope X, and K_X is a constant of proportionality.

The constant of proportionality depends not only upon the element of interest but also contains contributions from the analytical conditions used, for example primary ion beam conditions (ion energy, ion current, ion species), mass spectrometer transmission and secondary ion energy. In order to minimise the influence of the analytical conditions on the analysis

Dynamic secondary ion mass spectrometry

the secondary ion signal from element X is normalised to the secondary ion signal from a matrix element. Thus

$$C_X = \frac{K_{X/m} I_X}{f_X I_m},$$

where I_m is the secondary ion current of a matrix element and $K_{X/m}$ is the relative elemental sensitivity factor for element X. By this method the dependence on factors such as primary ion beam current and mass spectrometer transmission (analysed area) are removed. However, other experimental conditions: primary ion species, secondary ion energy, must be the same for the analysis of the standards and the unknown.

In general the relative sensitivity factor $K_{X/m}$ is constant over a wide range of values as long as the element X is present as a dilute solute in the matrix. Hence only one standard should be required. Once the concentration C_X exceeds a few atomic percent it is likely that a non-linear dependence of C_X on I_X will be found, i.e. $K_{X/m} = f(C_X)$. In this situation a range of standards are required to determine the functional dependence of $K_{X/m}$ on C_X.

Thus once the set of relative elemental sensitivity factors $K_{X/m}$s is known, the measured secondary ion signals I_Xs can be reduced to the relative atomic concentrations which, provided all the major elements in the sample are included, can be normalised to 100%. Using this approach quantitative analyses with accuracies of $\sim 10\%$ can be achieved. However, the method relies upon the existence of uniform homogeneous standards with accurately known compositions and can only be applied to samples where the analysed volume is uniform in composition.

This method of quantification has been successfully used for the analysis of glasses (Newbury & Heinrich, 1979) and minerals (Havette & Slodzian, 1980). It is particularly useful for the analysis of semiconductor materials since in this case there is a uniform well-defined matrix and the impurities are present at very low levels (much less than 1 atom %). In addition, well-characterised standards, in the form of bulk doped material or ion implanted samples, are available for the usual semiconductor dopant species.

4 Instrumentation

The basic instrumental requirements of any SIMS system are: an ion gun to provide a source of primary ions, a mass spectrometer with which to detect the secondary ions, a data acquisition system, and a vacuum environment in which to house the ion gun and mass spectrometer. A detailed description of the operation of the various components is outside the scope of this chapter; however their salient features are outlined in the following sections.

4.1 The ion gun

Ion guns used for dynamic SIMS applications consist of an ion source, an ion optical column to focus and deflect the primary ions, and a mass filter to ensure the purity of the primary ion beam.

The ion source most widely used in this application is the duoplasmatron which is a cold cathode discharge source capable of producing a range of primary ions from various gaseous species, generally O_2^+, O^- and Ar^+. As discussed in Section 2, oxygen ion bombardment enhances the secondary ion yields of the electropositive elements and also produces a more uniform erosion of the sample than inert gas bombardment. Two oxygen species are used in different analytical situations; the O_2^+ species provides the highest current density and is the 'general purpose' primary ion for analysis of metals and semiconductor samples where sample charging is not a problem. If the O_2^+ beam is used on an insulating sample the sample surface tends to become electrically charged and this charge has to be neutralised, usually with an electron flood gun. However, when the O^- species is used sample charging is not as severe and analyses of insulating samples can often be successfully achieved without the use of an external source of negative charge.

To complement the duoplasmatron a source of caesium ions is also required for optimum sensitivity to the electronegative elements. These are generally surface ionisation sources, although more recently liquid metal field ionisation sources have become available, which offer potential advantages because of their higher ion-optical brightness.

The ion optical column of the ion gun allows the primary ions to be accelerated up to energies of between 1 and 20 keV and focussed into spots of between 1 and 50 μm in diameter. Then by applying a raster to the primary ion beam a well-defined rectangular pit can be eroded in the sample surface. This is essential if high-quality analyses are to be performed. Typically, etch rates can be as high as $1 \, nm \, s^{-1}$ for O_2^+ or Cs^+ bombardment of silicon.

A primary beam filter of some form is generally incorporated into the primary ion column to remove any neutral species in the primary beam and also to remove any impurities in the beam which may be present in the ion source area. This is necessary because SIMS is such a sensitive analytical technique that trace level impurities in the primary ion beam may become implanted into the sample and then be detected as a false high background level of that species.

4.2 The mass spectrometer

Two types of mass spectrometer can be used for dynamic SIMS systems, namely the quadrupole mass spectrometer and the magnetic sector mass spectrometer. The quadrupole mass spectrometer is widely used for both static SIMS and dynamic SIMS applications and in applications where several surface analytical techniques are included in one vacuum system. The magnetic sector mass spectrometer, on the other hand, tends to be used only in dedicated dynamic SIMS systems.

4.3 The quadrupole mass spectrometer

The quadrupole mass spectrometer in its basic form comprises four rods to which DC and RF signals are applied; the resulting electric field will allow ions of a given charge to mass ratio entering the quadrupole to follow a stable oscillatory trajectory between the rods and so reach the detector. All other ions entering the quadrupole are forced to follow unstable trajectories of increasing amplitude and so will collide with the rods or housing and not reach the detector. The performance of the quadrupole is determined by the physical size of the rods and the frequency of the applied RF field. In order to have good mass resolution the energy of the transmitted ions should be as low as possible. However, the transmission of the quadrupole decreases with decreasing energy. Typical mass resolutions achievable with quadrupole mass spectrometers are in the range $M/\Delta M \simeq 50$ up to 1000 where M is the mass of the peak and ΔM is the width of the peak at 10% of the peak height.

The quadrupole mass spectrometer offers a high mass range, to greater than 1000 amu, freedom from magnetic fields, compactness and compatability with other surface analysis techniques in the same vacuum chamber. However, when compared with the magnetic sector mass spectrometer it has a lower transmission and is not capable of the high mass resolution that can be achieved with the magnetic sector instruments. A detailed review of the applications of quadrupole mass spectrometers to SIMS analyses has been given by Wittmaak (1982).

4.4 The magnetic sector mass spectrometer

In the magnetic sector mass spectrometer the ions of mass M and charge e are accelerated by a potential V and injected into a uniform magnetic field **B**. The ions will then follow circular trajectories of radius R, where

$$R \propto \frac{1}{\mathbf{B}}\left(\frac{MV}{e}\right)^{1/2},$$

thus for a given value of field, **B**, only ions of a certain value of e/M can follow the correct path to reach the detector and by varying **B** ions of different values of e/M can follow that trajectory. By changing the width of the entrance and exit slits to the magnetic sector the mass resolution of the mass filter can be adjusted allowing continuous variation of the mass resolution, $M/\Delta M$, for example, in the Cameca IMS3f it may be varied from ~ 200 to over 10 000. The magnetic sector based SIMS instruments offer a transmission that is greater than quadrupole based instruments and the higher mass resolution available is invaluable in resolving some spectral interferences. However, the mass range of the magnetic sector instruments is usually limited compared with the quadrupole instruments, but since dynamic SIMS is concerned with high-sensitivity elemental analyses a mass range which covers the periodic table is all that is necessary. The size and complexity of the magnetic sector mass spectrometer preclude its use as an addition to a multitechnique surface analysis system and it is only found in high-performance dedicated dynamic SIMS systems.

4.5 Ion energy analysis

Both types of mass spectrometer, the quadrupole and magnetic sector, are combined with some device to transmit ions selectively within a given kinetic energy range – the energy filter. This facility is essential for good dynamic SIMS analyses. Whilst the detailed design of the energy filter varies from instrument to instrument its function is common to all instruments. In quadrupole instruments, however, it has a two-fold role: as the peak shape in these mass filters is dependent upon the kinetic energy of the transmitted ion the energy filter is used as a narrow band pass device to give improved peak shape and consequently acceptable mass resolution.

The principal role of ion energy analysis in dynamic SIMS analysis is the separation of the required elemental signals from unwanted molecular signals. In Section 2 the ion energy distributions of elemental and molecular ions were shown to be markedly different, with the elemental ions having a much flatter distribution than the molecular species. By using the energy filter to allow only the higher-energy ions to reach the mass spectrometer, the transmission of the required elemental signal can be enhanced with respect to the unwanted molecular signal, the practical applications of this effect are illustrated in Section 5.

4.6 Ion imaging

Dynamic SIMS instruments can be used to study the spatial variations of the secondary ion currents across a surface (ion imaging) as well as with depth into the surface (depth profiling). Two distinct methods have been employed to produce ion images – the ion microprobe and the ion microscope.

4.7 The ion microprobe

The ion microprobe technique of ion imaging is an ion analogue of the electron microprobe and the scanning electron microscope. The primary ion beam is focussed to a fine spot and rastered over the surface of the specimen. The mass spectrometer output at a specific mass peak is then displayed in synchronism with the primary beam position to produce a map of secondary ion intensity across the surface. The spatial resolution in the final secondary ion image is then determined by the spot size of the primary ion beam. Spatial resolution of the order of a few microns can be achieved with duoplasmatron sources, and with finely focussed liquid metal ion sources now available submicron, resolution has been demonstrated and spatial resolutions of the order of 10 nm are predicted, Steiger *et al.* (1983).

However in decreasing the spot size of the primary ion beam, the current

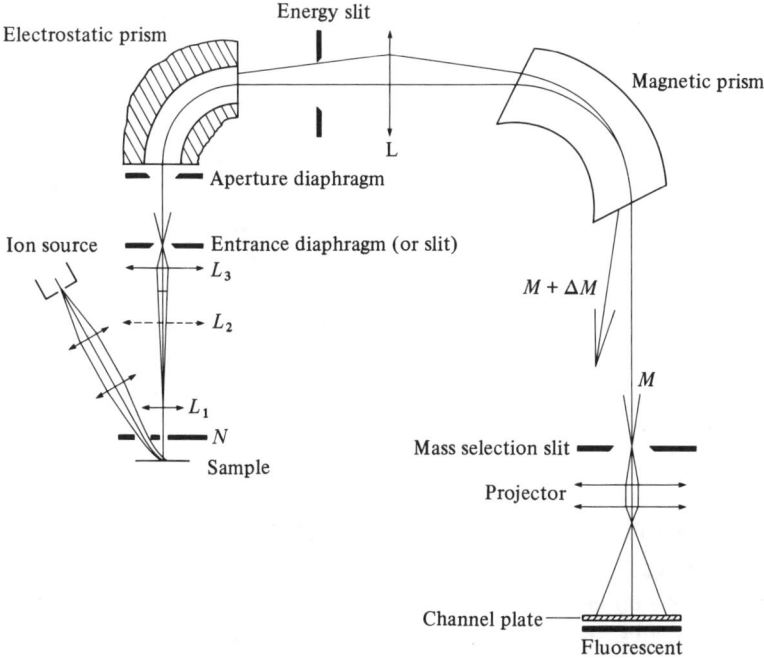

Fig. 7.6. Schematic diagram of a Cameca IMS 3f.

available in the beam is reduced, consequently the secondary ion signal and the rate of erosion are also reduced.

The ion microprobe technique for secondary ion imaging is the ion imaging method usually used with quadrupole-based mass spectrometers, but it can also be used with magnetic sector instruments. Its performance is dominated by the quality of the primary ion beam.

4.8   The ion microscope

The ion microscope is an ion analogue of the optical microscope and the transmission electron microscope. In this case the primary ion beam acts as a source of illumination and the ion image is formed using ion lenses to maintain the spatial distribution of the secondary ions as they are transmitted through the mass spectrometer. The secondary ion image is then formed by the impact of the spatially distributed secondary ions on an image intensifier/phosphor screen producing a real time display of the secondary ion distribution. This type of instrument is shown schematically in Fig. 7.6.

4.9   Additional facilities

The whole of the SIMS system is necessarily housed in a vacuum chamber of some description and, although the vacuum requirements are not as stringent as for the other surface analysis techniques, ultra-high vacuum (uhv) is still desirable, particularly around the sample. One obvious situation requiring uhv is in the detection of trace levels of C, O and H all species present in the residual vacuum, although this can be partly compensated for with high erosion rates. Another area where the vacuum conditions require consideration is in the attainment of good dynamic range/low backgrounds in ion-microprobe systems.

A useful facility is the oxygen flood which provides a directional jet of oxygen across the sample surface. Although this degrades the vacuum locally in the sample region it is used to beneficial effect to ensure complete saturation of the sample surface with oxygen, thus maximising the positive secondary ion yields when argon or oxygen at off-normal incidence is used as the primary ion species.

Electron flood guns are also added to SIMS instruments to overcome the problems of sample charging encountered during the analysis of insulating samples. Alternatively, a primary ion beam of O^- ions can be used to analyse insulating samples without the need for an external source of negative charge.

5 Some practical considerations

5.1 Interferences

The information produced during a SIMS analysis consists of a spectrum of number of counts vs mass and, as with most spectroscopic techniques, some peak overlaps occur where the required information is obscured by a peak from some other source. Considering the relative abundance of the naturally occurring isotopes, it is immediately obvious from this that different elements have isotopes with peaks at the same nominal mass, for example ^{50}Ti, ^{50}V and ^{50}Cr, ^{54}Cr and ^{54}Fe, ^{58}Fe and ^{58}Ni, etc. However, these do not normally present a problem as other isotopes can be used for interpretation. In addition to these elemental interferences, molecular interferences also occur where a combination of elements of low mass combine to form a molecular cluster which can appear at the same nominal mass as a higher mass element, for example at mass 56 one could have peaks from ^{56}Fe, ^{28}Si$_2$, ^{40}Ca^{16}O, ^{24}Mg^{16}O$_2$, ^{29}Si^{27}Al. A special type of molecular interference is the hydride molecule where an element combines with hydrogen either from the sample itself or from the residual vacuum of the SIMS instrument to produce a peak at the adjacent higher mass number, for example ^{30}SiH appears at mass 31 nominally ^{31}P, ^{31}PH and ^{32}S, ^{74}SeH and ^{75}As. Two methods are available by which the interfering species can be removed to allow the detection of the peaks of interest.

5.2 Energy filtering

In Section 2, the energy distribution of the secondary ions was shown to be a wide, flat distribution extending to high energies for atomic species, whilst molecular species had a narrow low-energy peak. By carefully selecting the energies of the secondary ions allowed to reach the detector the molecular species can be suppressed whilst the atomic species are transmitted, thus removing the unwanted molecular interferences from the required elemental mass spectrum. This effect is illustrated in Fig. 7.7 which show a mass spectrum recorded from a stainless steel sample using secondary ions in the energy interval 0–120 eV (Fig. 7.7(a)) and 100–220 eV (Fig. 7.7(b)).

An important semiconductor application of the energy filtering technique is in the analysis of arsenic in silicon. Arsenic is monoisotopic at mass 75 whilst silicon has three isotopes at masses 28, 29 and 30, so, when the oxygen primary beam is being used, a potential interference exists between ^{75}As and ^{29}Si^{30}Si ^{16}O. The effect of selecting ions of different energies on the measured implant profile of As ion-implanted into silicon

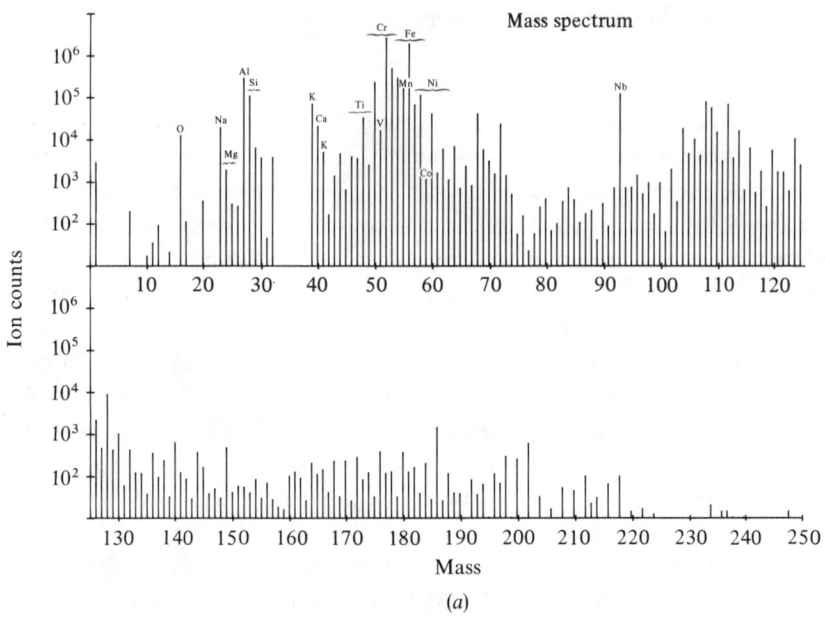

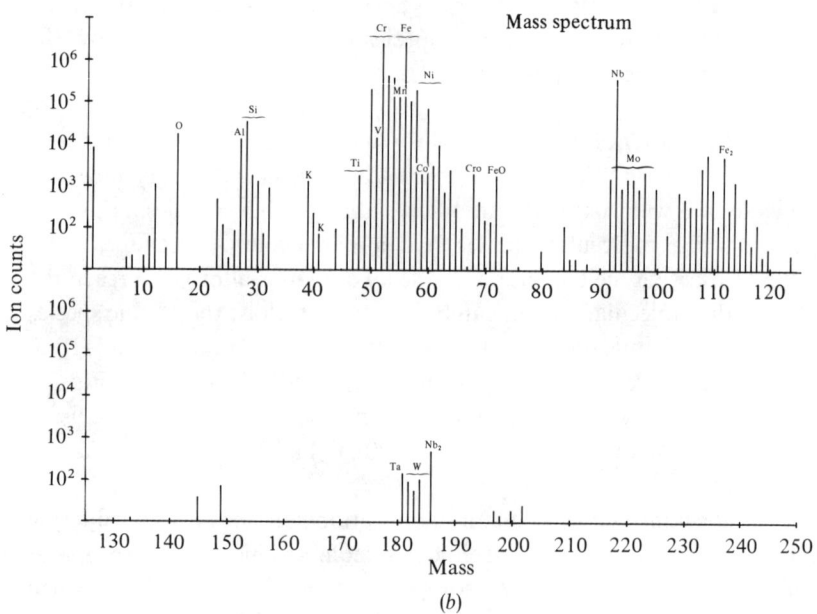

Fig. 7.7. Mass spectra recorded from a stainless steel sample with (a) energy window 0–120 eV, (b) energy window 100–220 eV, showing the ability to detect Ta present at 0.006%.

is shown in Fig. 7.8. If the secondary ions in the range 0–120 eV are collected the top of the implant profile is just discernible above the background Si_2O interference corresponding to a detection sensitivity for arsenic of only 10^{20} atoms cm^{-3}. If the energy interval of the secondary ions is charged to 20–140 eV, 40–160 eV and 60–180 eV, the peak of the arsenic signal is depressed slightly whilst the Si_2O background is dramatically reduced leading to a detection limit of better than 10^{17} atoms cm^{-3} for the conditions used in this analysis.

Energy filtering is a technique that lends itself to the suppression of the molecular interferences and can be implemented quite simply on either the quadrupole systems or the magnetic sector instruments. Unfortunately, it is not the solution to all interference problems as the hydride species have atomic like energy distributions.

5.3 High mass resolution

An alternative method to energy filtering for resolving interferences is the use of high-mass resolution. Although the interferences have

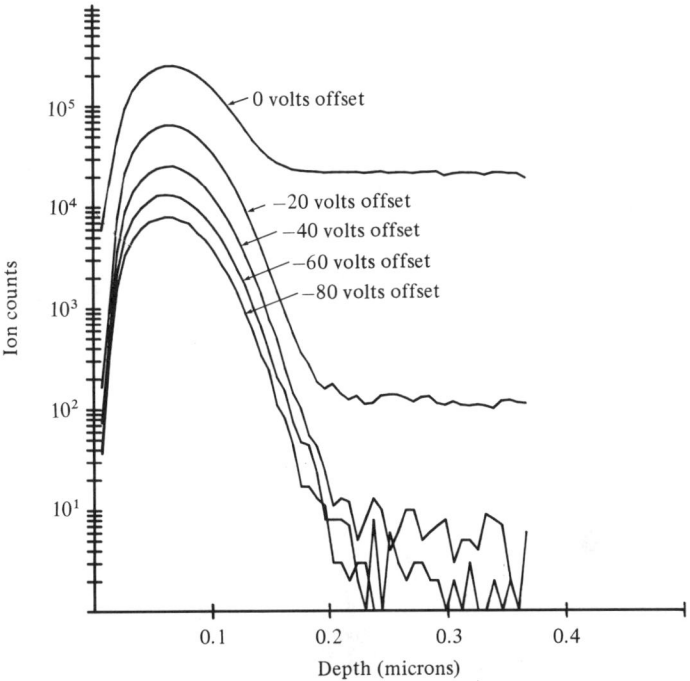

Fig. 7.8. Depth profile of As implanted in Si showing the effect of energy filtering.

the same nominal mass as the required analytical ion small-mass differences exist between the species. By using high-mass resolution the different masses can be resolved and so the interference can be overcome. This is the only method that can be employed to overcome the hydride type of interference and can only be implemented in the magnetic sector type of

Table 7.1. *Some typical interferences occurring in SIMS and the mass resolution required to resolve the interfering species*

Mass number	Required ion	Interference	Mass resolution to resolve peaks
14	^{14}N	$^{28}Si^{++}$	950
31	^{31}P	^{30}SiH	3961
32	^{32}S	^{31}PH	3362
56	^{56}Fe	^{55}MnH	5118
56	^{56}Fe	$^{28}Si_2$	2956
56	^{56}Fe	$^{40}Ca^{16}O$	2479
75	^{75}As	$^{29}Si^{30}Si^{16}O$	3250

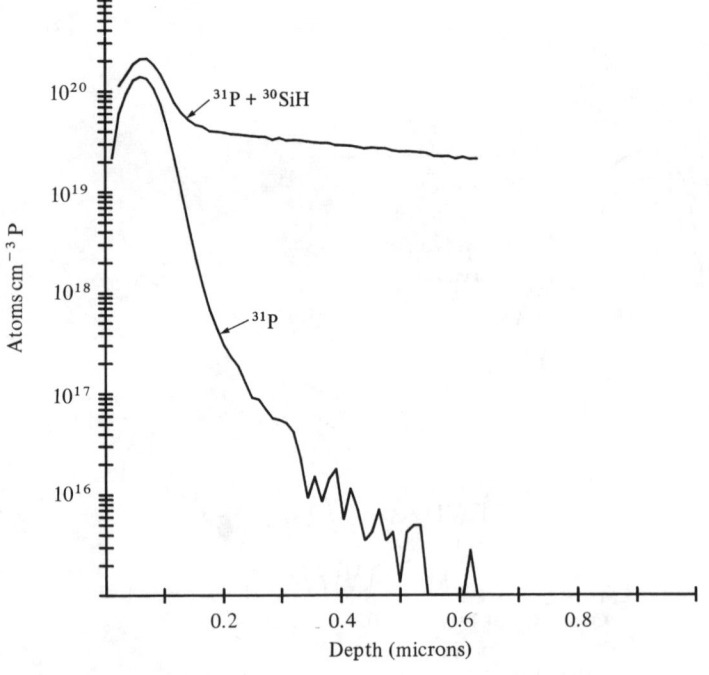

Fig. 7.9. Depth profile of P implanted in Si.

instrument. Some typical interferences are shown in Table 7.1. In general, a mass resolution of several thousand is required to resolve the analytical ion and the interfering species.

The use of high-mass resolution is illustrated in Fig. 7.9 which shows a 10^{15} atom cm^{-2} phosphorus ion-implant in silicon. Phosphorus is monoisotopic at mass 31 whilst silicon has a minor isotope at mass 30 (3.1% of the matrix) leading to a hydride type interference ^{30}SiH at mass 31. By using a mass resolution in excess of 4000 the required phosphorus signal and the interfering ^{30}SiH can be resolved. Figure 7.9 shows both a high-mass resolution profile and a low-mass resolution profile, i.e. the phosphorus profile and the phosphorus + silicon hydride profile, indicating that without high-mass resolution the detection limit for phosphorus in silicon would be 10^{19} atoms cm^{-3} whilst with high-mass resolution a detection limit of $\sim 10^{15}$ atoms cm^{-3} is achievable.

5.4 Dynamic range

An important concept which applies particularly to dynamic SIMS rather than the other surface analytical techniques is that of dynamic

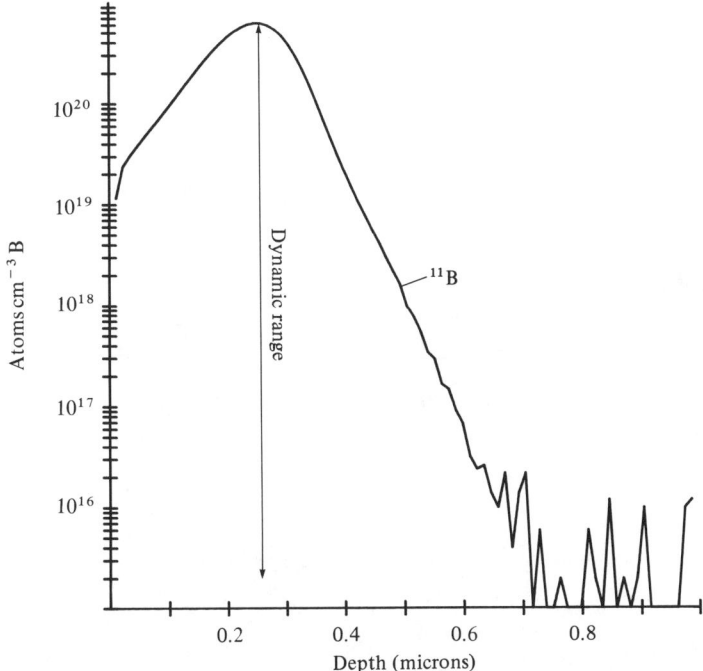

Fig. 7.10. Depth profile of B implanted in Si.

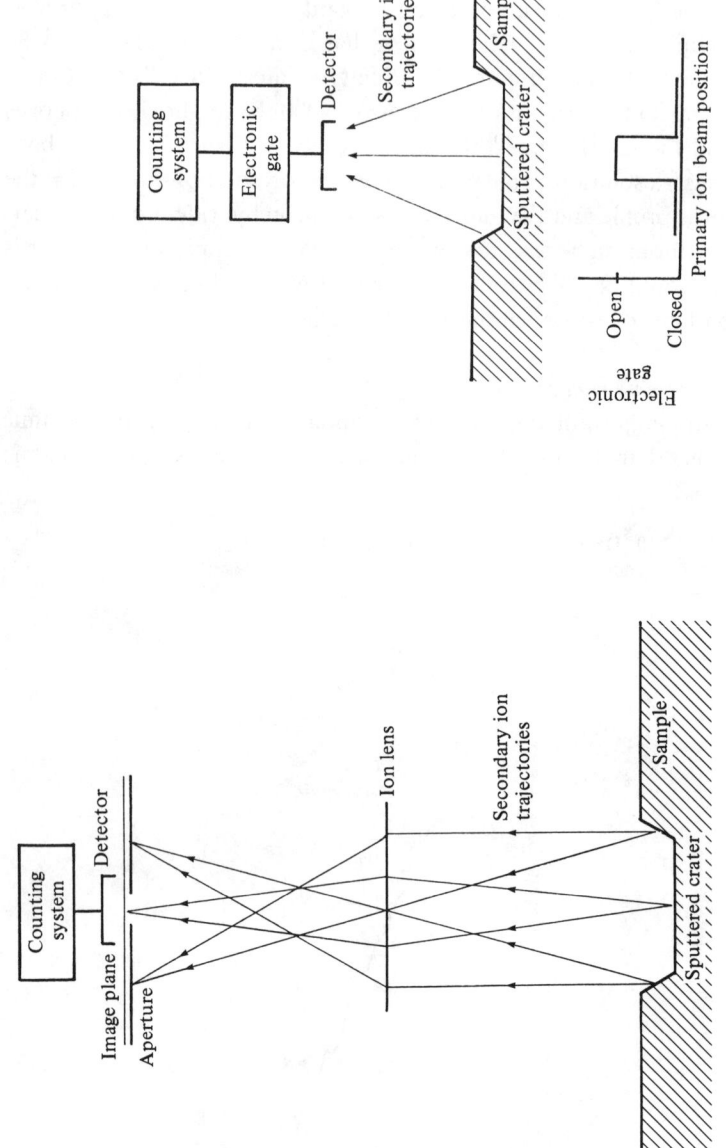

Fig. 7.11. Schematic diagrams of the imaging principle of (*a*) the ion microscope, (*b*) the ion microprobe.

range. It can be used to assess the quality of the instrumentation or the analysis performed. The dynamic range is usually expressed as the number of orders of magnitude between the highest level and the lowest level of the signal from a particular element in a given depth profile, usually that of an ion-implanted sample. For example, in Fig. 7.10, which shows the boron profile recorded from a 70 keV 10^{16} atoms cm^{-2} boron in silicon implant, a dynamic range of ~ 5 decades is achieved between the maximum signal of $\sim 6 \times 10^{20}$ atoms cm^{-3}, and the minimum, $\sim 6 \times 10^{15}$ atoms cm^{-3}. The dynamic range of an instrument can be limited for high signals by the inability of the detection system to respond to all the ions arriving at the detector (saturation), although this is not normally a problem. At the low-signal end of the profile the dynamic range is usually limited by some background signal. In the absence of any interference contributing to the background, this signal arises from a small amount of material from the high-concentration region of the sample entering the mass spectrometer. In order to understand how this comes about, it is necessary to consider how these large changes in signal are followed.

The primary ion beam is rastered across the sample surface so as to uniformly erode a square-shaped pit with a flat bottom. In order to achieve a good dynamic range it is essential that only those secondary ions produced from the flat-bottomed part of the crater are allowed to reach the detector, and ions originating from the sides of the crater, where the concentrations of elements are quite different to those at the bottom, are suppressed. This is achieved by different means in the two types of instrument used for dynamic SIMS measurements, as shown in Fig. 7.11(a) and (b). In the ion imaging system an ion lens is used to focus an image of the central part of the crater onto the detector, the size of this image can be further reduced by using apertures to select only the central portion of this image for analysis, Fig. 7.11(a). In the quadrupole-based instruments the discrimination is carried out electronically using an electronic gate which only allows the detected signal produced whilst the primary ion beam is being swept across the central portion of the crater to contribute to the measured output signal. This raster gating arrangement is shown schematically in Fig. 7.11(b).

Even with a perfectly focussed ion imaging system or an ideally focussed primary ion beam and optimum raster gating there is still a background signal produced which restricts the dynamic range that can be achieved. The mechanism by which this background is produced is different for the two types of instrument. In the raster-gated quadrupole system the ion detection system is opened to count secondary ions whilst the primary ion beam is crossing the central part of the crater. However, a small number of energetic neutral species may be present in the primary ion beam, or the primary ion beam itself may interact with the residual gas molecules in the vacuum system producing a flux of bombarding species which strike the sample in areas of high elemental concentrations whilst the gated detection system is open. As the quadrupole has no spatial discrimination it cannot differentiate between these ions and those produced from the bottom of the crater. Thus for this type of system a neutral free ion beam and uhv sample environment are essential for good dynamic range.

With the ion imaging systems the background problem arises from a quite different source. This type of instrument has an ion lens positioned close to the sample surface; in the course of analysis material from the sample is sputtered onto the surface of this lens. A small fraction of this material can then be redeposited onto the sample surface in the area being analysed, thus producing the background signal, the so-called memory effect. Quadrupole systems can also suffer from memory effect.

It has been shown by Clegg *et al.* (1984) that in the case of boron in silicon, a dynamic range of 5×10^5 can be achieved almost routinely in both types of instrument. Whilst Migeon & Le Goux (1982) have demonstrated a dynamic range of 10^7 for boron in silicon using an ion microscope type of system. In this case a large area crater was eroded through the sample over the first 3–4 decades of concentration covering the high-concentration regime. The rastered area of the primary ion beam was then reduced, whilst keeping the analysed area constant, and the profile continued, thus recording a second profile from a low-concentration sample. In this way, the memory effect can be suppressed.

The dynamic range should not be confused with sensitivity or detection limits. For example, in an ion microscope system where a dynamic range of 10^5 is routinely achievable, this implies a background limited detection limit of 10^{15} atoms cm^{-3} for an ion implant, peaking at 10^{20} atoms cm^{-3}. If, however, the implant peak was only 10^{18} atoms cm^{-3} then a background of 10^{13} atoms cm^{-3} could be reached.

5.5 *Depth resolution*

The depth resolution obtained in a SIMS depth profile is similar in concept to that obtained in depth profiles using the other surface analysis techniques relying upon sputtering to remove successive surface layers. The depth resolution in SIMS is not the same as dynamic range; indeed a profile showing good dynamic range can have poor depth resolution. Dynamic range is only the difference between the maximum and minimum signals, whilst depth resolution is a measure of the rate at which that change takes place with respect to depth.

As with the other sputter profiling techniques the depth resolution of a SIMS profile is degraded by poor instrumental alignment and sample roughness. However, even if perfectly flat samples are analysed with an ideally aligned system, some deviations from the true profile can occur. The precise mechanisms of the profile-broadening effects observed are outside the scope of this discussion, but generally result from the collision cascade induced in the sample surface by the bombarding primary ions. As long as the bombardment does not produce any microtopography – roughness – the depth resolution should remain a function of the primary ion energy only. If primary ion beam effects are causing a degradation of depth resolution this can be simply checked by repeating the profile under different primary beam conditions. For example, Fig. 7.12 shows the Al profile through a GaAlAs/GaAs superlattice with a repeat depth of 13 nm

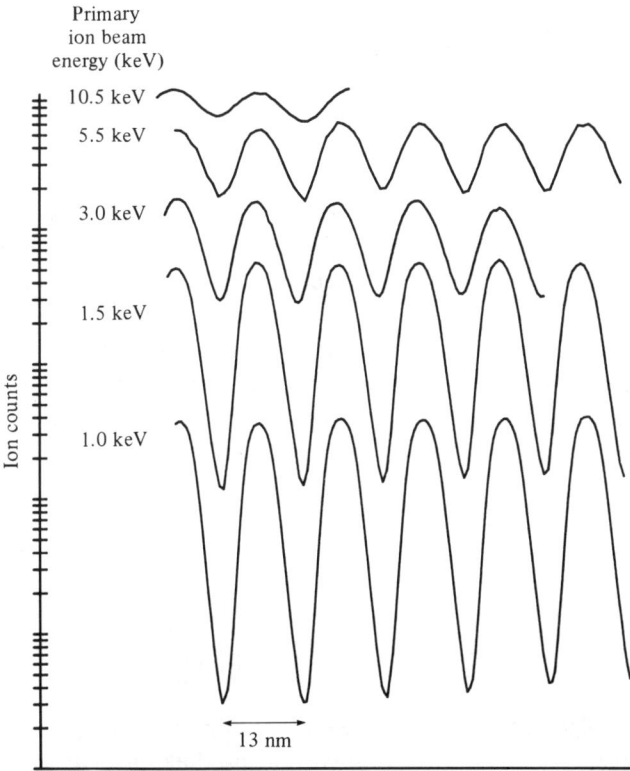

Fig. 7.12. Al profile from a GaAs/GaAlAs superlattice structure with a repeat depth of 13 nm.

recorded for a range of primary ion beam energies. Clearly, a reduction in primary ion beam energy increases the depth resolution and in this case the dynamic range. (In this particular example the angle of incidence of the primary ion beam is increased with a reduction of primary ion energy.) Although the profile shows the raw data only, no corrections for secondary ion yields or change in sputter rate with change in composition have been made, this result indicates that a depth resolution of the order of 2 nm can be achieved in dynamic SIMS. On the other hand, in the depth profiling of boron in silicon Clegg et al. (1984) have shown that for the relatively slowly varying implant distribution the depth profiles have the same depth resolution independent of primary ion beam conditions.

5.6 The secondary ion current

The measured value of the secondary ion current arriving at the detector in any SIMS system can be expressed in the following form:

$$I_m^+ = J\rho \cdot A \cdot S \cdot \beta_m^+ \cdot f \cdot c_m,$$

where I_m^+ is the measured secondary ion current for element m, J_ρ is the current density of the primary ion beam, A is the area of analysis, S is the sputter yield, β_m^+ is the degree of ionisation of element m, f is the instrument transfer function, and c_m is the concentration of element m.

So in order to have the highest sensitivity, i.e. the highest I_m^+ for a given c_m, the various terms in the equation must be maximised. An increase of the primary ion current density or the analysed area will increase I_m^+; these terms are directly proportional to the amount of material consumed in the analysis. When depth profiling is being carried out it is important to appreciate the volume of material removed per data point as a balance must be maintained between sensitivity and sampling interval commensurate with reasonable depth resolution. Also it should be noted that a reduction in analysed area A produces a reduction in ion current I_m^+ and hence an increase in the minimum detectable concentration for a given instrument transfer function. For example, if the minimum detectable concentration of an element is 10^{15} atoms cm^{-3}, for an analysed area of 60 μm in diameter, if the analysed area is reduced to 5 μm in diameter, all other parameters being equal, the minimum detectable concentration will be only 1.4 to 10^{17} atoms cm^{-3}.

The sputter yield, S, may be slightly increased by increasing the energy of the primary ion beam; however, this may produce other undesirable effects.

The degree of ionisation β_m^+ can be maximised by saturating the surface of the sample with oxygen for positive ion production or caesium for negative-ion production.

Dynamic secondary ion mass spectrometry

The instrument transfer function f depends upon the instrument design.
The product of S, β_m^+ and f can be readily assessed for a particular instrument configuration through the analysis of ion-implanted semiconductor samples. With an ion-implanted sample a known dose (number of ions/cm^2) N of a particular species is implanted into the surface of a semiconductor substrate. So by carrying out a SIMS profile of the implant distribution and integrating the total number of secondary ions detected through the distribution and correcting this for the analysed area the ratio N_m^+/N can be determined.

$$\frac{N_m^+}{N} = S \cdot \beta_m^+ \cdot f.$$

This quantity, the useful ion yield has been determined by Clegg et al. (1984) for a variety of instruments currently available and ranges from 3×10^{-4} for a good quadrupole based system to 3×10^{-3} for a magnetic sector type of instrument of the stigmatic imaging design.

6   Applications

SIMS is such a versatile analytical technique that it can be applied to almost any problem requiring the analysis of a solid surface. However, there are certain areas in which the extremely high sensitivity of the technique and its depth profiling capability make it the only choice available. By far the largest area of use of dynamic SIMS is in semiconductor analysis where the determination of the depth variation of trace-element distributions is a prime requirement. The technique can also provide valuable information in the areas of biology, geology, and materials science; some typical examples of the type of analyses that can be carried out are presented in the following examples:

6.1   Semiconductor applications

Ion implantation is a method widely used in the semiconductor industry for doping the surface of a semiconductor with an electrically active impurity. The as-implanted sample is damaged by the implantation and the dopant species are not located on lattice sites and so are not electrically active. In order to restore the crystallinity of the semiconductor surface and to activate the implant it is necessary to carry out an annealing process. SIMS is the ideal technique for studying the implant distribution in the as-implanted and annealed states and the chemical profile produced can be compared with the electrical profiles to determine the degree of activation, in addition other dopant species can be monitored at the same time. A typical example of this type of analysis is shown in Fig. 7.13(*a*)–(*d*) which shows the behaviour of Mg implanted into InP as the sample is

Fig. 7.13. Depth profiles of Mg implanted into Fe doped InP showing the effects of furnace annealing.

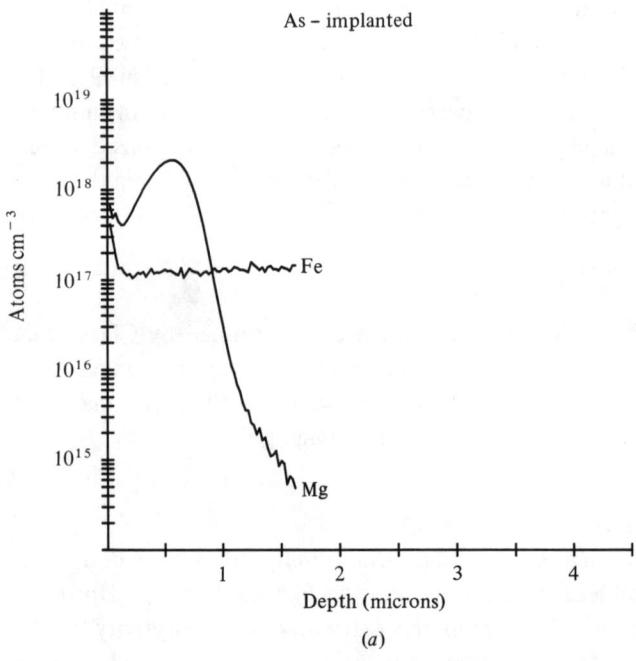

(a)

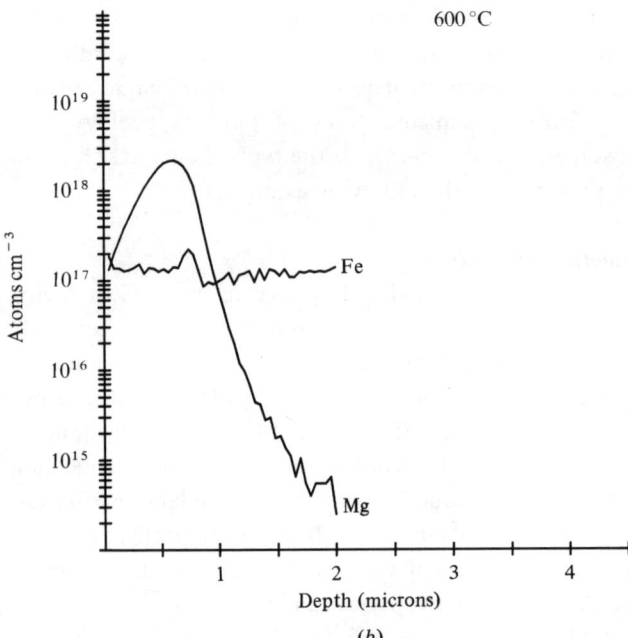

(b)

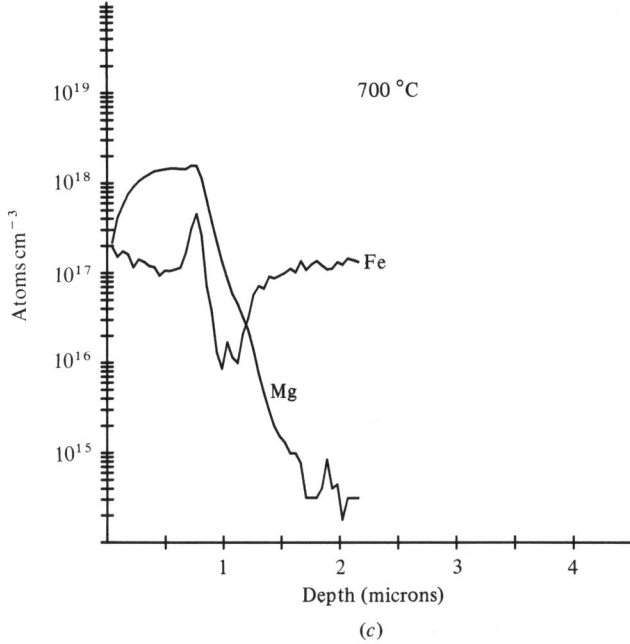

(c)

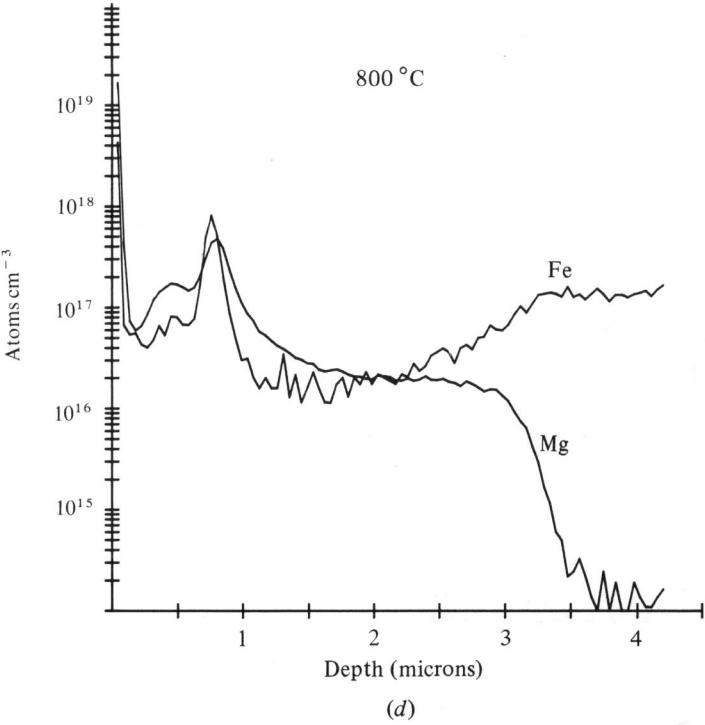

(d)

furnace annealed at various temperatures. Mg is a p-type dopant in InP, whilst Fe is present in the semi-insulating substrate material. The as-implanted sample shows the characteristic implant distribution of Mg together with a uniform Fe doping level. Annealing at 600 °C produces little movement of the Mg, but there is a slight redistribution of the Fe; by 700 °C the Mg is beginning to diffuse further into the InP and also back towards the surface; at the same time the Fe is gettered on the damage sites and is displaced from the bulk material by the diffusion of the Mg; at 800 °C the Mg has diffused 3.5 μm into the substrate and a substantial redistribution of the Fe has occurred.

In the characterisation of ion implants SIMS can also be used to study the unintentionally incorporated impurity species co-implanted with the required species. For example, Fig. 7.14(a) shows the relative abundance of the various Se isotopes implanted into GaAs in a nominal ^{78}Se implant. In this case all the Se impurity species show the same implant profile as the intentionally implanted ^{78}Se isotope, indicating a mass resolution problem in the ion implanter. On Fig. 7.14(b) a ^{77}Se implant from another implanter again shows the presence of the other Se isotopes, but in this case with a distinctly different depth distribution to the ^{77}Se implant indicating a

Fig. 7.14. Isotope impurities co-implanted with (a) ^{78}Se, (b) ^{77}Se, (c) ^{30}Si.

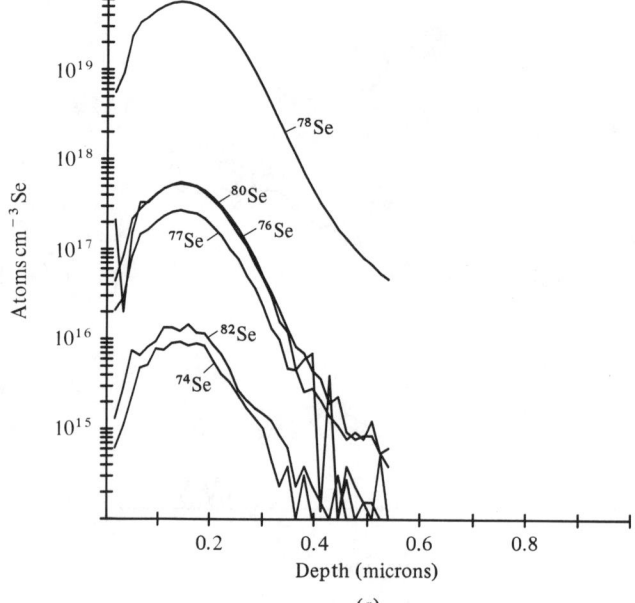

(a)

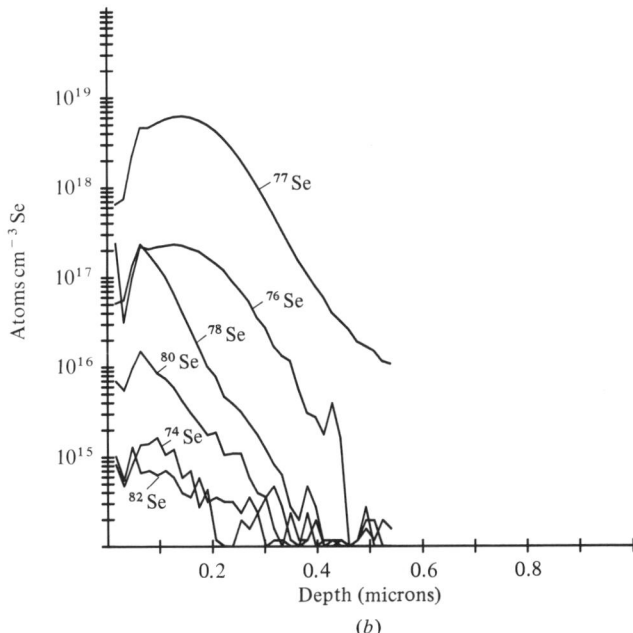

(b)

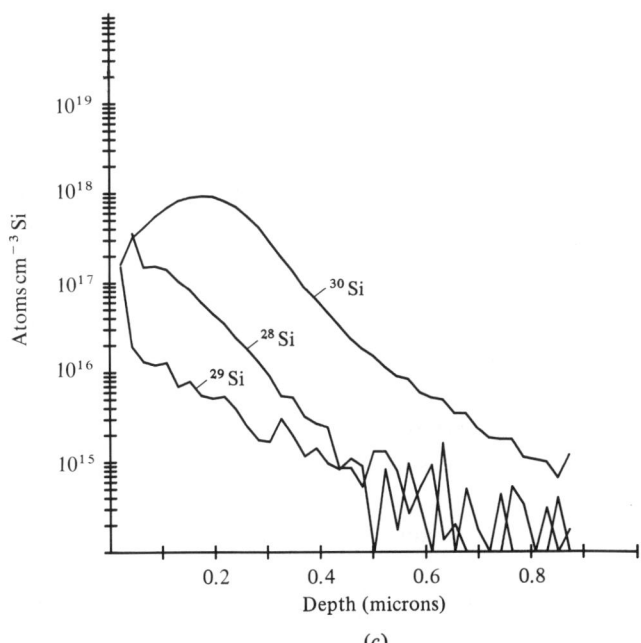

(c)

Table 7.2. *The relative concentration of ^{28}Si found in ten ^{29}Si implants from seven different ion implanters*

Implanter	Beam filter	Implant species	% ^{28}Si
A	OFF	$^{29}Si^{++}$	7
A	ON	$^{29}Si^{++}$	6
B	NOT AVAILABLE	$^{29}Si^{++}$	13
C	NOT AVAILABLE	$^{29}Si^{++}$	2
D	NOT AVAILABLE	$^{29}Si^{++}$	6
E	NOT AVAILABLE	$^{29}Si^{++}$	11
E	NOT AVAILABLE	$^{29}Si^{+}$	5
F	NOT AVAILABLE	$^{29}Si^{++}$	9
G	OFF	$^{29}Si^{++}$	14
G	ON	$^{29}Si^{++}$	10

problem other than mass resolution. Fig. 7.14(c) shows a ^{30}Si implant which shows similar characteristics to the ^{77}Se implant with the ^{28}Si and ^{29}Si showing a lower energy implantation than the desired ^{30}Si implant. This effect has been observed in all ^{29}Si and ^{30}Si implants analysed, and Table 7.2 shows the relative concentration of ^{28}Si found in ^{29}Si implants from seven different implants. It is thought that this effect arises from a scattering mechanism inside the ion implanter after mass separation, resulting in contamination of the chosen isotope with the other isotopes originally in the beam.

In addition to the ability to analyse the levels of trace impurities in semiconductors, SIMS can equally well be used to study the matrix element distributions. This is of particular value in the study of epitaxial layers of compound semiconductors where the matrix element profiles can reveal the presence of growth defects. Fig. 7.15 shows the ^{69}Ga and ^{27}Al profiles recorded through three GaAs/GaAlAs/GaAs structures grown using MOCVD to investigate the effectiveness of aluminium alkyl turn-off in different designs of epi-reactors.

A low primary ion beam energy was used to reduce atomic mixing effects and so give the sharpest interfaces commensurate with a reasonable erosion rate.

The samples consisted of a 200 nm GaAs layer on 200 nm GaAlAs on GaAs and the interface of interest is that between the GaAlAs layer and the outer GaAs layer. The interface corresponds to the turn-off of the aluminium alkyl supply to the reactor and is a sensitive indicator to the effectiveness of the turn-off.

Samples A and B (Fig. 7.15(a) and (b)) were grown in epi-reactors

Dynamic secondary ion mass spectrometry

Fig. 7.15. Ga and Al matrix element profiles from three MOCVD grown GaAs/GaAlAs structures.

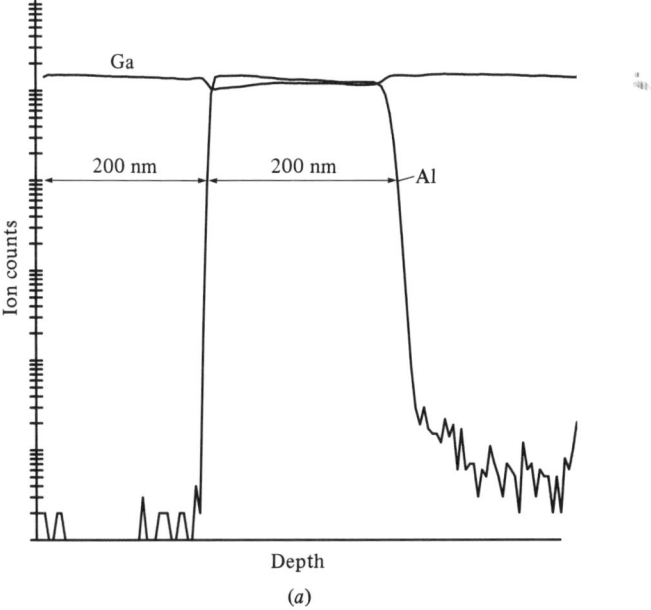

(a)

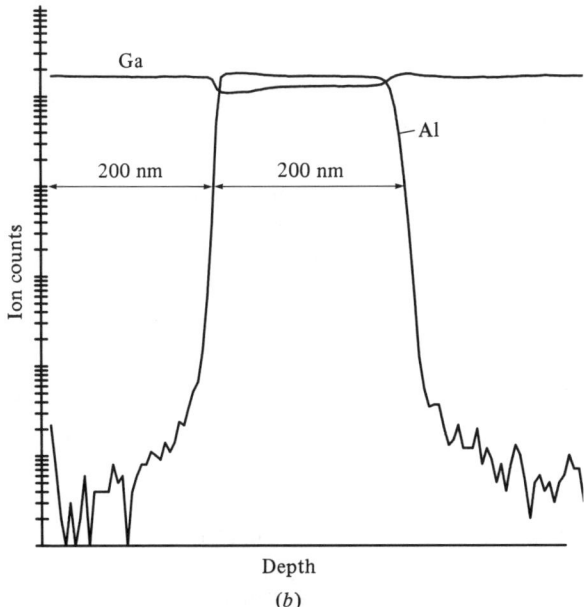

(b)

Fig. 7.15 (contd.)

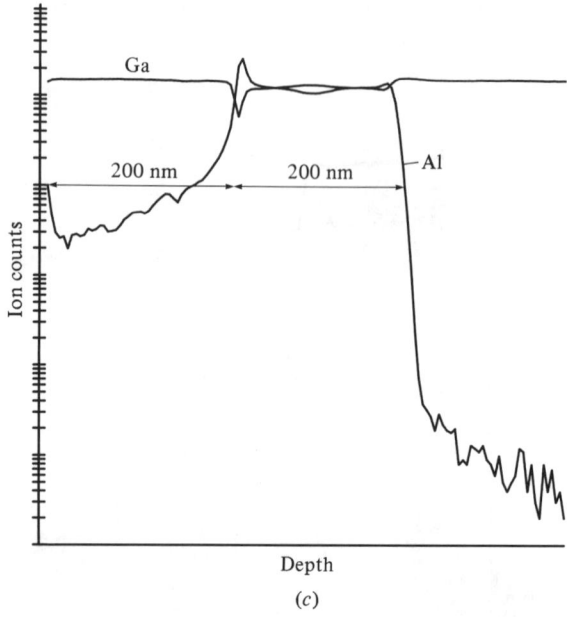

(c)

employing gas handling manifolds, whilst sample C (Figure 15c) was grown in a non-manifold system. The aluminium profiles show a distinctive difference between the manifold systems and the non-manifold system. In samples A and B the aluminium signal drops very rapidly on turn-off of the aluminium alkyl, by five decades in less than 20 nm for sample A and at a similar rate in sample B for the first three decades, followed by a slight tail. Sample C, on the other hand, shows a less rapid decrease on turn-off and a long tail extending all the way through the outer 200 nm of GaAs with the aluminium signal never falling below 1% of that in the GaAlAs layer. The magnitude of the tail signal is some 1000 times greater than that seen in sample B.

Sample C shows another interesting feature: there is a noticeable 'spike' in the aluminium signal at the turn-off point and a corresponding dip in the gallium signal representing quite a large compositional fluctuation. The width of this spike effect is of the order of 10 nm. Compositional spikes of this kind in MOCVD grown material are characteristic of pressure or flow transients in the reactor causing an imbalance in growth conditions. In this case the spike is associated with the turn-off of the aluminium alkyl and has previously been identified as resulting from a volume change caused by the valve in the aluminium alkyl line closing.

Returning again to the identification of impurity species in semi-conductor systems, by providing an elemental profile, SIMS can be used to identify unknown impurity species in epitaxial layers whose presence is suspected from electrical activity measurements. An example of this type of analysis is the identification of the major residual donor in MBE grown InP (Martin et al., 1985). Epitaxial layers of InP grown by MBE have a residual donor concentration of $\sim 10^{16}$ atoms cm^{-2}. By exploiting the high sensitivity of SIMS, with Cs$^+$ bombardment to give the maximum sensitivity to n-type impurities, the presence of possible impurity species can be determined.

Early studies on uniform layers showed that the principal residual n-type impurities present in the MBE grown InP layers were S at $\sim 10^{16}$ atom cm^{-3} and Si at $\sim 10^{15}$ atom cm^{-3}; whilst Se was less than the detection limit of 5×10^{14} atom cm^{-3}. A specially grown sample with step-like changes in the growth conditions was prepared to investigate the source of the contamination. An electrical profile from this sample is shown in Fig. 7.16(a). The features of interest are A the layer/substrate interface, A–B the region of uniform growth, B–C P flux reduced, C–D as A–B, D–E In flux reduced, E return to original conditions as in A–B. This evidence suggests the phosphorus as the source of contamination. SIMS profiles for S and Si through this layer (Fig. 7.16(b) and (c)) show that the elemental S profile correlates extremely well with the measured carrier concentration whilst the Si level is a decade below that of the electrically active species and shows no variations through the layer. This correlation positively identifies S as the residual donor species in MBE grown InP.

In comparing the SIMS profile and the electrical profile, it is interesting to note the better depth resolution of SIMS and the ability to provide a profile right up to the surface of the sample. Also, as it is the elemental profile that is monitored, the signals can be followed beyond the epi-layer and into the semi-insulating substrate.

6.2 Geology

Dynamic SIMS (or ion microprobe) has long been used by geologists as a small-area trace-analysis tool with the added advantage that it can be used to perform isotope ratio measurements on very small volumes of material. Isotope ratios are used as a method of dating rock samples and studying their evolution; with SIMS it is possible to compare isotope ratios in the centre and at the edge of a grain of rock a few tens of microns across. As, in general, the rock samples are insulating, the O$^-$ primary ion beam is most frequently used. Fig. 7.17(a) and (b) shows

Fig. 7.16. (a) C–V profile of InP epitaxial layer, (b) SIMS S profile, (c) SIMS Si profile.

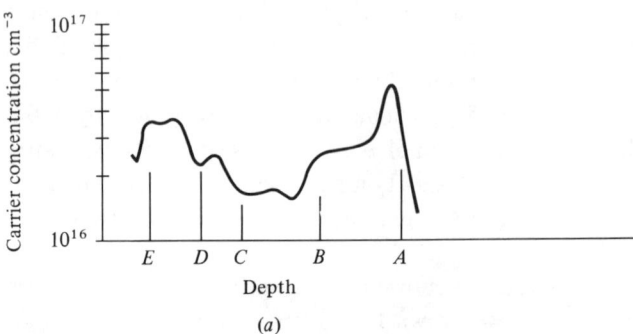

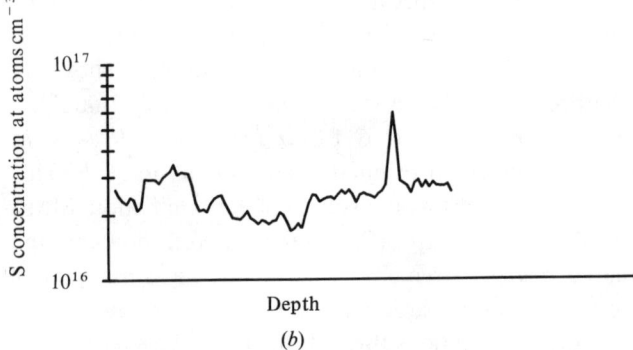

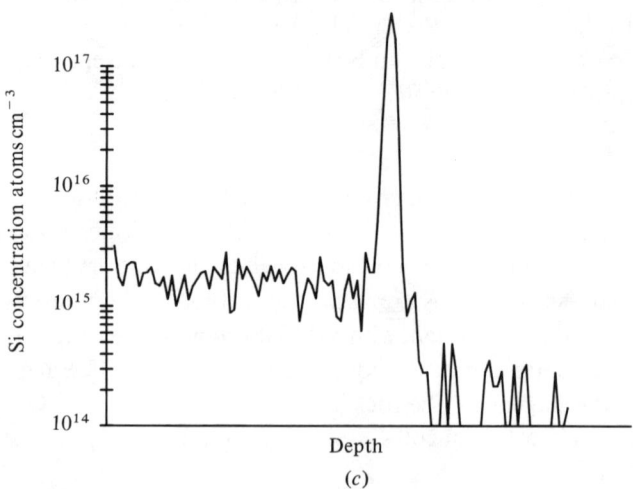

Dynamic secondary ion mass spectrometry 253

Fig. 7.17. Mass spectra recorded from a sample of pink granite: (*a*) feldspar phase, (*b*) quartz phase.

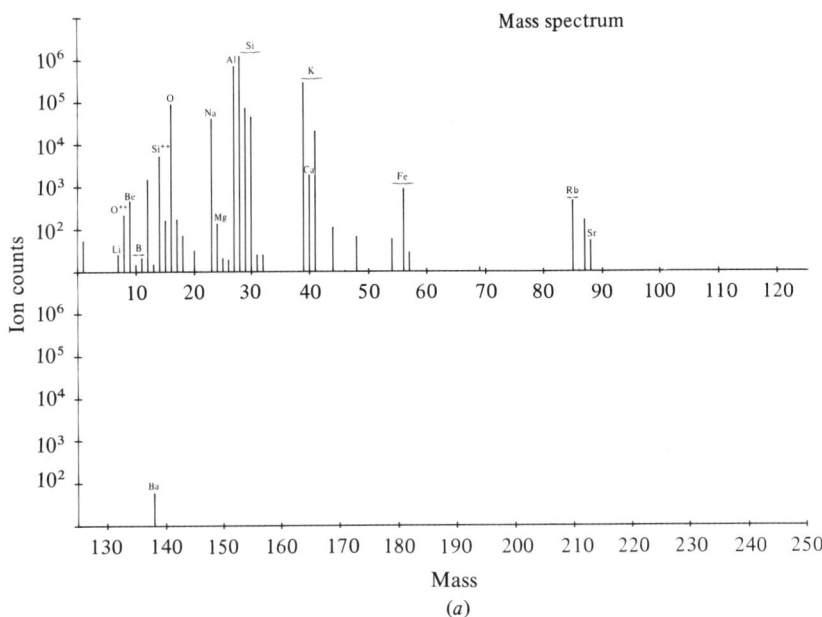

(*a*)

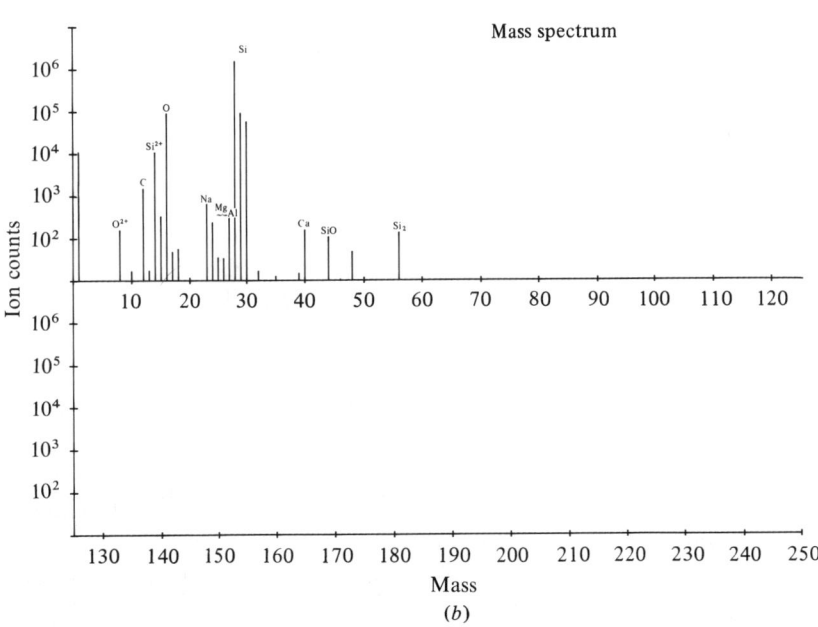

(*b*)

mass spectra recorded from the feldspar and quartz regions of a piece of pink granite. In the clear quartz region the dominant signal is silicon with traces of other elements at the few ppm level, whilst in the pink coloured region in addition to Al and Si most of the group one and two elements are found together with significant traces of iron. The much higher level of potassium than sodium identifies the region as a K-feldspar. The two peaks at masses 85 and 87 show the ease of detection of rubidium.

6.3 Biology

SIMS can be used in various ways for the characterisation of biological samples, ranging from determining the distributions of elements in samples of hard and soft tissue to the measurement of trace element levels in body fluids. The following example illustrates the applicability of SIMS for determining trace element distributions in soft tissue samples (Larras & Siami, 1985):

Fig. 7.18(a)–(d) show a series of secondary ion images recorded from a thin sample prepared from a section of the thyroid gland of a mouse. The main element of interest in this case was iodine with a secondary interest in the distributions of sulphur and phosphorus. Consequently, Cs^+ primary ion bombardment and negative secondary ion detection were chosen to give the maximum sensitivity to these electronegative species. The analysed area throughout the series of diagrams is the same 150 μm diameter site.

The basic structure of the sample is revealed in the CN^- image which shows the network of cells surrounding the follicles. The phosphorus image highlights the individual cells as it is found in the cell structures and not elsewhere, whilst sulphur is found in varying amounts through the section with slightly lower levels in the cells. The mass 127 image shows that the iodine is concentrated in the follicles. This particular animal had been treated with a preparation containing a radioactive isotope of iodine, ^{129}I, used in the treatment of thyroid conditions. (Unfortunately iodine has only the one stable isotope, ^{127}I, otherwise it would not be necessary to resort to radioactive isotopes, at least for the SIMS analysis.) The difference between the two iodine images $^{127}I^-$ and $^{129}I^-$ is quite striking with some follicles containing both isotopes with uniform distributions whilst in other cases the follicles are filled uniformly with ^{127}I and only have a thin outer layer with high concentrations of ^{129}I and no ^{129}I inside whilst one follicle shows a uniform ^{127}I and a high concentration of ^{129}I in an outer layer with a lower uniform level of ^{129}I inside. Fig. 7.18(f) shows the ^{127}I image from a specimen taken from a younger animal which reveals

Fig. 7.18. Secondary ion micrographs 150 μm diameter of mouse thyroid. (a) $^{26}CN^-$, (b) $^{32}S^-$, (c) $^{31}P^-$, (d) $^{127}I^-$, (e) $^{129}I^-$, and (f) $^{127}I^-$ from a younger animal.

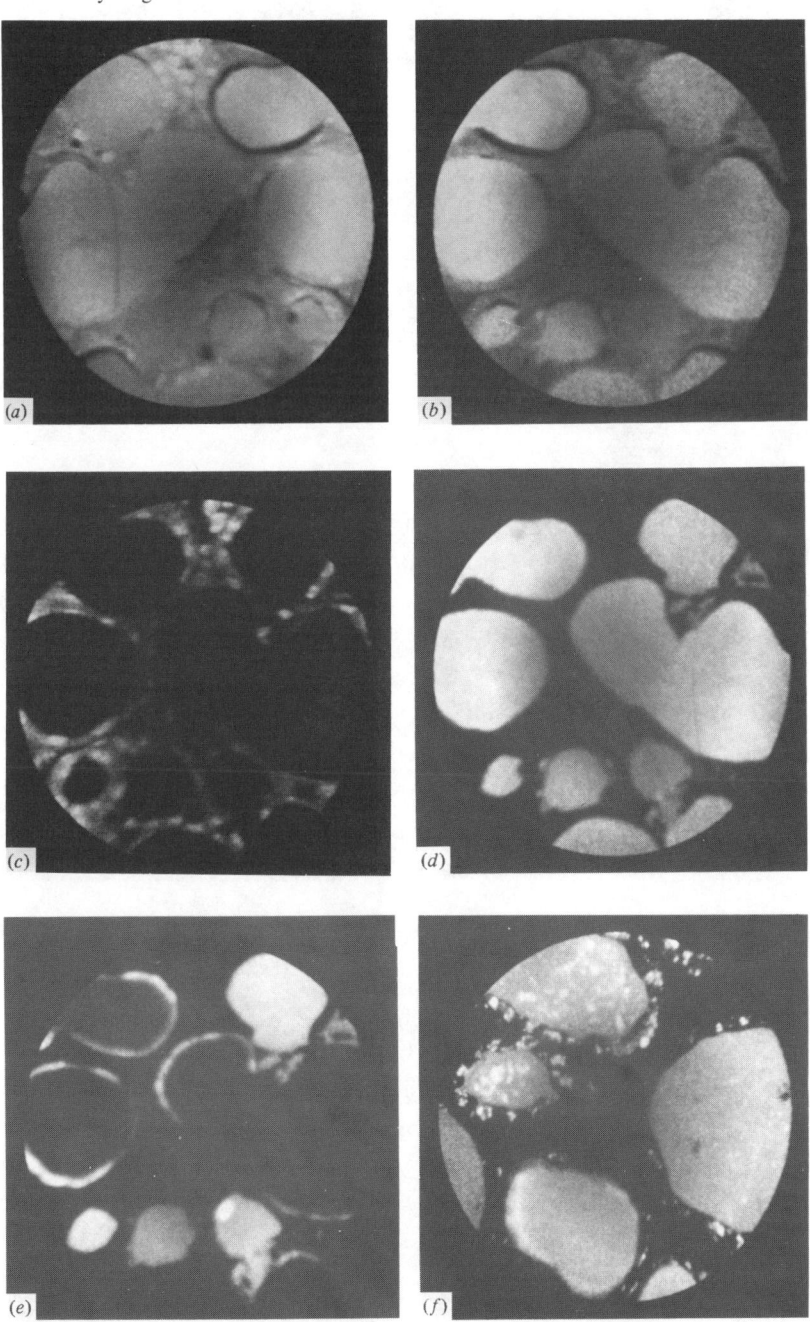

Fig. 7.19. Secondary ion micrographs of section of Fe/Ni lamp ferrule: (a) ^{56}Fe, (b) ^{63}Cu, (c) ^{64}Zn, (d) ^{107}Ag, (e) ^{23}Na, (f) ^{39}K.

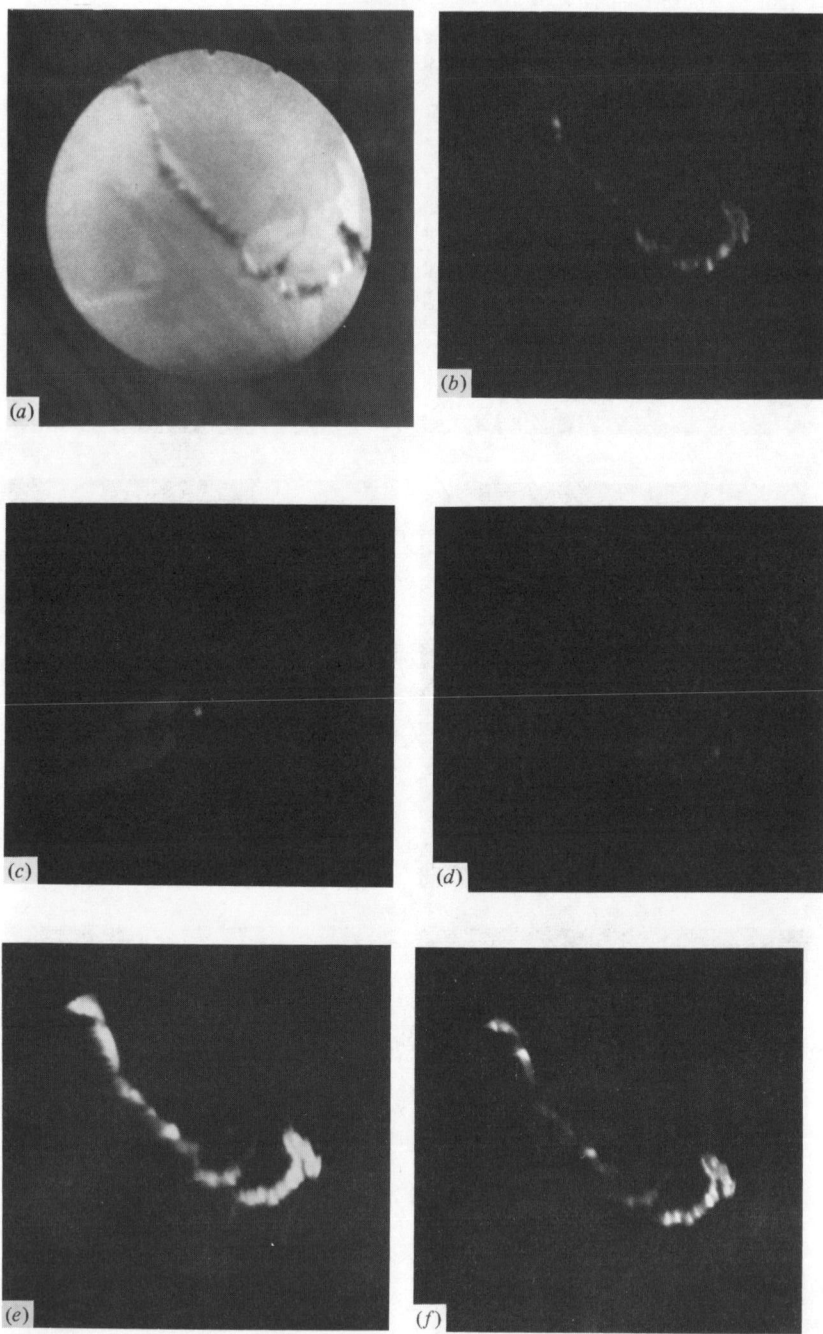

the presence of iodine outside the follicles in locally high concentration.

Using the isotope specificity of SIMS the biologist can study both elemental distributions and differences in distribution between isotopes of the same element. Spatial distribution in the ion imaging mode is adequate for resolving individual cells but as yet is not sufficient to investigate structure within the cells. Probably the greatest value to the biologist of using SIMS as an investigative tool is that non-radioactive tracer experiments can be performed, at least for the majority of elements, on very thin samples and with high lateral resolution.

6.4 Materials science

SIMS offers the materials scientist greater sensitivity, better lateral resolution and faster data acquisition than do the conventional electron optical techniques of energy and wavelength dispersive, X-ray analysis. However, the price paid for this is the partial consumption of the sample and less refined quantitative analysis. In the first example, Fig. 7.19 shows secondary ion images recorded from a polished cross section of an Fe/Ni alloy used as part of a lamp ferrule. The component is braized into place and in this case the braize has penetrated a grain boundary and cracked the metal. The ^{56}Fe image shows the grain structure of the alloy with the cracked region running diagonally through the image. (No special etches are required to show up the grain structure, this results from orientation contrast in the SIMS image). The braize elements, ^{63}Cu, ^{107}Ag and ^{64}Zn, show up clearly in the region of the braize penetration whilst the elements ^{23}Na and ^{39}K reveal residual traces of the flux left behind in the braized region.

Fig. 7.20(a) shows the mass spectrum recorded from the outside of the envelope of a high-pressure sodium lamp. The lamp envelope is made from sintered alumina to which yttrium is added to modify the grain growth during the sintering process to give the desired grain size. As in this case the sample is insulating the O^- primary ion beam was used for the analysis to minimise any charging problems. The spectrum shows an intense ^{27}Al signal with other Al-related species at masses 43 (AlO) 54 (Al$_2$) 70(Al$_2$O) a small Y peak is apparent at 89 together with its oxide at 105. The secondary ion image of mass 27 is shown in Fig.7.21(a) which shows up the grain structure of the aluminium with the grain boundaries appearing bright. This is because the lamp has been run and the outside of the envelope has become thermally etched. No spatial variation of the yttrium signal could be detected. Fig. 7.20(b) shows the mass spectrum from the inside of the lamp envelope. Now high levels of sodium and yttrium can be seen together with traces of barium (from the electrodes).

258   D.E. Sykes

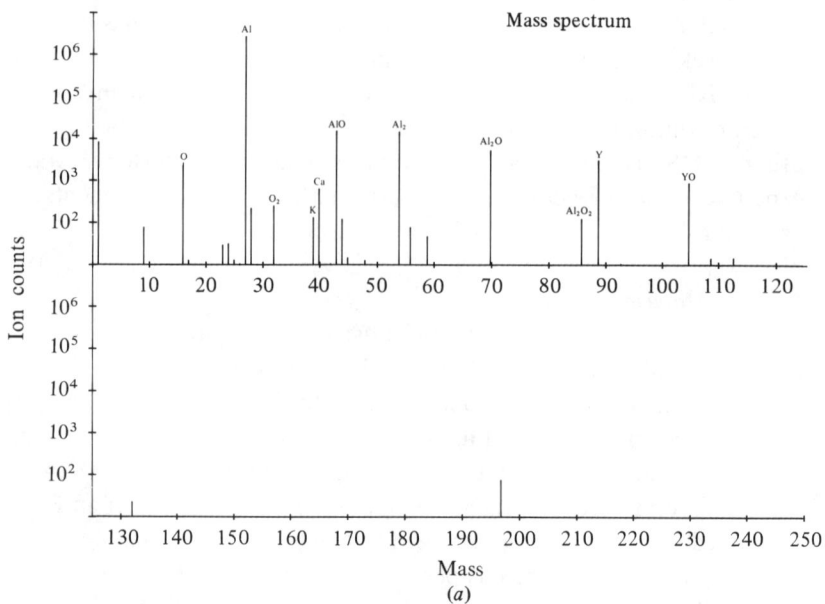

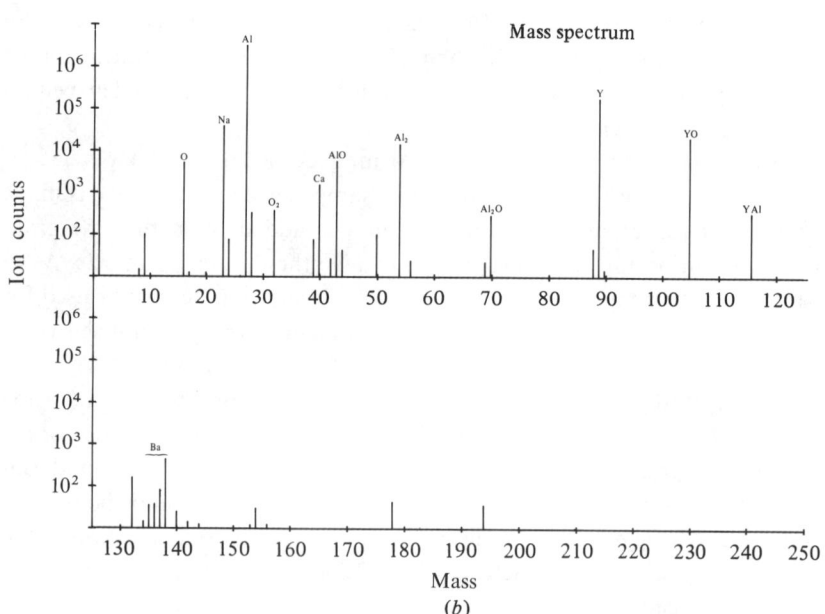

Fig. 7.20. Mass spectra recorded from the outside (*a*) and inside (*b*) of an alumina lamp envelope.

Secondary ion images shown in Fig. 7.21(b), (c) and (d) show the Al$^+$, Na$^+$ and Y$^+$ distributions. The ^{27}Al$^+$ image shows the grain structure, this time with the grain boundaries dark – no thermal etching. Sodium is found to have penetrated only in some of the grains and not in others whilst the yttrium is concentrated in the grain boundary network.

In the production of cast irons the addition of small quantities of certain elements can modify the morphology of the graphite phase and thus control the physical properties of the iron. The addition of trace levels of cerium, for example, can produce a spherical graphite morphology. The precise mechanism whereby these nodularising elements influence the graphite growth is uncertain and SIMS has been used to investigate their role (Franklin & Stark, 1984).

Fig. 7.21. Secondary ion micrographs from the outside (a) ^{27}Al$^+$ and inside (b) ^{27}Al$^+$, (c) ^{23}Na$^+$ and (d) ^{89}Y$^+$ of an alumina lamp evelope.

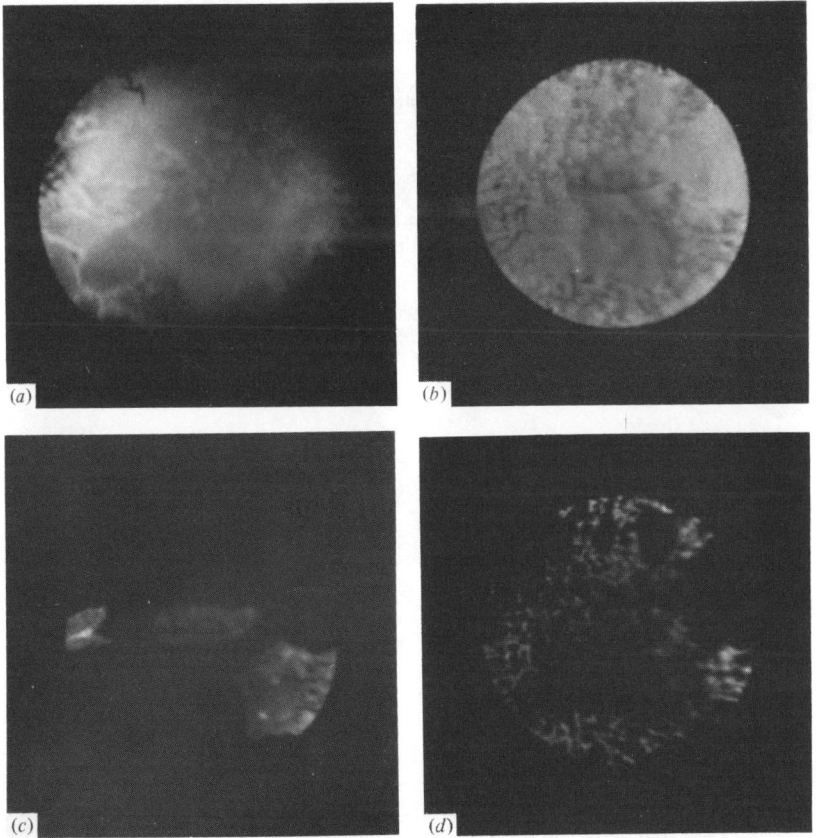

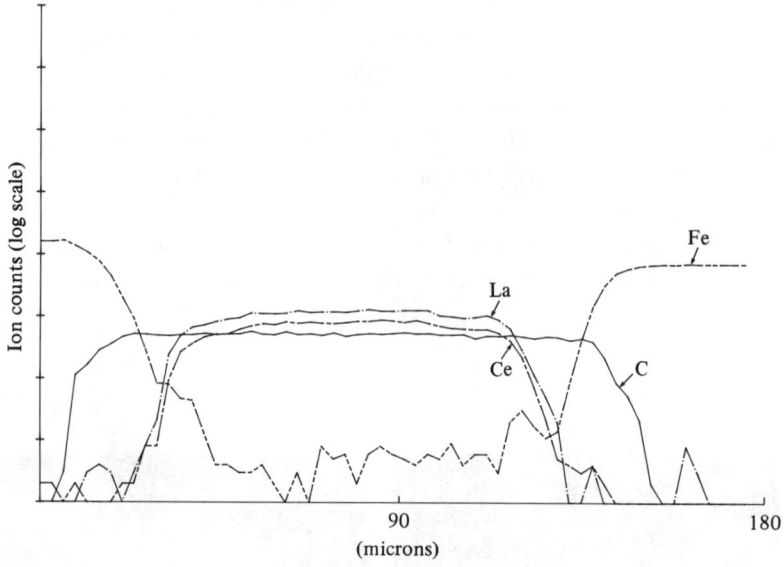

Fig. 7.22. Step scan for ^{12}C, ^{56}Fe, ^{139}La and ^{140}Ce across a duplex graphite nodule in cast iron. Total scan length 180 μm (60 × 3 μm steps).

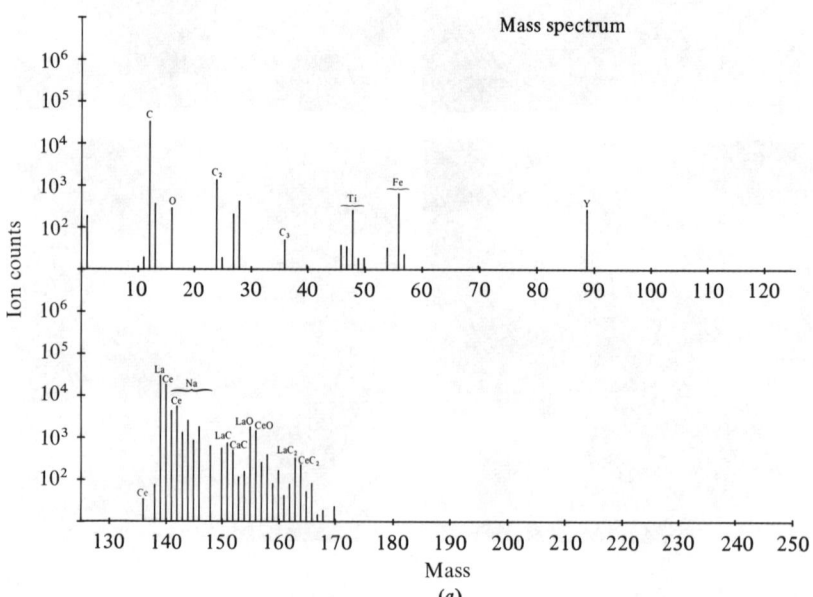

Fig. 7.23. Mass spectra from the centre of the duplex graphite nodule (a) and the surrounding iron (b) showing the presence of Ce which is present in the bulk at 0.08%.

Fig. 7.22 shows a line scan across the diameter of a spheroidal graphite nodule produced by stepping the sample in 3 μm steps across the analysed area 8 μm in diameter. The sample was prepared from a hypereutectic composition containing 0.19% of cerium mischmetall, and was metallurgically polished to reveal the duplex structure of the graphite nodules. The line scan shows the abrupt interface between the graphite and the iron in the Fe and C traces whilst the Ce and La are only detected in the central region of the duplex nodule. The Fe trace also shows a step in the region of this interface.

These duplex graphite nodules arise as a result of a two-stage growth process. The central part grows from the melt, whilst the outer halo is produced in the solid phase with the nodule surrounded by austenite. The rare earth elements are only found concentrated in the inner part of the nodule which grew from the liquid phase and not at all in the surrounding halo. Mass spectra recorded from the centre of the nodule, Fig. 7.23(a), and the surrounding iron, Fig. 7.23(b), show the concentration of the rare earths in the graphite and their absence in the iron.

7 Conclusion

In conclusion, an attempt has been made to illustrate the wide range of applications in which dynamic SIMS can be successfully used. The major modes of operation have been covered and it is hoped that, although the examples may not directly address the readers' own analytical

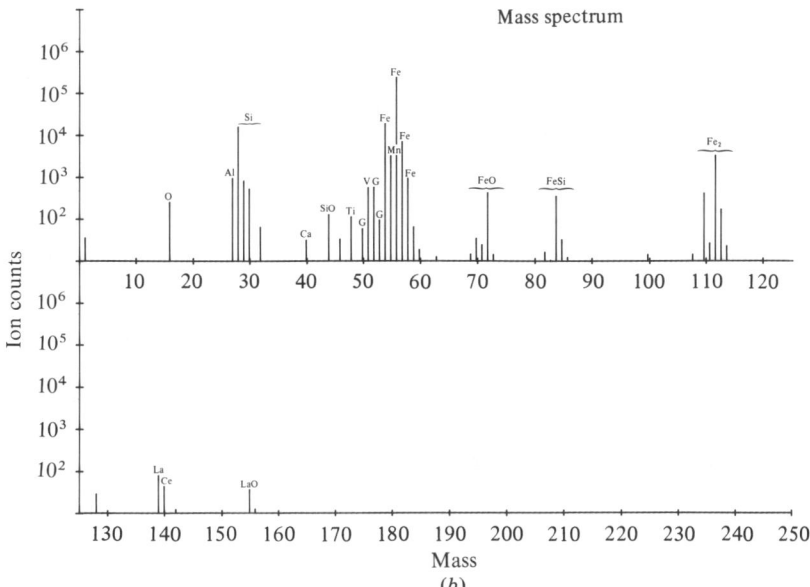

(b)

requirements, they may suggest how SIMS can be usefully used in other investigations. As the instrumentation develops to give higher sensitivity and better spatial resolution, the range of problems to which SIMS can be applied will increase.

References

Anderson, C.A. & Hinthorne (1973). J.R., *Anal. Chem.*, **45**, 1421.
Benninghoven, A. (1975). *Surface Science*, **53**, 596.
Clegg, J.B., Morgan, A.E., de Grefte, H.A.M., Simondet, F., Huber, A., Blackmore, G., Dowsett, M.G., Sykes, D.E., Magee, C.W. & Deline, V.R. (1984). *Surface and Interface Analysis*, **6**, 162.
Franklin, S.E. & Stark, R.A. (1984). *3rd International Symposium on the Physical Metallurgy of Cast Iron*, Stockholm, 1984.
Havette, A. & Slodzian, G. (1980). *J. Microsc. Spectrosc. Electron* **5**, 231.
Holland, R. & Blackmore, G.W. (1983). *Int. Journal of Mass Spectrometry and Ion Physics*, **46**, 527.
Larras-Regard, E. & Siami, K. (1985). *C.R. Acad. Sc. Paris*, t. 301, Serie III, **6**, 329.
Martin, J., Stanley, C.R., Iliadis, A., Whitehouse, C.R. & Sykes, D.E. (1985). *Appl. Phys. Lett.*, **46**, 994.
Migeon, H.N. & Le Goux, J.J.M. (1982). *Secondary Ion Mass Spectroscopy SIMS III*, eds. A. Benninghoven *et al.*, Springer-Verlag, p. 52.
Morgan, A.E. & Werner, H.W. (1976). *Anal. Chem.*, **48**, 699.
Morgan, A.E. & Werner, H.W. (1978). *Mikrochimica Acta (Wien)*, **11**, 31.
Newbury, D.E. & Heinrich, K.F.J. (1979). *Mikrochimica Acta (Wien) Suppl.*, **8**, 3.
Rudenauer, F. (1977). *Mikrochimica Acta (Wien), Suppl.*, **7**, 85.
Steiger, W., Rudenauer, F., Gnaser, H., Pollinger, P. & Studnicka, H. (1983). *Mikrochimica Acta (Wien), Suppl.*, **10**, 111.
Storms, H.A., Brown, K.F. & Stein, J.J. (1977). *Anal. Chem.*, **49**, 2023.
Takedoum, J., Pivin, J.C., Pons-Corbeau, J., Berneron, R. & Charbonnier, J.C. (1984). *Surface and Interface Analysis*, **6**, 174.
Werner, H.W. (1980). *Surface and Interface Analysis*, **2**, 56.
Wittmaak, K. (1980). *Nuclear Instruments and Methods*, **168**, 343.
Wittmaak, K. (1982). *Vacuum*, **32**, 65.
Yin, Shaiw-Yih (1980). *Microbeam Analysis*, **289**.
Yin, Shaiw-Yih (1981). *Microbeam Analysis*, **342**.

8

Ion scattering spectroscopy

D.G. ARMOUR

1 Introduction

Ion scattering spectroscopy (ISS) or low-energy ion scattering (LEIS) as the technique is also known, is used to obtain information about the composition and atomic arrangement of solid surfaces. In common with the other ion scattering techniques, Rutherford backscattering (RBS) and medium-energy ion scattering (MEIS), this information is deduced from measurements of the energy distributions of the particles scattered from the surface in a specified direction with respect to that of the primary beam. The basic experimental arrangement required for these measurements is conceptually simple, comprising a source of ions of well-defined mass, energy and direction, a target mounted in an environment which can be reliably controlled from the point of view of gas composition and an energy analyser and detector system for the scattered particles. A schematic diagram of a basic ISS system is shown in Fig. 8.1 where it is clear that by providing target and detector manipulation facilities and an ion beam system capable of delivering well-defined fluxes of different species at various energies, a wide choice of experimental scattering conditions may be employed. The conditions selected are of critical importance, since, by determining the precise nature of the scattering processes which contribute to the features observed in the measured energy spectrum, they specify the relationship between this spectrum and the mass and arrangement of the surface-layer atoms.

In practice, the mode of operation of an ion scattering system and the complexity of the apparatus used depend very strongly on the nature of the information to be obtained and on the requirements in terms of mass range and resolution, sensitivity, lateral and depth resolution, analysis time and quantitative reliability. Consequently, even though it is possible to quote figures to specify these performance parameters, their values will depend on the experimental scattering geometry and conditions and the capabilities of the ISS technique can only be properly described in terms of specific applications. However, a survey of the systems available shows that when used for straightforward composition analysis, the technique is capable of providing elemental (rather than chemical) information over the whole periodic table of the elements (including hydrogen using special techniques) with a depth resolution of 0.3 nm, essentially single atomic layer, lateral resolution of the order of 10^4 nm and a detection limit of about 10^{-3} ML for elements such as oxygen and sulphur adsorbed on metal surfaces.

One of the most important features of ISS is its inherent ability to provide information concerning only the outermost surface layer of the solid under investigation. This unique surface specificity, which is only achieved when rare gas ions at low energies are used as the probe in systems which detect only the ionised particle emission, is the result of a combination of elastic and inelastic scattering processes which lead to the virtual elimination of scattered ion emission from below the surface. The elastic scattering contribution to this subsurface yield reduction is caused by the rapid loss of beam particles in large angle scattering collisions

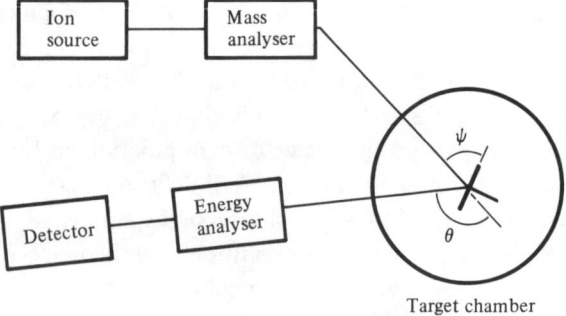

Fig. 8.1. Schematic diagram of basic ISS system.

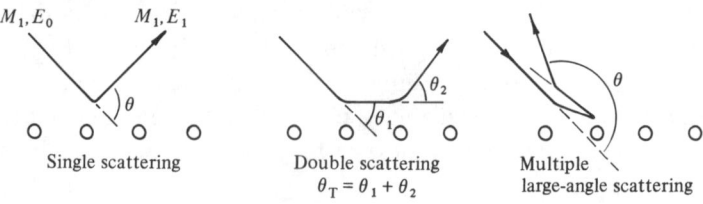

Fig. 8.2. Scattering processes at solid surfaces.

which are highly probable events at low projectile energies (i.e. the scattering cross-sections are large at low energies).

The significance of the large scattering cross-sections is not limited to the surface specificity feature of low-energy ion scattering. For small scattering angles and, in fact, at scattering angles close to 180° when small angles of incidence with respect to the surface are used, multiple scattering processes involving successive interaction of the primary beam particles with surface atoms occur. For primary energies above about 100 eV, depending on the atomic species involved in the collision, these multiple collision processes appear to involve sequences of binary collisions in which the impact parameter, p, at one collision depends on the preceding scattering angle. Consequently, as illustrated in Fig. 8.2 for a number of different geometries, these processes lead to measured energy distributions which depend on the geometrical arrangement of the surface atoms and hence to the ability of ISS to provide information on the short range order of solid surfaces.

By careful selection of the experimental conditions, it is often possible to effectively separate the contributions of single and multiple collision processes to the observed spectra and hence to optimise the operation of the system for either compositional (single collision) or structural (multiple collision) analysis. It is clear, however, that the choice of conditions and the interpretation of the data relies on a detailed understanding of the physical basis of the scattering process.

The ability of ISS measurements to yield highly surface specific composition and structure information has led to a steady increase in the range of applications of the technique and the interpretation of measurements on complex surfaces has created an increasing need to understand and accurately simulate the fundamental elastic and inelastic processes occurring during the interaction of the projectile ions with the surface. Consequently the following discussion of ISS as an analytical tool in surface science and technology applications is based on consideration of the physical basis of the technique in sufficient detail to explain the major practical features of measuring systems and the performance capabilities and limitations in terms of elemental range, sensitivity and detection limits, mass resolution, depth and lateral resolution and quantitative potential.

2 The physical basis of ion scattering
2.1 Introduction

The analytical capabilities of ion scattering are dependent on an ability to relate the energy of the scattered ions to the masses and, in structural analysis, the relative locations of the target surface atoms, and to assess the abundance of the atomic species or structural features present from the scattered ion yields. The required relationships are determined by consideration of the details of the atomic collision processes involved and, although the processes are understood sufficiently well to enable ion scattering data to provide valuable quantitative information, their overall complexity still precludes the universal application of the technique for quantitative measurements.

The bombardment of a solid surface with energetic ions (in all the following discussions, primary energies above about 100 eV are considered) results in a complex sequence of interactions during which the primary ion may penetrate and become trapped or be back-scattered with reduced energy, having transferred energy to the atoms of the solid via elastic and inelastic collision processes. In this context, elastic collisions are those in which the kinetic energy lost by the projectile is acquired by the target atoms without any change in the internal energy of the colliding particles. Inelastic energy losses occur when the kinetic energy of the projectile is converted into electronic energy via electron transfer, ionisation and excitation during the collision.

The extent to which these elastic and inelastic processes affect the final energy and charge state of the scattered particles is dependent on the

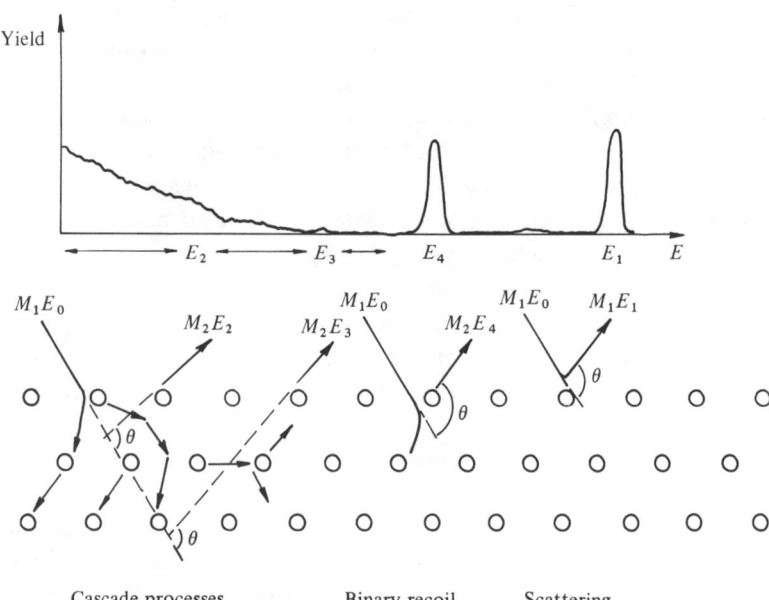

Fig. 8.3. Particle ejection processes during ion bombardment and the corresponding energy distribution.

primary energy, the identities of the collision partners and the scattering geometry.

Considering the situation depicted in Fig. 8.3, in which ions of mass, M_1 and energy, E_0, are impinging on a surface composed of atoms of mass, M_2, it can be seen that particles leaving the surface at an angle θ, to the incident beam direction may be scattered primaries which have suffered various different energy losses or target or secondary particles which have recoiled from the surface after a single collision with a primary ion or after interacting with other target particles involved in the collision cascade induced by the energy deposited by the primary ion impact. A collision cascade develops when a target atom is hit by an energetic projectile and recoils to transfer part of its energy to other atoms in the solid. In this way, when the energy deposition density is sufficiently high, large numbers of target atoms are momentarily set in motion. When this cascade of moving atoms interacts the surface and surface atoms receive sufficient energy to overcome the binding energy, sputtering occurs and those particles which are ejected as ions contribute to the very low energy region of ISS spectra and are, in fact, the particles mass analysed in SIMS.

The relative contributions of all the ejected particles to the measured ion energy distributions determine the characteristics of low-energy ion scattering as an analytical technique. The mass resolution depends not only on the details of the individual collision processes but on an ability to relate yields at different energies to scattering from different mass atoms rather than from atoms of the same mass but with different trajectories. The sensitivity and detection limits are dependent on the cross-sections for the different scattering processes, the probability of the particles being scattered as ions and the background count levels due to sputtering. The removal of target atoms in the sputtering process and the displacement of atoms from their original lattice positions represent perturbation effects which significantly affect the analytical capabilities of the technique since the possibilities of removing different constituents at different rates and changing the atomic arrangement place restrictions on the acceptable analysing fluence and hence on the counting times and ultimate sensitivity.

The dependence of the capabilities of ion scattering on the details of the atomic collision processes occurring is reflected in the close relationship between the basic collision physics and the design and performance of experimental systems. The following consideration of the basic theory of atomic collisions enables the major design features of scattering systems to be explained and indicates the way in which systems may be optimised for either composition or structure analysis.

2.2 Basic theory of atomic collisions in solids

Energetic ions (or atoms) incident on a solid surface may be simply back-scattered at the surface or may penetrate and slow down by transferring energy in elastic and inelastic processes. For primary energies above about 100 eV, the vast majority of the incident particles are trapped within the solid, typically less than 10% being reflected for light particles incident on heavy substrate, and the reflection coefficient decreases as energy is increased. The energy loss processes which occur during the interaction of the primary ions with the target atoms depend on the energy, the identity of the collision partners and the scattering geometry. Interpretation of measured energy distributions therefore depends on an ability to accurately model the collision processes and to specify the experimental conditions for which any particular model is valid.

On the basis of a detailed analysis carried out by Bohr (1948) it can be shown that classical mechanics may be used to accurately describe the elastic features of the collision processes which occur in any practical ion scattering system. This does not mean that basically quantum mechanical effects such as ionisation and electronic excitation are negligible since they determine the charge state of the backscattered projectiles and contribute significantly in some cases, to the observed energy losses. However, to a first approximation, elastic and inelastic processes may be assumed to be uncorrelated and hence considered as separated processes.

In terms of classical mechanics, collisions between the projectile and target atoms are described in terms of the interatomic potential between them which arises from the mutual forces (attractive and repulsive) between the atomic nuclei and the electrons. For the conditions relevant to ion scattering analysis, the particles approach so closely that the attractive part of the potential can be neglected and the interaction may be described in terms of a totally repulsive potential, the precise form of which determines the differential scattering cross-section and hence, from the point of view of elastic processes, the yield of back-scattered particles. In the ISS regime, the potential is generally described by a Coulomb potential with a screening function to account for the effect of the electron cloud between the nuclei. These screening electrons reduce the Coulomb repulsion and give a potential of the form

$$V(r) = \frac{Z_1 Z_2 e^2}{r} \varphi(r/a),$$

where r is the interatomic spacing, Z_1 and Z_2 are the atomic numbers of the colliding particles, e is the electronic charge, $\varphi(r/a)$ is the screening function and a is known as the screening length.

Various analytical expressions for the screening function have been used to describe the collisions occurring in ISS and scattering yields in specific situations can be predicted fairly accurately. In contrast to the situation for RBS, however, a generally applicable formalism for calculating cross-sections for any projectile–target combination is not available and the technique is essentially non-quantitative unless proper calibration can be carried out for the case of interest. The potentials which have been found to accurately describe low-energy ion scattering data are the Molière, which is based on the Thomas–Fermi model of the atom and the Biersack-Ziegler (1982) approach which is based on the free electron model and accounts for exchange and correlation energies. For the Molière potential which has been particularly widely used in ISS calculations the screening function is approximated by an analytical expression of the form (Molière, 1947)

$$\varphi(r/a) = 0.1 \exp(-6r/a) + 0.55 \exp(-1.2r/a) + 0.35 \exp(-0.3r/a),$$

where a is the Firsov screening length modified, in many ion scattering calculations to

$$a = 0.468 C(Z_1^{2/3} + Z_2^{2/3})^{1/2},$$

where C is a 'fitting' constant in the range $0.6 \leq C \leq 0.8$. The effect of variations in C and the general form of these potentials are shown in Fig. 8.4 for the case of argon–copper (Poelsema et al., 1974) and argon–magnesium collisions (Cruz et al., 1982).

A common feature of all these potentials is that they decrease very rapidly with atomic separation such that, during any particular collision

Fig. 8.4. Interaction potentials for (a) Ar^+—Cu and (b) Ar^+—Au.

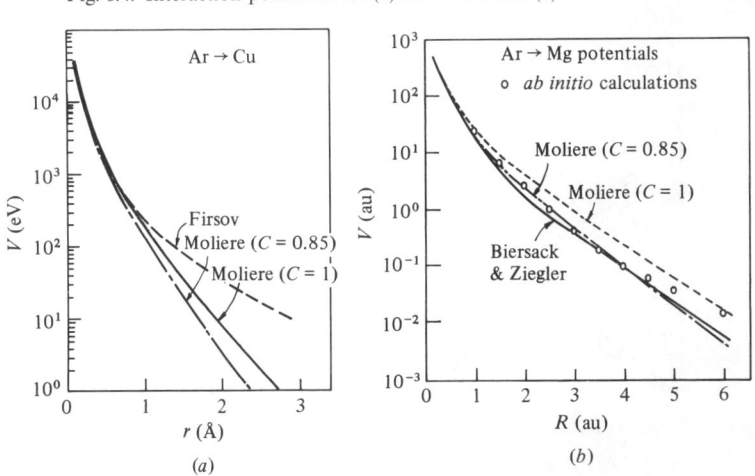

between the projectile and a target atom, the interaction with neighbouring atoms, which are typically a few tenths of a nanometre away, is negligible. This is an extremely important feature of the collision process since it means that the interaction of the projectile ion with the solid can be described in terms of simple binary collision theory. Additionally, since the time duration of such a collision is of the order of $10^{-15} - 10^{-16}$ s compared with lattice vibration periods of approximately 10^{-13} s and the energy transfer to the target atom is, in general, very much greater than the binding energy in the solid, it is reasonable to assume that the target atom in the binary collision will behave like a free atom. Under these circumstances the overall scattering process will consist of a single binary collision with a surface atom or a sequence of such collisions and it is on this basis that the qualitative analysis of ion scattering data becomes particularly straightforward.

The validity of this form of analysis depends on the assumption that elastic collision processes are dominant and a convenient comparison of the relative elastic and inelastic energy loss contributions, expressed in terms of the so-called stopping powers, can be obtained in terms of the reduced energy, ε, which is the energy expressed in Lindhard's dimensionless Thomas–Fermi units (Lindhard et al., 1963). It is the ratio of the Thomas–Fermi screening length to the distance of closest approach in an

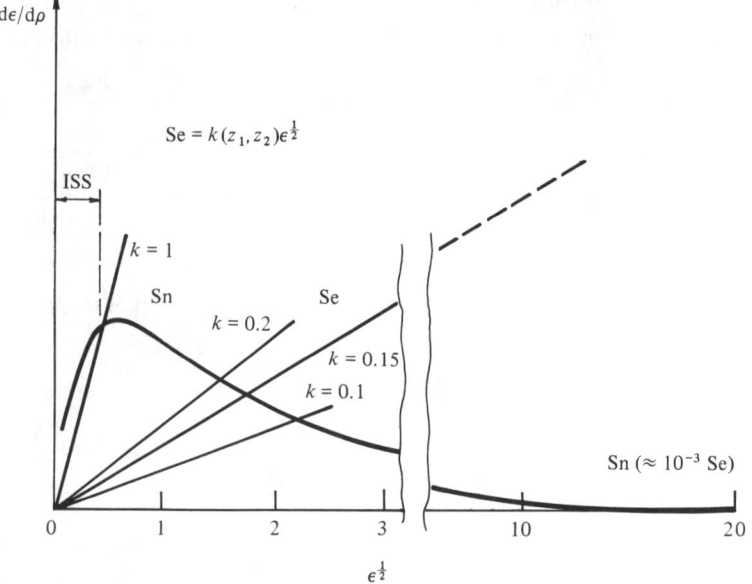

Fig. 8.5. Comparison of elastic and inelastic stopping powers as functions of reduced energy.

unscreened collision. The advantages of this energy scaling are that it allows the elastic energy loss or stopping power ($S_n = -dE/dx$) to be expressed in terms of a universal curve for all combinations of Z_1 and Z_2 and it enables the different scattering regions, i.e. low, medium and high (RBS) energy scattering, to be reliably defined. The curves shown in Fig. 8.5 for S_n and S_e against ε for the region where S_e is proportional to $\varepsilon^{1/2}$ indicate that for ε values below about 0.3 the elastic energy loss processes are dominant. Below this limit, which corresponds, for example, to 2 keV He$^+$ or 30 keV Ar$^+$ collisions with Cu atoms, the elastic collision cross-sections are comparatively large and the scattering is characterised by single, elastic collisions with surface atoms and it is in this regime that ISS measurements can be interpreted reliably in terms of simple, elastic binary collision theory.

For the case of a single binary collision, the energy retained by the scattered ion can be calculated on the basis of energy and momentum conservation laws without reference to interatomic potentials and, following the nomenclature in Fig. 8.6 (Lindhard et al., 1963), which illustrates such a collision.

$$E_1 = E_0 \left(\frac{\cos\theta \pm (A^2 - \sin^2\theta)^{1/2}}{(1+A)} \right)^2, \tag{1}$$

where A is the ratio of the target and projectile masses ($A = M_2/M_1$). The positive sign applies for $A > 1$, i.e. light particles incident on heavier substrates, and both signs are possible for $A \leqq 1$. In the latter case, the

Fig. 8.6. Schematic diagram of a binary collision process.

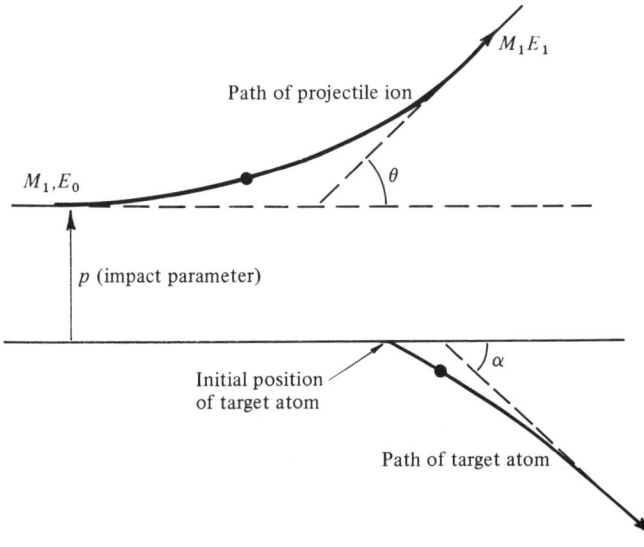

scattering angle is clearly limited by the requirement that

$$\theta_{max} = \sin^{-1} A. \qquad (2)$$

The energy transferred to the struck atom, E_2, is given by the expression

$$E_2 = E_0 \left(\frac{4A}{(1+A)^2} \right) \cos^2 \alpha \qquad (3)$$

in which α, the angle between the path of the recoiling particle and the incident projectile direction is limited to values $\leq 90°$.

The accuracy with which these simple equations predict the experimentally observed peak energies is illustrated in Fig. 8.7 (Hagstrum, 1954, 1977) for 2 keV He$^+$ scattering at a copper surface and 1.5 keV Ar$^+$ scattering from a copper surface. The high, low-energy yield in the Ar$^+$–Cu spectrum is the result of sputtering. For appropriate scattering conditions, the predictions of the simple binary scattering model agree with experimental observations to within about 1–2% for incident energies down to a few hundred eV.

The validity of this model forms the basis of the use of ISS for surface elemental analysis since it allows the mass of the target atom, M_2, to be

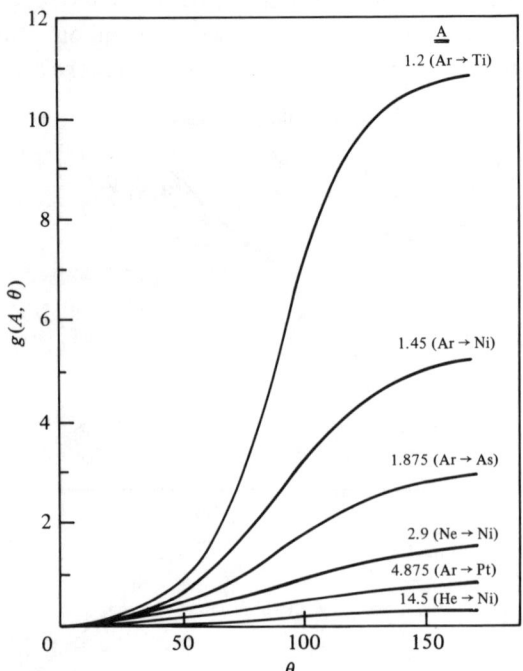

Fig. 8.7. Experimentally measured energy distributions for (a) 2 keV He$^+$—Cu and (b) 1.5 keV Ar$^+$—Au.

defined or measured. This means that the experimentally measured energy scale can be directly converted into a target atom mass scale without any reference to complex interatomic potential functions. When using a light projectile species such as He^+ as the probe, it is particularly convenient to use a scattering angle of 90° for which equation (1) reduces to determined in terms of E_0, E_1, M_1 and θ, all of which can be accurately

$$E_1 = E_0 \left[\frac{M_2 - M_1}{M_2 + M_1} \right] = E_0 \left[\frac{M_2 - 4}{M_2 + 4} \right]. \tag{4}$$

The arithmetic convenience with which the energy scale can be converted into a mass scale is not, however, the major factor which determines the choice of experimental conditions for elemental analysis. Sensitivity, mass resolution, multiple scattering effects, inelastic energy losses and neutralisation processes are all important factors in determining the optimum scattering conditions.

From the point of view of mass resolution, the simple binary scattering relationship between E_1 and M_2 makes it possible to specify the mass resolution of the system in terms of the energy resolution of the analysing system employed. For situations where $A > 1$, the mass resolution is given by

$$\frac{M_2}{\Delta M_2} = g(A, \theta) \frac{E_1}{\Delta E_1}, \tag{5}$$

Fig. 8.8. Dependence of the mass resolution factor on scattering angle and mass ratio.

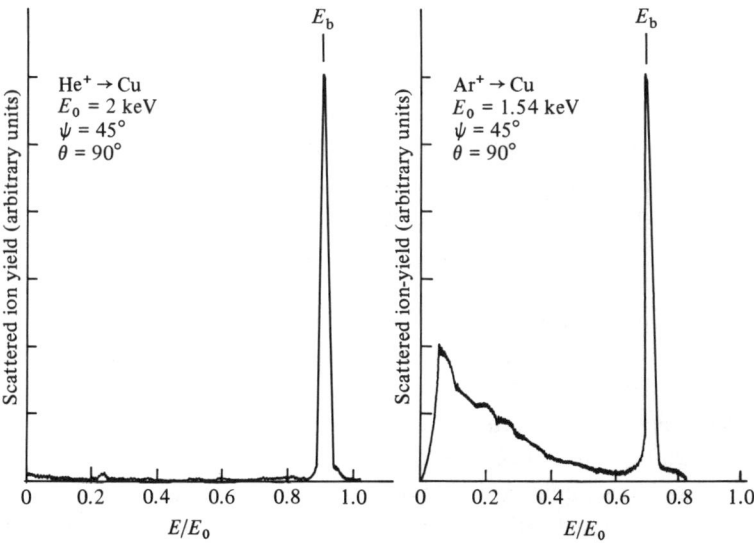

where $E_1/\Delta E_1$ is the resolution of the energy analyser and $g(A,\theta)$, obtained by differentiation of equation (1), is

$$g(A,\theta) = \frac{2A}{(1+A)} \cdot \frac{A + \sin^2\theta - \cos\theta(A^2 - \sin^2\theta)^{1/2}}{A^2 - \sin^2\theta + \cos\theta(A^2 - \sin^2\theta)^{1/2}}. \tag{6}$$

The dependence of $g(A,\theta)$ on θ and A is illustrated in Fig. 8.8 where it can be seen that the mass resolution increases for large scattering angles and small values of A, i.e. heavy projectile ions. It is interesting to note that these conditions are those for which the scattering cross-sections are reduced and, in common with other types of mass spectrometry, ISS, when used for surface elemental analysis, exhibits a trade-off between sensitivity and resolution.

In practice, in order to emphasise the single scattering contribution to the ion yield, low-mass primary ions, usually He^+ and large scattering angles must be employed.

2.3 Ion yield considerations

Having established the basic capability of ISS to provide a mass spectrum of a solid surface, it is necessary from the point of view of equipment design and applications to consider the relationship between the ion yields at different energies and the atomic concentration of the scattering centres (atoms m^{-2}). Quantification of the technique in this way is governed by factors associated with the basic collision processes and the transmission characteristics of the analysing and detection system.

The scattered ion yield is determined by the differential scattering cross-section, $d\sigma/d\Omega$, the probability of the scattered particle reaching the analyser in an ionised state, $(1 - P_n)$ where P_n is the neutralisation probability, and the extent to which, for the scattering geometry and primary energy and species employed, geometrical and inelastic shadowing effects reduce the yield for certain atoms on the surface. Geometrical shadowing occurs when the primary ion is prevented from scattering through the prescribed scattering angle from one atom by the presence of neighbouring atoms, while inelastic shadowing, which is partly responsible for the absence of an ion yield from below the surface when rare gas ions are used, is caused by the preferential neutralisation of ions which pass close to other surfaces. This form of shadowing is strongly dependent on the atomic species present on the surface and represents an important 'matrix' effect in ISS.

Considering bombardment of an area A of a surface comprised of atoms of species i and j with a probe beam of current density, J_0 and detection of the scattered ions at a scattering angle θ with an analyser–detector system having acceptance solid angle $\Delta\Omega$ and a transmission factor $T(E,\theta)$, which takes into account not only the characteristics of the analyser but also the extent to which the area 'seen' by the analyser–detector system overlaps the bombarded area, the yield of ions scattered from the surface atoms of species i is given by

$$Y_i = (J_0 A \, \Delta\Omega T_i(E, \theta)) \cdot \left(\left(\frac{d\sigma}{d\Omega} \right)_i (1 - P_n)_i (N_i - \alpha N_j) \right). \tag{7}$$

In this expression, the first term describes the instrumental dependence and the second term the matrix dependence of the yield. N_i and N_j are the surface atomic densities of species i and j and α is a shadowing coefficient which describes the extent to which atoms of species j shadow those of species i. For any given experimental system, the instrumental function can be accurately specified and quantitative interpretation of the yield measurements in terms of the number densities of the surface atoms depends on reliable evaluation of the differential scattering cross-section and the neutralisation probability. If they cannot be eliminated by the choice of an appropriate scattering geometry, an assessment of the role of shadowing effects is also required.

Although uncertainties in the precise form of the interatomic potential for ISS energies preclude totally reliable evaluation of the differential scattering cross-section, scattering simulations using potentials such as the modified Molière and Biersack–Ziegler are capable of providing reliable yield–energy data for a wide range of target–projectile combinations. In fact, uncertainties in the absolute value of the cross-sections now make a comparatively small contribution to the overall uncertainty in the quantitative interpretation of yield data. The main difficulties concern the neutralisation effects and the way in which they depend on the target matrix and scattering geometry. The latter has been found to be of prime importance in most cases since pure chemical effects appear to have only a small effect on the neutralisation coefficient such that calibration on the basis of pure element yields can be used.

The scattering geometry or trajectory dependence of the neutralisation probability stems from the complex interaction of the projectile with both the broad band delocalised electrons in the solid as well as the localised core electrons. The major features of the neutralisation probability are that rare gas ions are very effectively neutralised while alkali ions have a very high probability of escape in the ionised form (typically $\approx 90\%$) (Hagstrum, 1954, 1977). An overall reduction in the neutralisation probability as the velocity of the departing ion perpendicular to the surface increases is also observed and in certain cases, specifically low energy He^+ scattering from Pb, Ge, Bi and In, the probability has been found to exhibit an oscillatory behaviour (Erickson & Smith, 1975).

The implications of the theoretical models which have explained these observations are that the probability of neutralisation is strongly dependent on the electronic structure of the solid in the vicinity of the ion trajectory. Consequently, in ISS measurements, there will be a dependence of P_n, and hence ion yield, on the atomic arrangement of the surface atoms which cannot be uncoupled from the structural dependence of the scattering. This combination of inelastic and geometrical effects emphasises the importance of selecting single scattering geometries in ISS composition measurements and highlights the difficulty of interpreting multiple scattering spectra in a quantitative way.

2.4 Shadowing and multiple scattering effects

Shadowing and multiple scattering effects, which have already been shown to introduce problems in the interpretation of ion scattering data in composition analysis applications, form the basis of surface structure analysis using low energy ion beams. These structure related effects result from the large scattering cross-sections for low-energy projectiles and are characterised by the formation of wide shadow cones behind surface atoms and the occurrence of sequences of binary collisions with neighbouring surface atoms leading to back-scattered particle energy distributions which are dependent on the relative positions of the atoms. Although the two effects are based on the same fundamental processes, they have been exploited in two distinctly different ways to provide surface structure information.

In the shadowing technique, the basic principles of which are illustrated in Fig. 8.9, the angle of incidence of the primary beam or the target azimuth is varied, usually with a constant (and large) scattering angle, in order to determine the circumstances for which atoms are located within the shadow cones created by their neighbours. In Fig. 8.9(a), surface atom B and sub-surface atoms C and D are shadowed by atom A while in Fig. 8.9(b), atom B is visible to the incident beam. When rare gas ions are used, the subsurface atom shadowing is irrelevant since neutralisation processes dramatically reduce the ion yield for particles scattered from below the surface and the technique yields information on the relative atomic positions in only the outermost layer of the crystal. Analysis of the data is simplified if light primary ions, usually He^+, and large scattering angles are employed such that multiple scattering, blocking effects on the outward trajectory, and beam-induced damage are minimised. Under these circumstances the shadowing is essentially a straightforward geometrical effect and qualitative information about the surface atomic arrangement, for example whether or not a surface is reconstructed, can be obtained without detailed calculations involving interatomic potential and thermal vibration considerations. Quantitative analysis, however, does involve these factors since the screening function determines the effective size of the atoms as far as ion scattering is concerned and thermal vibrations inevitably affect the extent to which shadowing occurs.

The precision with which the relative positions of the surface atoms can be determined can be enhanced by increasing the scattering angle to close to 180°. Under these circumstances, the back-scattered ions undergo

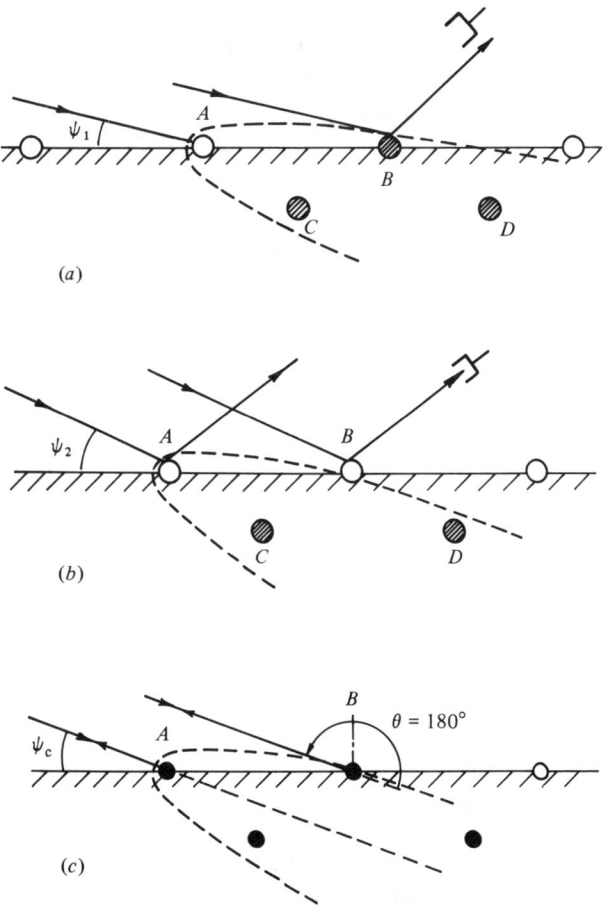

Fig. 8.9. Principles of the shadowing technique for surface structure analysis.

zero impact parameter (head-on) collisions with surface atoms such that, as illustrated in Fig. 8.9(c), the onset of shadowing of atom B by atom A occurs when the edge of the shadow cone coincides, to a good approximation, with the centre of atom B. Used in this mode, the technique is known as impact collision ISS or ICISS (Aono *et al.*, 1981) and the availability of reliable, universal expressions for the radius of the shadow cone as a function of distance behind the shadowing atom gives the technique accurate quantitative capabilities (Oen, 1983).

In contrast to the shadowing method, the multiple scattering technique is designed to provide structural information by making specific use of the multiple scattering processes which have a dominant influence on the scattering of heavy ions through small angles. The main features of multiple scattering are determined by shadowing effects which inhibit small impact parameter collisions to an extent which depends critically on the interatomic spacing. The basis of the technique is illustrated in Fig. 8.10 where it can be seen that the scattering along a single row of atoms is dominated by single and double collisions which are modified by additional small deflections on the inward and outward paths. These collision sequences are known as quasisingle, QS, and quasidouble, QD, collisions and they give rise to peaks in the energy distribution at energies above the single

Fig. 8.10. Quasisingle (QS) and quasidouble (QD) scattering processes along a row of surface atoms.

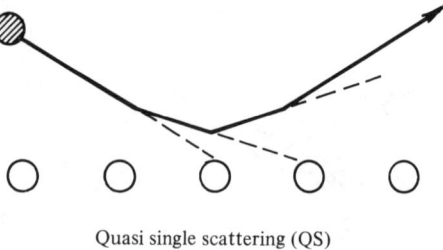

Quasi single scattering (QS)

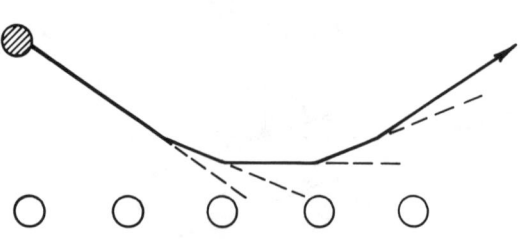

Quasi double scattering (QD)

Ion scattering spectroscopy

binary scattering energy. The energy at which the QS peak is observed and the relative intensities of the two peaks give a measure of the interatomic spacing in the row of atoms involved. Although computer simulations using both single atom row (so-called chain) and full surface models have shown that a significant fraction of the incident ions undergo more multiple collisions, efficient neutralisation of these particles when rare gases are used leads to only a small contribution to the measured ion energy distributions. As a consequence, the simpler, chain calculations are generally adequate for rare gas ions but are unable to accurately simulate alkali ion scattering spectra since these contain major contributions from more complex collision sequences involving not only the outermost surface rows but also atoms in the second layer. Even for the comparatively simple case of QS and QD scattering of rare gas ions along single rows of atoms, interpretation of the data requires the use of computer simulation if quantitative information about the surface atomic structure is to be obtained. This requirement, which contrasts strongly with the simplicity of other types of ISS measurement, results from a need to know the precise position of the trajectories with respect to the surface rather than simply the scattering angle.

A direct consequence of the occurrence of multiple collision processes is that for any specified angle of incidence of the primary beam, only a limited range of total scattering angles is available. This effect is illustrated in Fig. 8.11(a) where it can be seen that scattering through angles greater than θ_{max} is eliminated by the steering effect of atom B which prevents

Fig. 8.11. (a) Schematic illustration of the scattering angle limitations imposed by multiple scattering effects. (b) Energy versus scattering angle ($E \sim \theta$) loops for different angles of incidence.

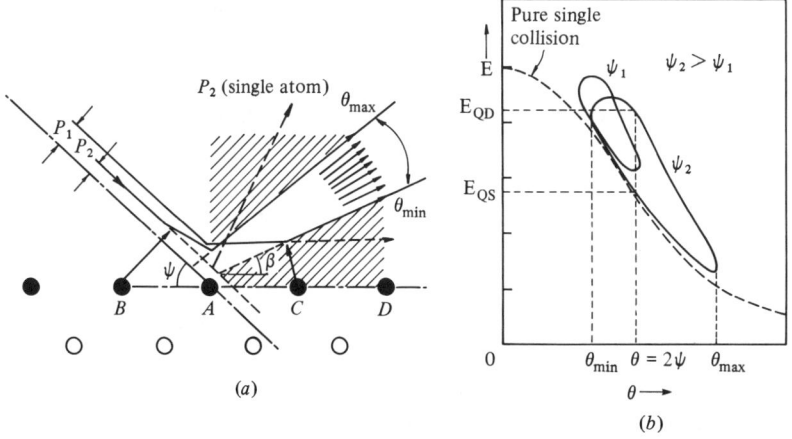

small impact parameter collisions with atom A. Similarly, scattering through angles less than θ_{min} is prevented by the additional deflections caused by interaction with atoms C and D. The scattered ion energy versus total scattering angle curves shown in Fig. 8.11(b) summarises these multiple scattering effects and shows, as already indicated, that the multiple scattered ions leave the surface with energies above that corresponding to single binary scattering through the same total angle. Computer simulation of these chain scattering phenomena indicate that the contribution of multiple scattering is enhanced for heavy particles at low energies, small interatomic spacings and small scattering angles (Parilis & Turaev, 1965). Since it is desirable to maximise multiple scattering, for structural analysis purposes heavy primary ions, low scattering angles and low primary ion energies are generally employed. In view of the fact that the range of allowed scattering angles decreases towards the specular condition as the angle of incidence is reduced (Fig. 8.11(b)), specular or close to specular scattering geometries are also required. The high probability of neutralisation for rare gas ions, however, limits the primary energies used in practice to values above about 1 keV and scattering angles to greater than about 20°. While these scattering conditions give energy spectra which are dominated by multiple scattering, the keV heavy ion beams produce significant surface damage and this technique is considerably more perturbing than the light ion shadowing method.

In multiple scattering studies, the presence of surface damage, e.g. vacant sites or steps, manifests itself in the form of scattered ion yields at scattering angles outside the allowed range and hence, by the choice of appropriate, non-specular geometries, surface damage can be studied. An indication of the rate at which damage production occurs is given by the observation of damage on an initially undamaged Ni(110) surface after a fluence of 10^{12} ions cm^{-2} of 6 keV Ar$^+$ at an angle of incidence of 15° with respect to the surface (Verheij *et al.*, 1982).

3 ISS instrumentation and modes of operation
3.1 Introduction

Consideration of the physical processes occurring during ion scattering from solid surfaces shows that the choice of all the main experimental parameters depends critically on the type of information required and on a series of compromises between sensitivity, resolution and ease of interpretation. The optimum conditions for different applications of ISS, deduced on the basis of theoretical considerations are summarised in a qualitative way, in Table 8.1. An indication is also given of the basic features of many practical systems.

Table 8.1. Modes of Operation of ISS

Application	Primary beam				Target		Detection
	Species	Energy	Fluence	Angle of incidence	Azimuthal angle		Scattering angle
Composition analysis low mass probe	He^+, Li^+	low 0.2–2 keV	not critical but $< 10^{15}$ cm^{-2}	fixed large 45–90°	fixed		fixed, large 90° – ≈ 180°
high mass probe	Ne^+, Ar^+, K^+	low 1–2 keV	critical ≃ 10^{13} ion cm^{-2}	fixed limited by θ_{max}	fixed		limited by A and θ_{max}
Structure analysis ISS	He^+, Li^+	low 0.2–2 keV	not critical $< 10^{15}$ cm^{-2}	variable 20–90°	variable ±180°		fixed 90° – ≈ 180°
ICISS	He^+, Li^+, Na^+	1–2 keV	not critical $< 10^{15}$ cm^{-2}	variable 0–90°	variable ±180°		fixed ≈ 180°
Multiple scattering	Ne^+, Ar^+, K^+	2–6 keV	low	variable 10–45°	variable ±180°		variable 20–90°

In terms of detailed instrumentation design, it is necessary to specify not only the probe beam species and energy and the overall scattering geometry, but also the probe beam intensity and energy and angular spread, the energy resolution and acceptance solid angle of the detector, the positional accuracy of the target and detector manipulations, the tolerable background gas pressure in the scattering chamber and the allowed analysing fluence. The last two features are of particular importance in view of the extreme surface specificity of the technique.

To some extent, all these instrumental factors are interdependent and the specification of practical systems in terms of them is based, in the first instance, on considerations of scattered ion yields and natural peak widths. The latter are determined by thermal vibrations, inelastic energy losses and small angle multiple deflection contributions and for the conditions of interest in most ISS applications, the energy widths of 'single' scattering peaks are of the order of 1–2% of the peak energy. For alkali ions, values two to three times wider are observed. To ensure that the natural peak widths are the main mass resolution limitation in a typical 90° scattering situation, energy analysis systems with resolutions, $E/\Delta E$, of 100–200 and acceptance angles of 0.5°–2° are required.

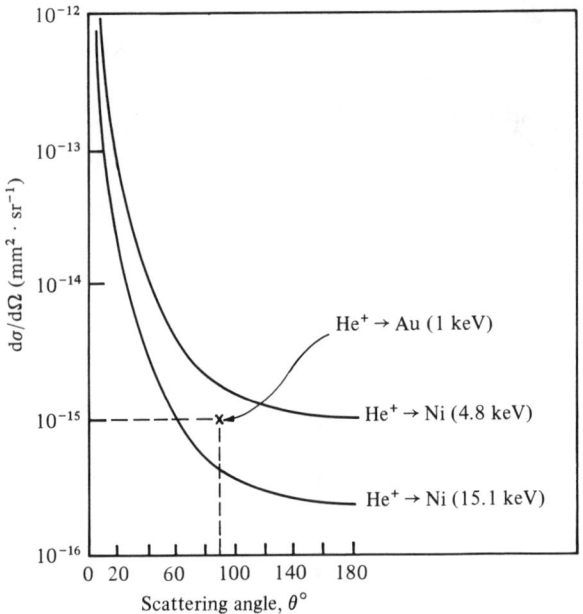

Fig. 8.12. Dependence of the differential scattering cross-section on scattering angle.

From the point of view of measured ion yields the value of the differential scattering cross-section and its dependence on energy and scattering angle plays a major role in determining the detailed design features of ISS apparatus. The curve and points shown in Fig. 8.12 illustrate the generally observed cross-section properties for the collision conditions of interest in ion scattering. The important features are that the cross-section falls dramatically (by a factor of the order of 10^2) over the scattering angle range from 20° to 90° and then only slowly at larger angles, reducing by a further factor of about two between 90° and 180° (Eckstein *et al.*, 1974; Taglauer *et al.*, 1975).

For the typical case of 1 keV He^+ scattering through 90° from a surface comprised of heavy atoms ($A \gg 1$), the differential cross-section is approximately 10^{-15} mm² sr⁻¹ and the probability of being scattered as an ion $(1 - P_n)$ is of the order of 10^{-3}. In apparatus complying with the energy resolution and acceptance angle criteria specified above, an acceptance solid angle of 10^{-3} sr and an analyser transmission efficiency of 50% are typical. For a bombarded area of 1 mm² and a surface atom number density of $\approx 10^{13}$ mm⁻², the ion yield is given by

$$Y_i = 5 \times 10^{-9} J_0 A.$$

With a primary beam current density of 10^{-8} A mm⁻², which is readily achievable with a wide range of ion sources and beam transport systems and is not too high from the point of view of bombardment fluence considerations, a yield of 10^{-17} A would be expected. This result immediately emphasises the need for electron multipliers and low-noise counting systems in ISS and indicates that detection limits of 10^{-2}–10^{-3} monolayers are to be expected using rare gas scattering. It also illustrates the lateral resolution limitations of the technique since a reduction in beam diameter at constant current density, which is required if the lateral resolution is to be increased without excessive perturbation of the bombarded area, inevitably leads to an unacceptable loss in sensitivity. This fact essentially precludes the use of ISS for high-resolution elemental mapping, even though sensitivity increases of factors up to 10^4 have been obtained using alkali ion probes and highly efficient cylindrical mirror analysers which accept particles scattered into the entire cone of solid angles specified by the geometry. These calculations combined with the theoretical considerations summarised in Table 8.1, from the basis of the design of several different types of ISS system.

3.2 *Practical systems for ISS measurements*

Surface analysis by ISS, whether composition or structure information is of interest, requires vacuum and target preparation conditions which are, in general, similar to those for other surface sensitive techniques such as low-energy electron diffraction, Auger electron spectroscopy and SIMS. To some extent the vacuum requirements are more exacting in view of the extreme surface specificity of the technique and ultra-high vacuum is essential if reproducible measurements are to be carried out. In common with other techniques, *in-situ* cleaning by ion bombardment followed by annealing forms the final, essential, stage of target preparation. In this respect the inherent availability of a suitable ion gun is particularly convenient and, in most cases, the optimum cleaning procedure is to bombard with 500 eV–1keV Ar^+ ions at an angle of incidence of 20°–30° with respect to the surface. For these conditions, ion impact desorption cross-section for contaminants such as oxygen on metal surfaces are in the range 10^{-12}–10^{-13} mm². Removal of surface damage caused by the clean-up bombardment generally requires annealing at a temperature of at least 20 °C but the need to use much higher temperatures for materials such as silicon has led to the general use of electron bombardment heating systems capable of giving target temperature in excess of 1200 °C.

ISS systems for composition and structure analysis have basically similar specifications in terms of energy resolution, acceptance solid angle and sensitivity and the two systems shown in Fig. 8.13 illustrate the main features of two approaches to system optimisation. Both systems may employ rare gas or alkali ion sources and mass analysis of the primary beam is essential. Because beam current densities required to be low, of the order of 10^{-8} A mm^{-2}, simple electron bombardment ion sources are suitable for the production of rare gas ions and thermionic (surface ionisation) sources are generally used for alkali ions. In the case of rare gas ion beams, electron bombardment sources of the Nier or Bayard–Alpert type are characterised by low energy spreads, high stability, long life and low pressure operation, typically about 5×10^{-4} mbar in the ionisation region, such that differential pumping requirements can be easily met.

The ion gun shown in Fig. 8.13(*a*) is typical of the type used in ISS and is designed to deliver mass analysed beams over the energy range 100 eV–2 keV with an energy spread of 0.5 eV or 0.5% (whichever is the greater). The beam size and angular spread are typically set by apertures to be 1 mm FWHM and < 0.5° respectively. Currents in the range 1–100 nA can be obtained routinely with this type of system and pressure rises in the target chamber during operation of less than 10^{-9} mbar are generally observed. In applications such as adsorption measurements, where beam non-uniformities, particularly edge effects, may be important, simple physical aperturing of the primary beam is not always acceptable

Ion scattering spectroscopy

and electronic aperturing using a beam scan amplitude which is compatible with the angular resolution requirements of the system must be used. This technique, which is widely used in depth profiling in SIMS, involves raster scanning the primary beam and gating the counting electronics such that signal is recorded only when the beam is in the central region of the scan (see Chapter 7).

In addition to electron bombardment sources, various types of gas discharge source have also been used in ISS systems. Duoplasmatrons,

Fig. 8.13. Practical ISS systems: (a) deflection type ESA, (b) CMA arrangement.

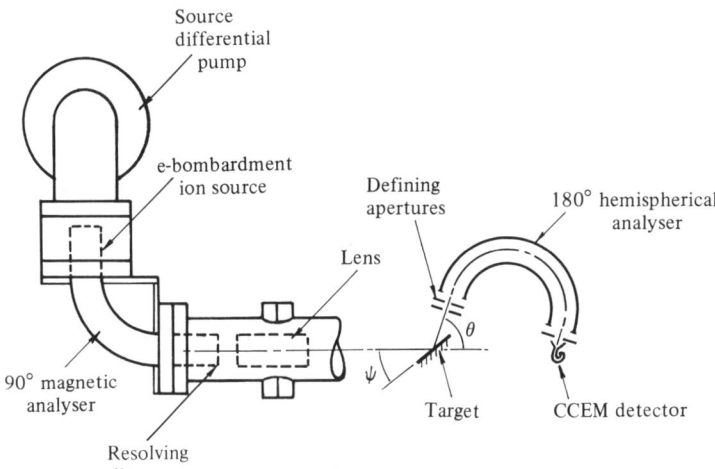

(a)

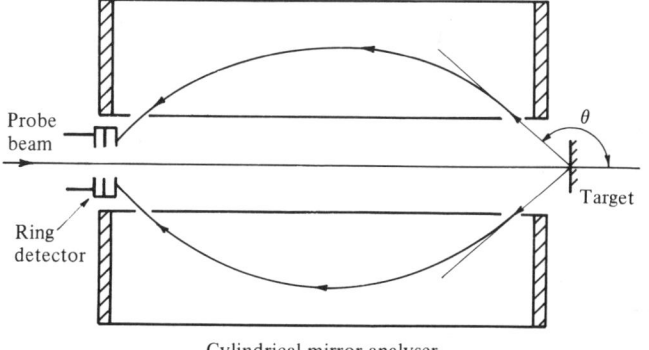

(b)

oscillating electron and radio frequency sources do not match the electron bombardment type for energy spread or stability but they do produce considerably more intense beams. While this is not directly useful in normal operation, it does enable target cleaning to be carried out more rapidly and has, in fact, been utilised in ISS systems operating in non-uhv conditions. With a sufficiently intense primary beam it is possible to maintain 'clean' surface conditions at high background gas pressures, e.g. 10^{-7} mbar, by sputtering adsorbates at a sufficiently high rate. Interpretation of data obtained in this way must take into account the possible matrix effects introduced by the adsorbed and recoil implanted impurities.

The energy analyser shown in Fig. 8.13(a) is a 180°, hemispherical sector instrument but 127° cylindrical analysers are also widely used. Typical systems use analysers with mean radii between 50 and 100 mm which are positioned with their entrance apertures up to 100 mm from the target to obtain the required angular acceptance conditions. (Taglauer et al., 1975; Andrew et al., 1972). The use of two apertures can make it possible to restrict the area 'seen' by the analyser to only the central region of the bombarded area such that edge effects are automatically eliminated.

In composition analysis applications, a fixed scattering geometry is generally adequate and specular scattering through 90° is widely used. However, it is not uncommon to observe peaks in the measured energy spectra which are difficult to identify and which may originate either from high-energy recoils or scattered primary ions. By providing facilities to vary the scattering geometry, the origin of such peaks can be identified via the relationships given in equations (1) and (3).

It is this variable geometry facility that is essential in structure analysis applications and the arrangement shown in Fig. 8.13(a) in which sophisticated, high precision target and detector manipulators are incorporated, has been widely used. The specifications of typical analysis and detection systems and the associated manipulators are indicated in Table 8.2. In contrast to the system shown in Fig. 8.13(a) the CMA arrangement in Fig. 8.13(b) is capable of accepting a major part of the complete cone of solid angles corresponding to the acceptance capability

Table 8.2. *Typical analyses and detection systems and associated manipulators*

	TARGET				ANALYSER AND DETECTOR		
Temperature	Incidence Angle	Azimuthal Angle	Tilt Angle	Angular acceptance (planar)	Energy Resolution	Horizontal Axis	Vertical Axis
RT-1200 °C	±90°	±180°	±10°	2–5°	0.4–3%	160°	±32°

of the analyser for the specified planar scattering angle. (Brongersma et al., 1978). This arrangement enables large scattering angles, and hence good mass resolution, to be combined with high sensitivity. It is typically about two orders of magnitude more sensitive than systems which use alternative analysers. The arrangements in use employ a scattering angle of $\approx 142°$ and are most suitable for operation with He^+, Li^+ or Na^+ probe ions in view of the restricted target mass range observable with heavier ions. The major drawback of this type of system is the fixed geometry but the very high sensitivity does present the prospect of using ISS for elemental mapping.

In addition to the systems described, energy analysis using time-of-flight systems in which both ionised and neutral backscattered particles can be detected (at energies above about 1 keV at which channeltron or channel plate multipliers have acceptable gain) and gas stripping cells in conjunction with normal electrostatic analysers, have also been employed in ISS systems. In both cases, major increases in sensitivity can be achieved but as for alkali ion scattering, there is some loss of surface specificity. The time-of-flight technique, which typically uses an average pulsed ion beam current of ~ 0.05 nA and a flight path of approximately 1 m, is between 10^2 and 10^3 times more sensitive than the standard hemispherical analyser method. When combined with electrostatic energy analysis, time-of--flight measurements allow mass/charge identification of secondary ions sputtered by the probe beam.

In all the experimental systems considered, the scattered ion yields are sufficiently low to require the use of electron multipliers and counting systems. Because their gain and background count rates are generally better than those obtainable with conventional multiple dynode multipliers, continuous channel electron multipliers are widely employed. These devices are small in size, stable against atmospheric exposure can operate at count rates as high as 10^6 cps with gains of the order of 10^6 and background counts as low as 0.1 cps.

An interesting recent development, which enables a range of energies to be measured simultaneously using concentric hemispherical energy analysers, is the position sensitive detector. Use of this type of detector can reduce the measuring time for a typical spectrum (or alternatively increase the signal-to-noise ratio) by a factor of about 30. The detector is essentially comprised of a microchannel plate assembly coupled in the simplest systems to a resistive strip anode. It is usually necessary to clamp the channel plates to form a doublet to achieve adequate gain. The spatial resolution is limited by the transverse energy of the electrons leaving the second channel plate and, for channel diameters of 15 μm, a value between 50 and 100 μm is obtained. Resistive strip anodes, which indicate the coordinates of the impinging electron pulse, and hence the energy of the scattered ion reaching the detector, by the ratio of the charge collected at probes located at their corners, can detect single pulses up to count rates of about 10^5 cps. Improved collectors based on arrays of charged coupled devices are currently becoming available.

3.3 Quantitative capabilities and calibration

The relationships between the measured signal and the concentrations of different elements or surface structural features are complicated in ISS by neutralisation and geometric shadowing effects. These factors are evident in the yield equation discussed earlier. Because the influence of these processes, particularly neutralisation, is dependent on the composition of the surface under investigation, there is no universally applicable relationship between scattered ion yield and the surface density of different atomic species or features. This means that the basic data obtained using ISS is qualitative and that the technique is most useful when combined with other methods.

The growing need for reliable quantitative information concerning the outermost surface layer has led, however, to major efforts to improve the quantitative capabilities of the technique either by studying the fundamental physics of the neutralisation and scattering processes or by the use of reference materials or *in situ* calibration against an alternative technique. Where multiple component materials have been used, e.g. alloys such as Cu—Pt (Brongersma *et al.*, 1976), it has generally been found that the surface composition can be determined with high accuracy by calibrating the ion yields for the alloy components against the relevant pure material yields. This indicates that pure chemical effects are small and that in many, but not necessarily all, cases, pure elements can be used for calibration of the elemental sensitivity.

Calibration of ISS against other techniques such as Auger electron spectroscopy, electron microprobe or neutron activation must take into account the different depths to which the techniques are sensitive. For submonolayer coverages of adsorbed species, however, this type of calibration is reliable and has allowed quantitative sensitivity factors to be evaluated.

In situations where information about the first few layers is of interest, the use of TOF detection techniques or alkali ion probe beams can reduce the problems caused by neutralisation processes. However, alkali ion yields have been found to depend on the work function of the surface and adsorption studies in which this parameter changes with composition still present quantitative problems.

3.4 Applications of ISS

The most interesting and useful applications of ISS are those which utilise its unique capability of providing information about the structure and composition of the outermost surface layer of solid materials. This ability, which has been particularly elegantly demonstrated by the complete suppression of He$^+$ scattering from Si after the adsorption of a single monolayer of Br[18], makes it necessary to carry out measurements in very clean vacuum systems since adsorbed overlayers obscure the real surface and the use of high probe ion fluxes to maintain low equilibrium coverage levels in high background pressure systems ($p > 10^{-8}$ mbar) causes significant recoil implantation and excessive damage build-up. This requirement has led to the use of the technique to investigate the more fundamental aspects of a wide range of technologically interesting phenomena and detailed studies of surface segregation and diffusion processes have proved to be of considerable value in areas such as catalysis and surface bonding. Adsorption, ion impact desorption and preferential sputtering studies for which ISS is particularly well suited, are becoming increasingly important in semiconductor technology as processes such as molecular beam epitaxy (MBE), reactive ion etching (RIE) and reactive ion beam etching (RIBE) assume growing roles in the fabrication of low-dimensional device structures.

In the study of surface segregation in alloys, the use of conventional rare gas ion scattering using an electrostatic analyser to obtain information about the outermost layer, combined with time-of-flight measurements which are dependent on the immediate subsurface layers, has provided valuable information on both the extent of, and the mechanisms responsible for, the segregation processes. ISS measurements at elevated temperatures on a wide range of alloys including Cu—Ni, Cu—Au, Ni—Au and Fe—Sn have been reported and comprehensively reviewed (Buck, 1982) and the results for a $Ni_{0.48}Cu_{0.52}$ (Brongersma et al., 1978) bulk alloy illustrate the nature of the information obtained. This alloy was found to exhibit marked Cu segregation at the surface as expected on the basis of a number of established segregation theories. The experimental data obtained using 3 keV Ne$^+$ ions as the probe and a high sensitivity, large scattering angle, cylindrical mirror analyser are shown in Fig. 8.14(a) and (b). The spectra in Fig. 8.14(a) illustrate the excellent mass resolution of this type of instrument since the Ni and Cu isotopes are partially resolved even with the low mass probe ion. The very low Ni yield in the higher temperature spectrum clearly demonstrates that the surface layer is composed almost entirely of Cu atoms. The plot of surface composition versus bulk composition in Fig. 8.14(b) also illustrates the strong Cu segregation for the three bulk compositions used since the straight diagonal line would be followed if no segregation occurred. Measurements such as these are playing an increasingly important role in explaining the nature of the driving forces responsible for surface segregation effects.

A similar type of measurement can be used to study the diffusion of one material through another, a process which is of interest in certain electronic devices in which multiple, thin film, layers of different materials are employed. The migration of Cr through Au and the subsequent formation of Cr_2O_5 on the gold surface has been found to interfere with the thermocompression bonding used in hybrid microcircuit technology. By depositing a 3400 Å Au layer on top of a 475 Å Cr layer on a glass substrate maintained at low temperature during deposition, and subsequently monitoring the build-up of Cr_2O_5 on the Au surface as a function of heat treatment time at different temperatures, Nelson and Holloway were able to investigate the kinetics of Cr grain boundary diffusion (Nelson & Holloway, 1976).

The surface layer sensitivity has also been utilised to study the mechanism of operation of barium activated, tungsten impregnated dispenser type cathodes (Baun, 1980). Comparison of the spectra obtained at room temperature and at a temperature at which electron emission is occurring, shows that activation of the cathode causes complete coverage of the W surface by Ba atoms. The fact that the high-temperature spectrum was obtained at a reduced probe ion energy is related to the very high ion impact desorption cross-section observed for Ba atoms on W. By

Fig. 8.14. (a) Energy spectra of 3 keV He^+ scattered from a Cu—Ni alloy surface at different temperatures. (b) Bulk versus surface composition of different Cu—Ni alloys at 520 °C.

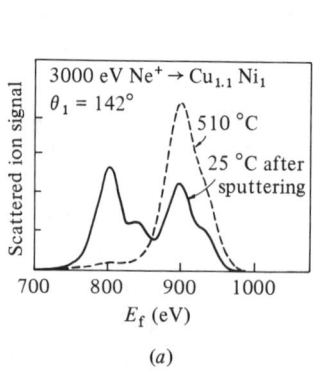
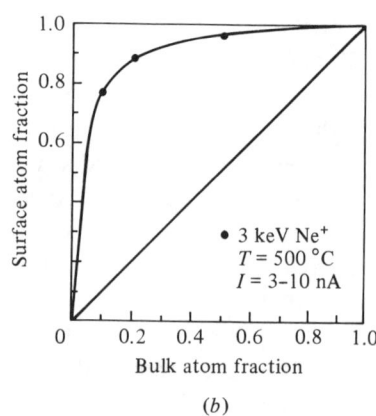

(a)

(b)

reducing the probe ion energy to 300 eV it was possible to obtain a spectrum with only a very small W peak.

Probably the major application of ISS has been into the investigation of adsorption processes, particularly on single crystal surfaces and extensive studies of the O—Ni (110) (Heiland & Taglauer, 1972), O and S—Ni (100) (Brongersma & Theeten, 1976), O—Ag (110) (Heiland et al., 1975), and O on W (110) and (100) systems as well as many others have been reported. (Niehus & Bauer, 1975; Prigge et al., 1977). In most cases these studies have been primarily concerned with the atomic arrangement of the surface atom but simple composition analysis has provided interesting and useful information about the adsorption or deposition processes. The deposition of metal films on the Si (111) surface has been studied using ISS and interesting contrasts observed for different materials (Oura et al., 1984). For Pb, 1 ML is sufficient to completely suppress the scattering yield from Si atoms while for Au, 2 ML is sufficient to uniformly cover the whole Si substrate with no reaction induced Si atom appearance on the outermost layer. In the case of Pb deposition, however, the ISS signal originating from the Si substrate persists until about 25 ML of Pd have been deposited. This observation, combined with LEED studies, provides evidence for the occurrence of significant intermixing or replacement processes between the Si and Pd atoms at room temperature.

One of the most important, and difficult to detect, atomic species is H and analysis of surfaces on which this gas is adsorbed or materials containing it cannot be carried out by conventional ion scattering, at either ISS or RBS energies, or Auger electron spectroscopy. It is possible, however, to detect the presence of H on a surface using an ISS system either by monitoring the reduction in scattering from the original surface during H exposure or by energy analysis and detection of the binary recoil particles. The former method, which has been used to investigate H adsorption on W (100) (Taglauer & Heiland, 1980), gives a qualitative indication of the presence of H, but the lack of a complete understanding of the yield reduction mechanisms, i.e. whether simple geometric shadowing or complex inelastic shadowing based on the trajectory dependence of the neutralisation processes is occurring, makes quantification difficult. The same problem applies, to some extent, when recoil monitoring is employed but the use of time-of-flight techniques to detect the binary recoils regardless of their charge state does improve the quantitative reliability. An interesting feature of binary recoil monitoring is that the ion fluence required to obtain acceptable peak intensities are typically about two orders of magnitude lower than that required to obtain a full ISS energy spectrum. This means that probe fluences of the order of 10^{11} to 10^{12} ion cm^{-2} may be used such that minimal perturbation of the surface occurs during analysis. The recoil technique has been used to detect H on the outermost surface layer of hydrogenated amorphous silicon (Oura et al., 1984) and to study the adsorption of H on a stepped Pt (997) surface (Koeleman et al., 1983). As in all studies of H on surfaces, these measurements have to be carried out in extremely clean, uhv systems since H is generally the dominant impurity in stainless steel systems at pressures below 10^{-9} mbar.

During adsorption or deposition onto single crystal surfaces, the positions taken up by the arriving species are of considerable interest and the ability of ion scattering to provide information about the relative positions of the surface atoms has been widely exploited. Since ISS measurements are concerned with the short-range structural features of the surface, they are most useful when combined with LEED studies since a knowledge of the overall crystallography of the surface enables the scattering geometries, particularly the target azimuthal directions, used on ISS to be optimised. The complementary nature of the two techniques is also reflected in the use of ISS data to give a rapid, qualitative indication of the surface atomic arrangement which is of potential value in reducing the number of trial structures in a full LEED analysis.

Structure analysis by means of simple shadowing observations using light projectile ions and large scattering angles is a particularly straight-

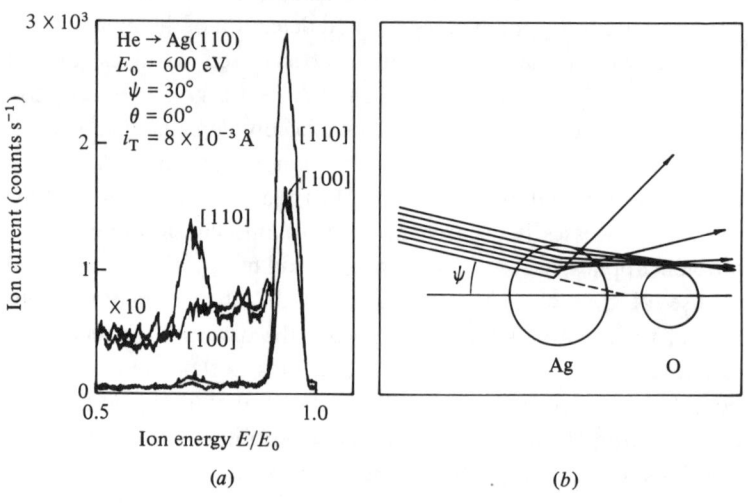

Fig. 8.15. (a) Energy spectra in different azimuthal surface directions for He$^+$ scattering from an oxygen contaminated Ag (110) surface. (b) Determination of the relative atom locations using shadowing calculations.

forward means of locating the positions of adsorbed atoms at submonolayer coverage levels. An example of this type of measurement is the study of O adsorption onto Ag(110)[25]. On this particular surface, a (2 × 1) LEED pattern is observed after the adsorption of 0.5 ML of oxygen. The ion scattering spectra for 600 eV He$^+$ taken at constant angles of incidence and scattering of 30° and 60°, respectively, for the $<\bar{1}10>$ and $<00\bar{1}>$ azimuths shown in Fig. 8.15(a) immediately suggest that the 0 atoms are adsorbed between the top $<\bar{1}10>$ rows of Ag atoms since virtually no 0 signal is observed when the scattering plane is oriented along the $<00\bar{1}>$ direction. The effectiveness of the 0 shadowing in this direction further suggests that the most probable position of the adsorbate atoms is between top layer Ag atoms rather than directly above Ag atoms in the second layer.

In contrast to the quite simple interpretation of the scattering data in terms of the qualitative surface atomic arrangement, quantitative estimation of the actual height of the O atoms relative to the Ag surface atoms is complicated. The geometrical shadowing argument requires the shadowed region behind the different atoms to be defined precisely and the effective size of the atoms to be specified. One approach to this calculation is to take the radii of the atoms in a simple two-atom surface model as the impact parameters for scattering through $\delta\Theta$ for the shadowing atom and Θ for the shadowed atom, where $\delta\Theta$ is the overall angular acceptance of the system (typically 1°–2°) and Θ is the total scattering angle. This model is valid when, as is usual, experimental conditions are chosen to maximise the simple binary scattering contribution to the measured ion yields. This situation is illustrated in figure 8.15(b) where the radii of the atoms and the trajectories are dependent on the interatomic potential. For the O—Ag (110) surface, using the Thomas–Fermi Molière potential, the most probable position for the O atoms is at the same level as the Ag atoms.

Taking this into account uncertainties in the potential parameters and the effects of thermal vibrations, the accuracy with which the height of adsorbate atoms with respect to the surface can be determined has been estimated to be ± 0.2 Å. However, the assumption that only geometrical shadowing occurs is not necessarily valid and trajectory-dependent neutralisation effects as well as changes in ion yield due to the changing surface composition may also be important. This is graphically illustrated for the case of O adsorption onto a Mo (001) surface when the O exposure dependence of He$^+$ and K$^+$ scattering yields for the $\langle 110 \rangle$ azimuth are compared (Overbury et al., 1984). The monatomic reduction in He$^+$ scattering from the Mo atoms as exposure increases cannot be explained in terms of simple geometrical shadowing since no comparable decrease in the K$^+$

scattering intensity is observed. Since, on the basis of scattering cross-section arguments the K^+ should actually decrease more rapidly due to shadowing, the decrease in He^+ scattering from Mo atoms must be caused by the increase in neutralisation probably due to the additional electron density contributed by the adsorbed O. These observations illustrate the usefulness of combining rare gas and alkali ion scattering and this procedure represents a growing trend in ISS analysis.

Although neutralisation problems remain, uncertainties in atomic radii and shadow cone dimensions can be minimised using the ICISS technique and quantitative measurements of surface atomic geometries to a precision of better than $\sim \pm 0.1$ Å can, in principle, be achieved. An additional feature of ICISS data is that the often complex azimuthal dependence of the scattered ion yields reflect the two-dimensional spatial distribution of the surface electrons. The technique has been applied to the study of a number of different surfaces such as Ti-C (111) (Aono et al., 1981), Si (001) (Aono et al., 1983), Si (111) (Aono et al., 1983), and Si (111) $-(\sqrt{3} \times \sqrt{3})$ R30° Ag (Aono, 1985) and both rare gas (He^+) and alkali (Li^+) ions have been employed. In addition to providing valuable information concerning the structure of such surfaces, measurements have made a useful contribution to an understanding of the charge exchange processes occurring during the interaction of the projectile with the surface. This understanding, combined with the basic experimental and analytical simplicity of the

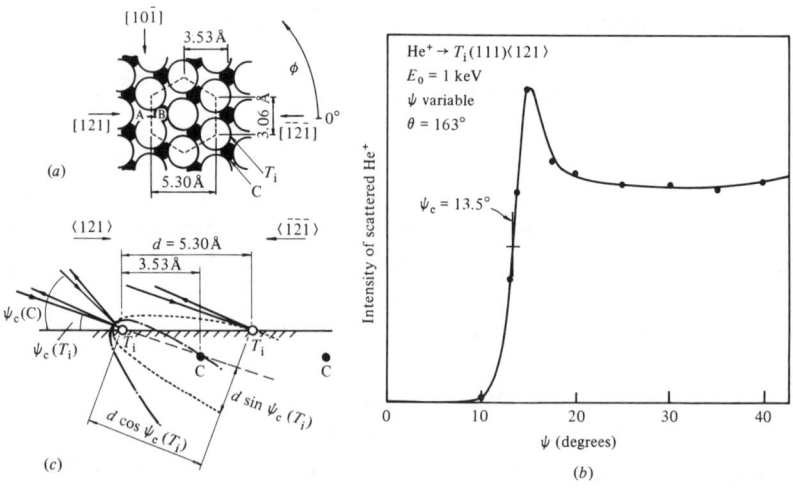

Fig. 8.16. ICISS studies of TiC: (a) surface atomic arrangement of TiC (111), (b) scattering intensity versus angle of incidence, (c) atom position calculations based on shadow cone dimensions.

technique, makes it one of the most powerful methods of investigating surface atomic arrangements.

An example of the application of this technique is the measurement of the Ti—C distance perpendicular to the TiC (111) surface. This compound is of current interest as a carbide catalyst and as a potential material for the first wall of fusion reactors. Sputter cleaned TiC (111) exhibits a sharp (1 × 1) LEED pattern and has a surface layer of Ti atoms as indicated in Fig. 8.16(a). By measuring the scattered He^+ intensity, I, along a specific surface azimuth, e.g. $\langle 121 \rangle$, as a function of angle of incidence, a critical angle, ψ_c, at which the intensity drops to zero is observed. This is illustrated in Fig. 8.16(b) where it can be seen that the I versus ψ curve is characterised by a sharp fall at ψ_c and a peak at angles slightly greater than ψ_c. This peak reflects the focussing or ion flux concentration at the edge of the shadow cone. For $\psi < \psi_c$ each Ti atom lies in the shadow cone of its neighbour. As shown in Fig. 8.16(c), for a surface azimuth with known interatomic spacing, d, evaluation of ψ_c enables the shadow cone radius at $d \cos \psi_c$ from the centre of the atom to be evaluated since it is simply given by $d \sin \psi_c$. Alternatively, if d is unknown and the shadow cone radius as a function of distance behind the atom is evaluated using, for example, the calculations published by Oen (1983), d can be determined with a precision dependent on the accuracy of the cone dimension calculations and the measurement of ψ_c. For this particular surface, the interatomic spacings in the different azimuthal directions are known and measurements of ψ_c for different values of d allows the shape of the cone to be determined experimentally. Using this knowledge and measuring ψ_c for scattering from the immediate subsurface C atoms in the $\langle 121 \rangle$ and $\langle \bar{1}2\bar{1} \rangle$ azimuths, the position of the C atoms below the surface can be determined. These measurements have yielded a value of 0.87 ± 0.08 Å for the Ti—C spacing perpendicular to the surface. This value is 30% smaller than the corresponding bulk value.

The straightforward nature of these measurements and their interpretation has made this technique particularly attractive for the study of complex surface structures and it is anticipated that a combination of rare gas and alkali ion scattering measurements will enable unambiguous determination of such structures without the sophisticated computer simulation analysis which is characteristic of LEED and low angle, multiple scattering ISS.

Although quantitative analysis of low angle, multiple scattering ISS measurements requires extensive computer simulation, qualitative information concerning surface atomic spacings can be obtained relatively easily by observation of the QS and QD peaks for different surface crystallographic directions. This is illustrated in Fig. 8.17(a) for 6 keV Ar^+ Ni (110) surface (Verheij et al., 1979). The general trends of reductions in QS peak energy and QD to QS peak intensity ratio as the atomic spacing increases are immediately evident.

The fact that multiple scattering spectra inherently contain information about both the identity of the chain atoms and their separation makes this type of measurement particularly useful for adsorption studies where changes in composition and structure may occur simultaneously. The adsorption of oxygen onto the Ni (110) surface illustrates the usefulness of this technique in that an increase in the atomic spacing in the $\langle\bar{1}10\rangle$ surface direction at 0 coverages greater than about 0.05 ML is readily detected by the appearance of a QS peak in the energy spectrum at an energy which, when analysed in detail, corresponds to a spacing which is double that of the clean surface. By using Ne^+ ions for which scattering from 0 to 30° is energetically possible, the location of the O atoms with respect to the Ni atoms on the reconstructing surface can be determined by observing the shadowing effects in the different azimuthal directions. This means that the energy spectra yield information on the Ni atoms spacings through the QS and QD peaks and the O atom positions simultaneously. The different behaviour of the O peak in the $\langle\bar{1}10\rangle, \langle\bar{1}1\bar{1}\rangle$ and $\langle00\bar{1}\rangle$ directions is shown in Fig. 8.17(b) and the onset of O shadowing

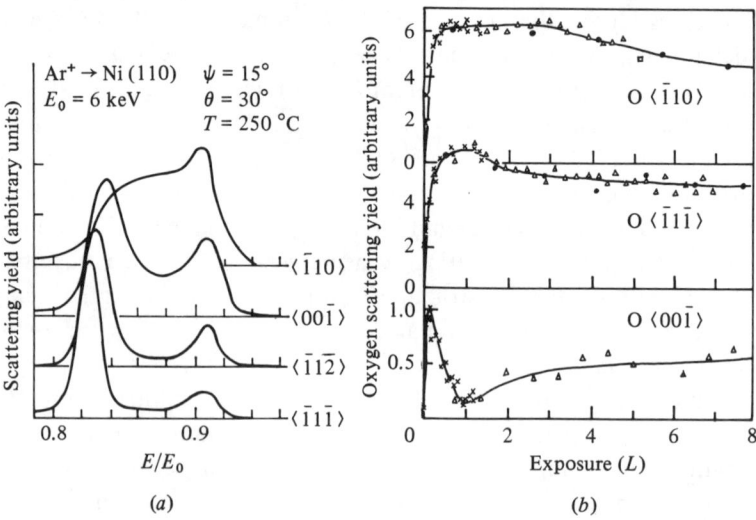

Fig. 8.17. (a) Azimuthal dependence of the multiple scattering spectra for 6 keV Ar^+ on Ni (110). (b) Azimuthal dependence of the oxygen scattering yield versus exposure curves for the O—Ni (110) system.

in the $\langle 00\bar{1}\rangle$ direction coincides with the growth of the QS peak associated with the change in Ni atom spacing in the $\langle \bar{1}10\rangle$ direction. This type of study, despite the difficulties in overall interpretation due to the need for computer simulation, represents a powerful method of observing the kinetics of adsorption and surface reconstruction processes (van den Berg et al., 1980).

4 Summary and future developments

Since the analytical potential of ISS was first demonstrated by Smith (1966), the capabilities of the technique have been extended by the exploitation of many different features of the scattering process. Although quantification difficulties still exist in that a universally applicable relationship between scattered ion yield and surface elemental or structural feature concentration is not available, calibration against standard materials or alternative techniques does allow quantitative information about specific surfaces to be obtained. Progress in this area and in the general development of the technique has been extensively reviewed (van den Berg & Armour, 1981; Brongersma & Buck, 1978).

The improved understanding of the neutralisation processes and the increasing use of TOF detectors and alkali probe ions have reduced quantitative uncertainties by removing the dominant neutralisation effects and have allowed both surface and immediate subsurface layers to be investigated. In addition, the increased yields resulting from the lack of neutralisation have enabled extremely low probe fluences to be employed. This feature is important when easily damaged surfaces such as semiconductors are to be studied and it is anticipated that ISS used in these modes will find increasing application in this area.

On the basis of present trends, the steady improvement in the theoretical understanding of the scattering processes and the availability of more sensitive multichannel detection systems will make ISS one of the most powerful surface analytical techniques available. Used in conjunction with LEED, which gives basic crystallographic information, and photo-electron spectroscopy, which provides data on the chemical binding of the surface atoms, ISS will undoubtedly contribute extensively to major improvements in our understanding of surfaces and surface processes.

References

Andrew, R., Riley, F.M., Armour, D.G. & Carter, G. (1972) *Vacuum*, **22**, 457.
Aono, M., Oshima, C., Zaima, S., Otani, S. & Ishizawa, Y. (1981) *Jap. J. Appl. Phys.*, **20**, L829.
Aono, M., Hou, Y., Oshima, C. & Ishizawa, Y. (1982) *Phys. Rev. Lett.*, **49**, 567.

Aono, M., Sauda, R., Oshima, C. & Ishizawa, Y. (1983). *Phys. Rev. Lett.*, **51**, 801.
Aono, M. (1985). *Proc. 9th Symp on Ion Sources and Ion Assisted Techniques, Tokyo (1985)*, p. 531.
Baun, W.L. (1980). *Applications of Surface Science*, **4**, 374.
Biersack, J.P. & Ziegler, J.F. (1982). *Nucl. Instr. and Meth.*, **194**, 93.
Bohr, N. (1948). *Mat Fys Medd Dan Vid Selsk*, **18**, 8.
Brongersma, H.H. & Mul, P.M. (1973). *Surf. Sci.*, **35**, 355.
Brongersma, H.H. & Theeten, J.B. (1976). *Surf. Sci.*, **54**, 519.
Brongersma, H.H., Buck, T.M. (1976). *Nucl. Instr. and Meth.*, **132**, 559.
Brongersma, H.H., Sparnaay, M.J. & Buck, T.M. (1978). *Surf. Sci.*, **71**, 657.
Brongersma, H.H., Hazewindus, N., van Nieuwland, J.M., Otten, A. & Smets, A.J. (1978). *Rev. Sci. Instr.*, **49**, 707.
Brongersma, H.H. & Buck, T.M. (1978). *Nucl. Instr. and Meth.*, **149**, 569.
Buck, T.M. (1982). *Chemistry and Physics of Solid Surfaces IV*, R Vanselow and R Howe (eds.), Springer-Verlag, Berlin, Heidelberg, New York, p. 435.
Cruz, S.A., Alonso, E.V., Walker, R.P., Martin, D.J. & Armour, D.G. (1982). *Nucl. Instr. and Meth.*, **194**, 659.
Eckstein, W., Schaffler, H.G. & Verbeek, H. (1974). *Report IPP9/16*, Garching.
Erickson, R.L. & Smith, D.P. (1975). *Phys. Rev. Lett.*, **34**, 297.
Hagstrum, H.D. (1954). *Phys. Rev.*, **96**, 336;
Hagstrum, H.D. in *Inelastic Ion-Surface Collisions*, N.H. Tolk, J.C. Tully, W. Heiland & C.W. White (eds.), Academic Press, New York (1977).
Heiland, W., Iberl, F., Taglauer, E. & Menzel, T. (1975). *Surf. Sci.*, **53**, 383.
Heiland, W. & Taglauer, E. (1972). *J. Vac. Sci. Technol.*, **9**, 620.
Koeleman, B.J.J., de Zwart, S.T., Boers, A.L., Poelsema, B. & Verheij, L.K. (1983). *Nucl. Instr. and Meth., in Phys. Res.*, **218**, 225.
Lindhard, J., Scharff, M. & Schiott, H.E. (1963). *Kgl Dan Vid Selsk Mat Fys Medd*, **33**, 14.
Molière, G. (1947). *Z. Naturforsch*, **2a**, 133.
Nelson, G.C. & Holloway, P.H. (1976). *Surface Analysis Techniques for Metallurgical Applications, ASTM STP*, **596**, 68.
Niehus, H. & Bauer, E. (1975). *Surf. Sci.*, **47**, 222.
Oen, O.S. (1983). *Surf. Sci.*, **131**, L407.
Oura, K., Yabuuchi, Y., Shoji, E., Hanawa, T. & Okada, S. (1983). *Nucl. Instr. and Meth.*, **218**, 253.
Oura, K., Shoji, F. & Hanawa, T. (1984). *Jap. J. Appl. Phys.*, **23**, L694.
Overbury, S.H., Dekoven, B.M. & Stair, P.C. (1984). *Proc. 10th Int. Conf. on Atomic Collisions in Solids, Bad Iburg, Germany. Nucl. Instr. and Meth.*, **82**, 384.
Parilis, E. & Turaev, N. Yu. (1965). *Sov. Phys. Sokl.*, **10**, 212.
Poelsema, B., Verheij, L.K. & Boers, A.L. (1974). *Surf. Sci.*, **64**, 554.
Prigge, S., Niehus, N. & Bauer, E. (1977). *Surf. Sci.*, **65**, 141.
Smith, D.P. (1966). *Bull. Am. Phys. Soc.*, **11**, 770.
Taglauer, E., Melchior, W., Schuster, F. & Heiland, W. (1975). *J. Phys. E, Scientific Instruments*, **8**, 772.
Taglauer, E. & Heiland, W. (1980). *Appl. Surf. Analysis. ASTM STP*, **699**, 111.
van den Berg, J.A., Verheij, L.K. & Armour, D.G. (1980). *Surf. Sci.*, **91**, 218.
van den Berg, J.A. & Armour, D.G. (1981). *Vacuum*, **31**, 259.
Verheij, L.K., van den Berg, J.A. & Armour, D.G. (1979). *Surf. Sci.*, **84**, 408.
Verheij, L.K., van den Berg, J.A. & Armour, D.G. (1982). *Surf. Sci.*, **122**, 216.

9

Rutherford back-scattering spectrometry

W.A. GRANT

1 Introduction

1.1 General concepts

Rutherford back-scattering is a widely used surface analysis technique in many branches of materials science. The basic concepts of Rutherford backscattering spectrometry (RBS) are simple to follow. A beam of energetic ions is directed at the solid under investigation. The ions collide elastically with lattice atoms within the sample and are scattered back into a suitable detector which counts the number of scattered particles and measures their energy. The information contained in the scattered particles can be interpreted to give, *inter alia*, data on the composition of the sample, the distribution of components within a sample and the sample thickness.

Rutherford back-scattering has long been used for surface analysis by nuclear physicists, since it provides a quick and simple method of examining target purity and thickness in nuclear physics experiments. In the last 20 years, however, the growing need for surface analysis in many areas of materials science has prompted rapid expansion in the development and use of RBS. Many accelerators in the 1–3 MeV energy regime, that were previously used for nuclear physics, are now used for solid state analysis by RBS. The electronic industries have provided the largest driving force for the development of RBS techniques, with their requirements for detailed information on shallow doped layers in semiconductors and on metal-semiconductor contacts. But RBS techniques are now extensively employed in a number of fields (see, for example, J.R. Bird & G.J. Clark, 1981).

A schematic RBS spectrum is given in Fig. 9.1. The analysing beam of energy, E_0, strikes a sample containing two elements of masses, A, and, B. Particles back-scattered from target atoms at the surface have an energy,

E, after collision of either E_A or E_B depending on the type of target atom from which they scatter. The energy after collision is related to the energy E_0 collision by a constant K which depends on the mass of the struck atom. Thus $E_A = K_A E_0$ and $E_B = K_B E_0$. The constants K_A and K_B can be calculated for a given projectile/target combination and for a given scattering geometry (see Section 2.2). Consequently the signals appearing on the energy scale of Fig. 9.1 at positions E_A and E_B can be identified as coming from masses A and B, respectively, and the energy scale is thus converted into a *mass scale*.

Only a small fraction of the beam will be scattered by surface atoms. Further scattering occurs from atoms located at successively deeper positions below the surface. In each case the collision can be treated as a binary collision and the energy *immediately after* such a collision is related to the energy *immediately before* the collision via the kinematic constant, K. In moving through a sample to reach the depth at which it elastically scatters, an ion loses energy *inelastically*. Passing back out through the sample the ion continues to lose energy inelastically. The rate

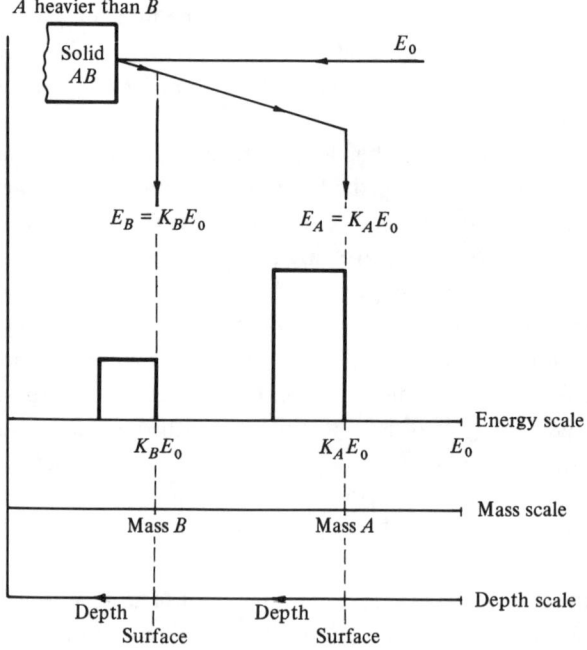

Fig. 9.1. Rutherford backscattering from a solid composed of elements of mass A and B.

of inelastic energy loss is known for many ion–target combinations. As shown in Fig. 9.1, scattering from A atoms which are below the surface will give signals at energies less than $K_A E_0$ to an extent that increases the deeper the scattering is below the surface. The energy scale of Fig. 9.1 can therefore also be converted to a depth scale. A single RBS spectrum consequently provides both mass identification and the depth distribution of these masses.

The number or yield of particles scattered depends principally on the Rutherford collision cross-section (see Section 2.3). The ion experiences simple Coulombic repulsion which is accurately and quantitatively described by the Rutherford cross-section. The yield of scattered particles in the RBS spectrum of Fig. 9.1 can thus be converted to give *quantitative* data on target composition.

By combining RBS analysis with channelling, further information on the state of surfaces and surface layers becomes possible (see Section 5). A projectile can be directed to enter a crystal lattice in a low index direction. Many of the ions entering in such a direction pass down the open channels of the crystal and do not come sufficiently close to lattice atoms to elastically collide and be back-scattered. The RBS yield for single crystals therefore depends on the angle between the analysing beam and the lattice of the sample. The combination of RBS and channelling can provide data on the degree of disorder within a single crystal and the lattice position of impurities within the lattice. In addition, for experiments conducted under UHV conditions, measurements can be made of the relaxation in position of crystal surface atoms and the reconstruction that can occur on certain crystal faces. These latter effects are discussed in Section 5.5.

The overall aim of the chapter is to present the basic concepts of Rutherford back-scattering in a manner that will allow the reader to appreciate the power and limitations of the technique. In the limited space available it is inevitable that the material presented is highly selective. However, interested readers can refer to the excellent book by Chu, Mayer & Nicolet (1978) which not only provides comprehensive coverage of RBS but has been the principal information source for the present author.

1.2   *Experimental instrumentation*

A schematic of the equipment used in RBS is shown in Fig. 9.2. A high-energy beam, typically 2 MeV $^4\text{He}^+$, is extracted from a suitable accelerator and passed through a magnetic analyser. The beam is then collimated and directed onto the target. Ions which are back-scattered into a particular solid angle are detected by a solid state detector. The number and energy of the particles striking the detector is stored electronically.

The commonest form of accelerator used for RBS is based on the electrostatic van der Graaff generator. The high voltage (typically several MeV) at the accelerator terminal is produced by means of a rapidly moving insulating belt that conveys charge between ground potential and the terminal. Ions are extracted from an RF ion source and accelerated down an insulating column that is maintained at $\sim 10^{-6}$ torr. The column and the high voltage terminal are surrounded by a tank which is pressurised to provide high-voltage insulation. The discharge current between a series of sharp points (corona points) and the high-voltage terminal is monitored and used as a controllable load to stabilise the terminal voltage.

RBS requires a highly energy-stablised (± 2 keV) ion beam. Fluctuations in the terminal voltage can be corrected by feedback systems. A common method is to sample the beam, after magnetic analysis, using high- and low-energy slits. A change in the terminal voltage causes the beam to shift its spatial position on leaving the magnet with a corresponding increase in the current to one of the two slits. The slit signals are used to control the terminal voltage via the corona discharge current. The energy of the accelerator can be calibrated using nuclear resonance reactions. Reactions such as ^{27}Al$(p, \gamma)^{28}$Si occur at precisely known beam energies and provide absolute energy calibration.

The ^{4}He$^+$ ion beam from an ion source is often accompanied by ^{16}O$^+$ ions. If these latter ions lose a further electron after acceleration down the column, but before magnetic analysis, they will not be separated from the ^{4}He$^+$ beam. This contamination interferes with RBS analysis by giving

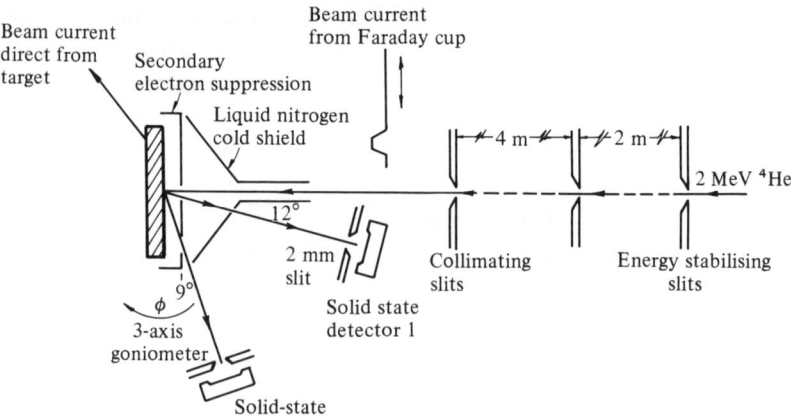

Fig. 9.2. Experimental arrangement for RBS.

spurious signals at the lower energies in spectra and by causing radiation damage in single crystals in channelling experiments. The ion current that strikes the target is measured by current integration. For accurate measurements care must be taken to suppress secondary electrons at slits and at the target. Alternatively, the current can be measured using a Faraday cup. In many RBS measurements, relative rather than absolute measurement of ion fluence is sufficient.

The scattered beam is energy analysed and counted. The highest energy resolution is provided by magnetic analysis but data accumulation is slow since only a single energy is measured for a specific magnetic field strength. Solid state detectors are more commonly used since they detect particles of all energies simultaneously. In a typical experimental arrangement a silicon surface barrier detector (area $\sim 25\,\text{mm}^2$) is placed at a scattering angle of $\sim 170°$ and a distance $\sim 15\,\text{cm}$ from the target. The angular spread of the beam at the detector can be reduced by placing slits in front of the detector.

The initial stage of signal amplification is a charge-sensitive preamplifier which provides a minimum of pulse shaping but preserves a maximum signal-to-noise ratio. The subsequent main amplifier creates a suitable pulse shape to optimise resolution and count-rate capability. The shaping produces individual pulses whose amplitudes contain the required (energy) information. The shaping helps to eliminate pulse-pile-up which occurs if the spacing between pulses is small such that they overlap and cause erroneous amplitude measurements. Data acquisition, storage and display is usually provided by a multichannel analyser (MCA) which is essentially a dedicated microcomputer. The commonest mode of operation of the MCA is pulse-height-analysis (PHA). The desired RBS spectrum is accumulated by measuring the amplitude of each signal (caused by a scattered particle striking the detector) and storing it at an appropriate channel-address (or channel number) that is proportional to the pulse height. RBA spectra commonly require 512 channels and so a memory of 4096 channels permits storage of eight spectra at one time. An alternative mode of operation for the MCA is the multi-channel-scaling mode (MCS). An individual channel counts all the incoming data for a predetermined time interval. During subsequent intervals counting is transferred to the next channel. This mode of operation is useful in channelling experiments in which the RBS yield is measured repeatedly at various angles of the analysing beam relative to a crystal channelling direction.

The pressure in the target chamber is typically 10^{-6} torr although surface studies of the type described in section 9.5 require UHV conditions, i.e. $< 10^{-9}$ torr. It is possible to produce a UHV environment by placing a cryo-shield around the target which is otherwise in a low ($\sim 10^{-6}$ torr) vacuum system. Low pressures help to reduce contamination on the target surface from hydrocarbons which are dissociated by the ^{4}He beam to form an adherent carbon layer.

For standard RBS measurements simple target holders which contain a number of samples that can be rotated into the analysing beam are sufficient. Channelling experiments require crystals to be oriented with respect to the beam and this involves use of a goniometer. A three-axis goniometer, providing tilt in two planes combined with 360° rotational movement, is commonly used. Stepping motors drive each of the three-axes in small, $\sim 0.01°$, angular steps. In channelling experiments the RBS yield is monitored as a function of sample orientation relative to the beam. Repeated measurements at different angles are used to orient the crystal to a position such that the beam is directed in a low index crystallographic direction.

2 Physical concepts
2.1 Introduction

In Rutherford back-scattering spectrometry we are concerned with projectiles that move through a target, losing energy along their path, and are scattered by collision with a target atom. The interaction between the projectile and target atom can be described as an elastic collision between two isolated particles. The energy of the projectile after the collision can be related to its energy before the collision by means of a *kinematic factor K*. The likelihood that scattering will occur, i.e. the probability of scattering, depends on the *scattering cross-section*. As the projectile passes through the scattering medium it suffers an average energy loss dE/dx and this leads to the concept of the *stopping cross-section*. Finally, since there are statistical fluctuations in the energy loss of a projectile as it penetrates a solid, particles entering a solid with a given energy will not have identical energies after travelling the same distance. This phenomenon is called *energy straggling*.

The above four basic physical concepts form the basis of RBS. The kinematic factor K leads to the ability for *mass analysis*. The scattering cross-section provides RBS with a *quantitative* capability. The stopping cross-section results in the capability for *depth analysis* and energy straggling sets limits on mass and depth resolution.

2.2 Kinematic factor K

The elastic collision between two particles can be solved by applying the principles of energy and momentum conservation (Goldstein,

1959). The collision is depicted in Fig. 9.3. The energy, E, of the projectile, M_1, after collision with the target atom, M_2, is related to its energy, E_0, before the collision by the kinematic factor, K, defined by

$$E = KE_0 \tag{1}$$

and

$$K = \left\{ \frac{M_1 \cos\theta + (M_2^2 - M_1^2 \sin^2\theta)^{1/2}}{M_1 + M_2} \right\}. \tag{2}$$

The kinematic factor, K, depends on the masses, M_1 and M_2, and the scattering angle, θ. In RBS a projectile of known mass, M_1, and known energy, E_0 is directed at the target of unknown mass, M_2. By measuring the energy of particles scattered at an angle, θ, the unknown M_2 can be found. If a target contains two masses that differ by a small amount, ΔM_2, the difference, ΔE_1, in the energy, E_1, of the projectile after collision is given by (Chu, Mayer & Nicolet, 1978),

$$\Delta E_1 = E_0 \left\{ \frac{dK}{dM_2} \right\} \Delta M_2. \tag{3}$$

The largest change in K for a given change, ΔM_2, occurs when $\theta = 180°$ and so large scattering angles are preferred experimentally in order to maximise the mass discrimination.

If the energy resolution of the analysing system is δE_1 then the mass resolution δM_2, for a scattering angle $\theta = 180°$ is given by (Chu, Mayer & Nicolet, 1978)

$$\delta M_2 = \frac{\delta E_1}{E_0} \frac{(M_2 + M_1)^3}{4M_1(M_2 - M_1)}. \tag{4}$$

The mass resolution for a 2.0 MeV and a 1.0 MeV ^{4}He beam and with a system resolution $\delta E_1 = 15$ keV is shown in Fig. 9.4. Isotopic resolution

Fig. 9.3. The trajectories and energies of particles scattered at the surface or at a depth x perpendicularly below the surface.

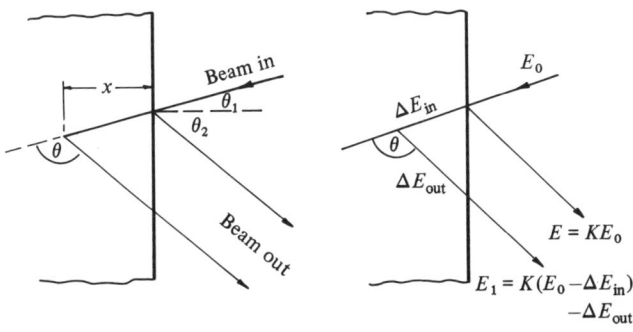

is possible up to about chlorine but the resolution degrades with increasing target mass. Thus heavy elements (e.g. Au) cannot be distinguished unless they differ by ~20 amu. Mass resolution in RBS can be improved by increasing the beam energy, improving the system resolution, and using a heavier analysing beam. For ^{4}He beam energies above 2 MeV care must be taken that the beam does not induce a nuclear reaction. Ions heavier than ^{4}He do not offer substantial advantages, partly because of the poorer energy resolution of solid state detectors with heavy ions and partly because heavy ions cause radiation damage at a faster rate (compared to ^{4}He) to both detectors and targets.

2.3 Scattering cross-section $d\sigma/d\Omega$

In Fig. 9.5 particles are scattered at an angle, θ, into a solid state detector that subtends a solid angle that is small (typically less than 10^{-2} sr). The number of counts, Y, registered by the detector is given by

$$Y = Q \cdot Nt \cdot \frac{d\sigma}{d\Omega} \Omega, \tag{5}$$

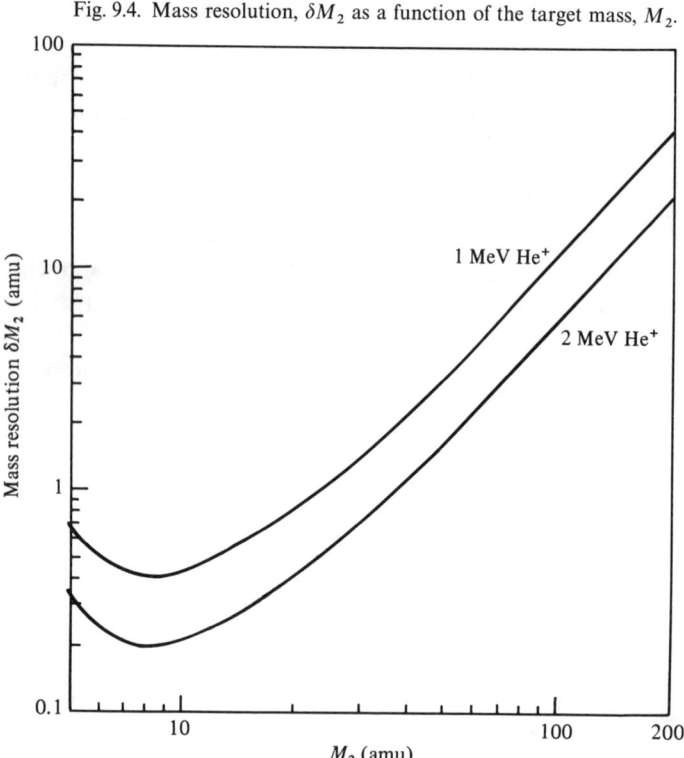

Fig. 9.4. Mass resolution, δM_2 as a function of the target mass, M_2.

where Q is the number of particles that strike the target, N is the volume density of target atoms, t is the target thickness and $d\sigma/d\Omega$ is the average differential scattering cross-section for scattering into a solid angle at a scattering angle θ. For small values of Ω the average $\langle d\sigma/d\Omega \rangle$ can be approximated by the differential scattering cross-section $d\sigma/d\Omega$ and this is usually referred to simply as the scattering cross-section and given the symbol σ. Thus the yield can be written

$$Y = Q \cdot Nt \cdot \sigma \cdot \Omega. \qquad (6)$$

If the number of particles striking the target and detector are counted and are both known then the number of atoms per unit area of the target,

Fig. 9.5. Scattering into a solid angle, $d\Omega$, at a scattering angle θ.

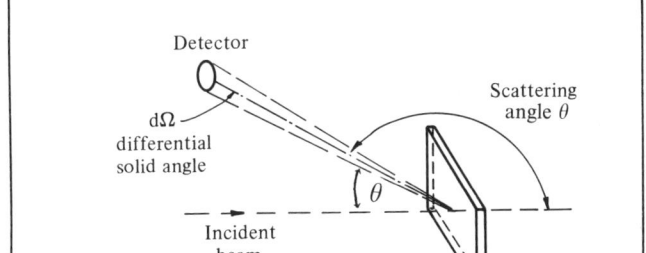

Fig. 9.6. Stopping of ^{4}He in Si as a function of the He energy.

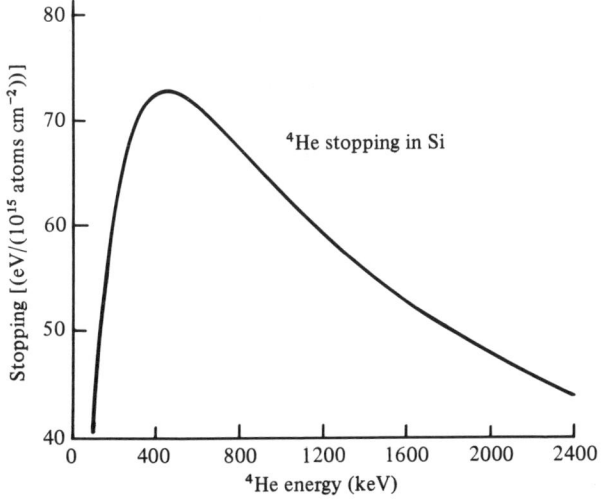

i.e. Nt can be determined. The differential cross-section for an elastic collision between two atoms in which the force of interaction is the Coulomb repulsion of the two nucleii is given by Rutherford's formula (Goldstein, 1959),

$$\frac{d\sigma}{d\Omega} = \left\{\frac{Z_1 Z_2 e^2}{4E}\right\} \frac{4}{\sin^4 \theta} \cdot \left\{\frac{[1-((M_1/M_2)\sin\theta)^2]^{1/2} + \cos\theta}{[1-((M_1/M_2)\sin\theta)^2]^{1/2}}\right\}^2 \quad (7)$$

where E is the energy of the projectile immediately before scattering and Z_1 and Z_2 are the atomic numbers of the projectile and target atom respectively. For $M_1 < M_2$ this expression for the cross-section can be rewritten:

$$\frac{d\sigma}{d\Omega} \sim \left\{\frac{Z_1 Z_2 e^2}{4E}\right\} \frac{1}{\sin^4 \theta/2}. \quad (8)$$

The magnitude of the Rutherford cross-section is dominated by the term $(Z_1 Z_2 e^2/4E)$. A beam of 1 MeV ^{4}He scattering from silicon at $\theta = 170°$ has a cross-section $\simeq 1.0 \times 10^{-24}$ cm^2 sr^{-1} (i.e. ~ 1.0 barn). The same beam scattering from gold has a cross-section $\sigma \simeq 32.8$ barns. The greater scattering yield from the heavier element is given, approximately, by

$$Y_{Au}/Y_{Si} = (Z_{Au}/Z_{Si})^2 = (79/14)^2 \simeq 32$$

and illustrates the sensitivity of RBS to heavy elements. Note also that the cross-section for scattering is inversely proportional to the square of the projectile energy. The yield from a 2 MeV ^{4}He beam is one-quarter of that for a 1 MeV ^{4}He beam.

2.4 Stopping cross-section

As an energetic particle passes through a solid it loses energy. For the light particles and the energies used in RBS the two dominant energy loss processes are:

(1) interactions with bound or free electrons in the target i.e. electronic stopping; and
(2) interactions with the screened or unscreened nuclei of the target atoms i.e. nuclear stopping. For beams of ^{1}H and ^{4}He nuclear stopping is negligible except at the very lowest energies (i.e. at the deepest penetration of the probe beam). The rate dE/dx at which a particle loses energy is typically 10–$100\,eV$Å^{-1} for ^{4}He and dE/dx depends on the energy, E, of the projectile as shown in Fig. 9.6 for ^{4}He penetrating silicon. The energy, E, of a particle at a depth x will be given by

$$E = E_0 - \int_0^x \frac{dE}{dx} \cdot dx. \quad (9)$$

In RBS our attention is confined to particles that travel very small distances in the surface layers of the target. Under these circumstances the rate of energy loss can be considered constant at the value $dE/dx|_{E_0}$ i.e. its value at the energy E_0 of the incident projectile.

Equation (9) can then be integrated to give

$$E = E_0 - \left.\frac{dE}{dx}\right|_{E_0} \cdot x. \tag{10}$$

This approximation is referred to as the *surface energy approximation* (Chu, Mayer & Nicolet, 1978).

In place of dE/dx it is common to use the *stopping cross-section* ε defined by

$$\varepsilon = \frac{1}{N} \cdot \frac{dE}{dx}, \tag{11}$$

where N is the target atomic density. Ziegler & Chu (1974) have tabulated semi-empirical values of the stopping cross-sections for ^{4}He in all elements and from 0.4 to 4 MeV. The stopping cross-section varies with energy in a similar fashion for all elements, showing a broad maximum at ~ 1 MeV. A 1 MeV ^{4}He beam scattering from silicon has a stopping cross-section $\varepsilon \simeq 66.3 \times 10^{-15}$ eV cm^{-2}.

When a beam slows down in a target composed of more than one element the energy loss can be calculated using *Bragg's rule* this states that the total energy loss ε^{AB} in a compound $A_m B_n$ is given by

$$\varepsilon^{AB} = m\varepsilon^A + n\varepsilon^B, \tag{12}$$

where ε^A and ε^B are the stopping cross-sections of the atomic constituents A and B. In a sample of SiO$_2$, for example, ε^{SiO_2} is given by

$$\varepsilon^{SiO_2} = \varepsilon^{Si} + 2\varepsilon^O.$$

2.5 Energy straggling

As a particle moves through a target it loses energy in a succession of individual collisions. The energy loss process is subject to statistical fluctuations and, as a result, identical energetic particles do not have the same energy after passing the same distance. The energy loss, E, is subject to fluctuations and this phenomenon is called *energy straggling*. The essence of RBS is to measure the energy of a scattered projectile beam and to calculate from this energy the depth and/or the mass from which scattering occurred. Any uncertainty in particle energy due to straggling leads to a reduction in the precision with which mass and depth analysis can be performed (except for mass analysis of atoms located *at* the surface). As an example a 2 MeV ^{4}He beam traversing an Al film of 4300 Å loses

125 keV and the standard deviation in this loss is 7.0 keV, i.e. 5.6% (Harris & Nicolet, 1975).

3 Back-scattering spectrometry

3.1 Depth scale

The energy, E_1, of a particle that strikes the detector can be related to the depth x (perpendicular to the surface) at which scattering occurred as shown in Fig. 9.3. A particle scattered at the sample surface has an energy KE_0. A particle that penetrates to a distance x has a path length $x/\cos\theta_1$ and loses energy ΔE_{IN} given by

$$\Delta E_{IN} = \int_0^{x/\cos\theta_1} \frac{dE}{dx} \cdot dx.$$

Assuming a constant value of $(dE/dx)|_{E_0}$ for this ingoing path gives an energy loss of

$$\Delta E_{IN} = \frac{x}{\cos\theta_1} \cdot \frac{dE}{dx}\bigg|_{E_0}.$$

If the particle energy *immediately* before the collision at a depth x is E, its energy immediately after the collision is KE and it then suffers a further energy loss ΔE_{OUT} as it travels a distance $x/\cos\theta_2$ to leave the target. The loss on the outward path is

$$\Delta E_{OUT} = \int_0^{x/\cos\theta_2} \frac{dE}{dx} \cdot dx.$$

Assuming a constant value of

$$\frac{dE}{dx}\bigg|_{KE_0}$$

for this outgoing path gives an energy loss

$$\Delta E_{OUT} = \frac{x}{\cos\theta_2} \frac{dE}{dx}\bigg|_{KE_0}.$$

The particle energy, E_1, as it leaves the target is then $KE - \Delta E_{OUT}$. Thus

$$E_1 = KE - \Delta E_{OUT}$$
$$= K(E_0 - \Delta E_{IN}) - \Delta E_{OUT}$$

or

$$KE_0 - E_1 = K \cdot \Delta E_{IN} + \Delta E_{OUT}. \tag{13}$$

The energy difference, ΔE, between particles scattered at the surface and particles scattered at a depth x is therefore

$$\Delta E = K \cdot \Delta E_{IN} + \Delta E_{OUT}$$
$$= x \left\{ \frac{K}{\cos \theta_1} \cdot \frac{dE}{dx}\bigg|_{E_0} + \frac{1}{\cos \theta_2} \frac{dE}{dx}\bigg|_{E_0} \right\}. \tag{14}$$

Any energy difference E can thus be converted into a depth x using equation (14), which is the surface energy approximation. If in place of dE/dx we use the stopping cross-section, equation (14) becomes

$$\Delta E = N \cdot x \cdot \left\{ \frac{K}{\cos \theta_1} \cdot \varepsilon(E_0) + \frac{1}{\cos \theta_2} \cdot \varepsilon(KE_0) \right\} \tag{15}$$
$$= N \cdot x \cdot [\varepsilon], \tag{16}$$

where

$$[\varepsilon] = \left\{ \frac{K}{\cos \theta_1} \cdot \varepsilon(E_0) + \frac{1}{\cos \theta_2} \cdot \varepsilon(KE_0) \right\}$$

and

$$\varepsilon(E_0) = \frac{1}{N} \frac{dE}{dx}\bigg|_{E_0} \quad \text{and } \varepsilon(KE_0) = \frac{1}{N} \frac{dE}{dx}\bigg|_{KE_0}$$

If the energy difference, ΔE, is chosen to be the energy width, δE, of a single channel the in multichannel analyser then each channel corresponds to a thickness, δx, given by

$$\delta x = \delta E/N \left\{ \frac{K}{\cos \theta_1} \cdot \varepsilon(E_0) + \frac{1}{\cos \theta_2} \cdot \varepsilon(KE_0) \right\} \tag{17}$$
$$= \delta E/N \cdot [\varepsilon]. \tag{18}$$

As a specific example for $168°$ scattering of 2 MeV ^{4}He from a silicon target and for a normal incidence beam we have the following parameters: e.g. $\theta_1 = 0°$, $\theta_2 = 12°$; $K = 0.5657$ (assuming a weighted average of the kinematic factors for the isotopes); $\varepsilon(E_0) = 47.87 \times 10^{-15}$ eV cm^2; $\varepsilon(KE_0) \simeq 57.3 \times 10^{-15}$ eV cm^2; $N = 5 \times 10^{22}$ atoms cm^{-3}; $\delta E = 5$ keV. Using these values in equation (18) gives $\delta x \times 115$ Å and, consequently, a depth scale can be drawn, as shown in Fig. 9.7. Note that the silicon surface (KE_0) is located at a position where the yield is 0.5 where H is the height of the back-scattering spectrum. The leading edge of the RBS spectrum is not vertical because of the finite energy resolution of the measuring system.

The system energy resolution can be calculated from the slope of the leading edge by measuring the energy spread from 12% to 88% of the step height H. In Fig. 9.7. this gives an energy resolution of 25 keV.

The depth resolution in RBS, calculated using the surface energy approximation outlined above, is typically between 100 Å and 300 Å for

312 W.A. Grant

Fig. 9.7. Random RBS spectrum for 2 MeV He scattering from Si.

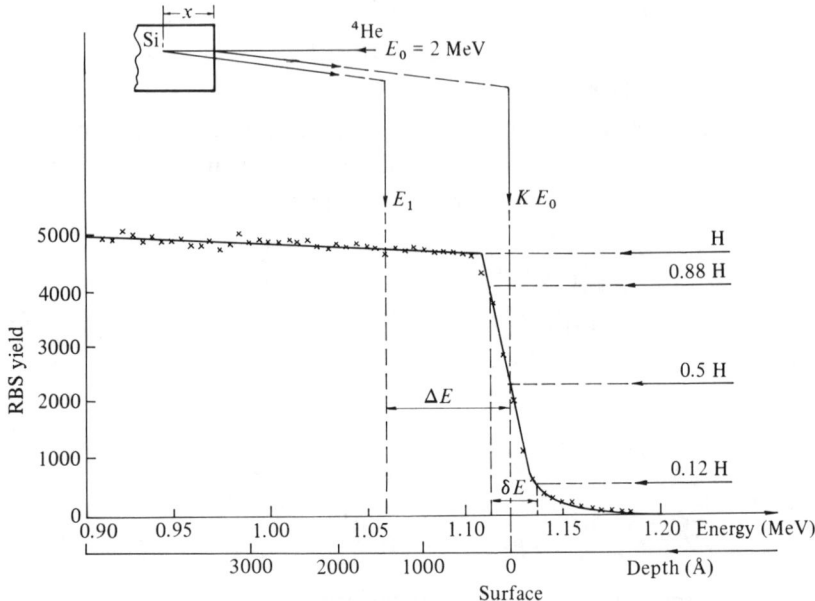

Fig. 9.8. The depth resolution δx (or $N\delta x$) as a function of the atomic number of the target.

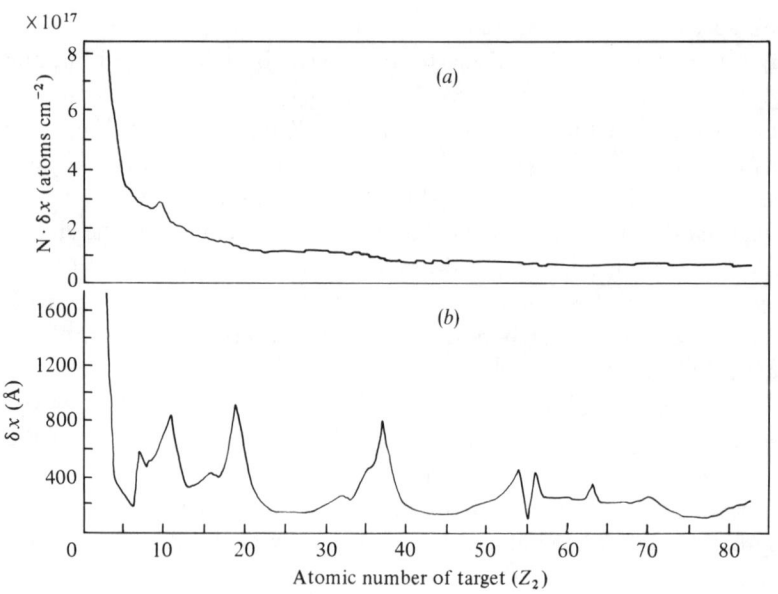

near-surface analysis using 1.0–2.0 MeV ^{4}He and with a system resolution $E = 15$ keV. Large variations in depth resolution occur between adjacent elements in the periodic table and this is caused by sharp variations in bulk densities. A low-density material requires a thicker layer for a given (15 keV) energy loss. An alternative formulation for the depth resolution in RBS is to specify the number of atoms per unit area, $N \delta x$, over the distance, δx. This is a more fundamental term since the energy loss of a particle is determined not by the distance, x, that it travels but by the number of atoms per unit area, $N \delta x$, with which it interacts to lose energy. The depth resolution, $N \delta x$, calculated using the surface energy approximation and for 2 MeV ^{4}He scattering at $\theta = 168°$ with a system resolution of $E = 15$ keV, is shown in Fig. 9.8(a) as a function of the target atomic number Z_2. Values for the depth resolution, δx, calculated using the same conditions, are shown in Fig. 9.8(b).

3.2 Height spectrum

The height H_0 of an RBS spectrum, i.e. the number of counts per channel, at the surface region is given by

$$H = \sigma(E_0) \cdot \Omega \cdot Q \cdot (Nt), \tag{19}$$

where $\sigma(E_0)$ is the average differential scattering cross-section at the energy E_0, Ω is the solid angle of the detector, Q is the total number of incident probe particles and (Nt) is the number of scattering centres per cm^2 in the thickness, t, over which the scattering occurs. If the analysing beam enters at an angle θ_1 it passes by a greater number of scattering centres as it passes through t; the number is increased by $1/\cos \theta_1$.

But a single channel of energy width E is equivalent to a thickness δx: thus

$$t = \delta x \text{ and } H = \sigma(E_0) \cdot \Omega \cdot Q \cdot \left(\frac{N \delta x}{\cos \theta_1} \right). \tag{20}$$

From equation (18)

$$N \cdot \delta x = \frac{\delta E}{[\varepsilon_0]},$$

and the expression for the height, H_0 of the RBS spectrum at the surface is

$$H = \sigma(E_0) \cdot \Omega \cdot Q \cdot \frac{\delta E}{[\varepsilon_0] \cos \theta_1}. \tag{21}$$

Note that the height does not depend on the atomic density, N, of the target. The yield, $H(E_1)$ of a spectrum at a channel corresponding to an energy E_1 is given by

$$H(E_1) = \sigma(E) \cdot \Omega \cdot Q \cdot \frac{\delta E}{[\varepsilon(E)] \cos \theta_1} \frac{\varepsilon(KE)}{\varepsilon(E_1)}, \tag{22}$$

where $\sigma(E)$ is the cross-section evaluated at the energy E of the projectile immediately before scattering at a depth x, $[\varepsilon(E)]$ is the stopping cross-section factor evaluated at the same energy, $\varepsilon(KE)$ is the stopping cross-section evaluated at KE, and $\varepsilon(E_1)$ is the stopping cross-section evaluated at the energy E_1. The ratio $\varepsilon(KE)/\varepsilon(E_1)$ in equation (22) corrects for the change with depth of the thickness δx corresponding to a single channel width δE.

For an energy E_1, the corresponding energy, E, immediately before scattering at a depth x can be found by an energy loss ratio method

$$E = (E_1 + \alpha E_0)/(K + \alpha), \qquad (23)$$

where $\alpha = \Delta E_{OUT}/\Delta E_{IN}$ is assumed to be constant independent of the depth at which scattering occurs.

In the surface energy approximation

$$\alpha \sim [\varepsilon(KE_0)/\varepsilon(E_0)]\beta \qquad (24)$$

and

$$\beta = \cos\theta_1/\cos\theta_2. \qquad (25)$$

Equations (23) and (24) allow a value for E to be found for a given E_1.

Equation (22) shows that the height, H, of the RBS spectrum is proportional to $\sigma(E)$ and inversely proportional to $\varepsilon(E)$. Both of these quantities vary with the energy E. In general the variation of $\sigma(E)$ with energy is greater than that of $\varepsilon(E)$ and so the yield $H(E)$ tends to increase with energy. The ratio $\varepsilon(KE)/\varepsilon(E_1)$ usually has a minor effect on the shape of the RBS spectrum.

3.3 Depth scale for a compound sample

If a target is composed of a homogeneous mixture of different atoms, the stopping cross-section must be evaluated using Bragg's rule. In a compound $A_m B_n$, the stopping power is given by equation 12

$$\varepsilon^{AB} = m\varepsilon^A + n\varepsilon^B.$$

A particle penetrating the sample $A_m B_n$ loses an energy E_{IN} along the inward path. The energy E_{OUT} after a scattering collision depends on whether scattering occurred with A-type atoms or B-type atoms. The conversion of a given energy loss to an equivalent depth thus results in two depth scales, one for each element.

If $[\varepsilon_0]_A^{AB}$ is the stopping cross-section factor for particles scattered from A-type atoms (using a surface energy approximation),

$$[\varepsilon_0]_A^{AB} = \frac{K_A}{\cos\theta_1} \cdot \varepsilon^{AB}(\varepsilon_0) + \frac{1}{\cos\theta_2} \cdot \varepsilon^{AB}(K_A E_0), \qquad (26)$$

where $\varepsilon^{AB}(E_0)$ is the stopping cross-section for the compound $A_m B_n$

evaluated at an energy E_0 using equation (12); K_A is the kinematic factor for scattering from A atoms. A similar equation can be written for $[\varepsilon_0]_B^{AB}$.

The two depth scales are usually less than 10% different for 2 MeV ^{4}He ions in most cases.

3.4 Height of spectrum for a compound sample

As mentioned earlier the energy of the probe beam after collision depends on the collision partner. Since the scattering cross-section depends on energy, the height, H, of the RBS spectrum will also depend on the collision partner. At the surface the height, H_0, for scattering from A atoms is

$$H_0^A = \sigma_A(E_0) \cdot \Omega \cdot Q \cdot m \cdot \frac{\delta E}{[\varepsilon_0]_A^{AB} \cos \theta_1} \tag{27}$$

and

$$H_0^B = \sigma_B(E_0) \Omega \cdot Q \cdot n \cdot \frac{\delta E}{[\varepsilon_0]_A^{AB} \cos \theta_1}$$

and the ratio is

$$\frac{H_0^A}{H_0^B} = \frac{\sigma_A(E_0)}{\sigma_B(E_0)} \cdot \frac{m}{n} \cdot \frac{[\varepsilon_0]_B^{AB}}{[\varepsilon_0]_A^{AB}}. \tag{28}$$

To determine the composition $A_m B_n$ from an RBS spectrum the ratio $[\varepsilon_0]_B^{AB}/[\varepsilon_0]_A^{AB}$ can be taken as unity to a first approximation (it is ~ 1 within 10% for most ^{4}He scattering at 2 MeV).

At any depth x the height $H(E_1)$ is the sum of the signals $H_A(E_1)$ and $H_B(E_1)$ from the atoms A and B, respectively, where

$$H(E_1) = H_A(E_1) + H_B(E_1) \tag{29}$$

and

$$H_A(E_1) = \sigma_A(E_A) \cdot \Omega \cdot Q \cdot n \cdot \frac{\delta E}{[\varepsilon(E_A)]_A^{AB} \cos \theta_1} \cdot \frac{\varepsilon^{AB}(K_A E)}{\varepsilon^{AB}(E_1)} \tag{30}$$

and

$$H_B(E_1) = \sigma_B(E_B) \cdot \Omega \cdot Q \cdot n \cdot \frac{\delta E}{[\varepsilon(E_B)]_B^{AB} \cos \theta_1} \cdot \frac{\varepsilon^{AB}(K_B E_B)}{\varepsilon^{AB}(E_1)}. \tag{31}$$

3.5 Thin film samples

Consider next a thin film (say 1000 Å thick) of a metal deposited on a lower mass substrate. The RBS spectrum is shown in Fig. 9.9. The energy width, E, of the particles scattered at the front and back surfaces of the film is related to the number (Nt) of atoms per unit area in the film.

$$\Delta E = [\varepsilon_0] N t, \tag{32}$$

where $[\varepsilon_0]$ is the surface approximation for the stopping cross-section factor and N is the atomic density. Equation (32) gives the thickness, t, of the film if a value of N is assumed. The area, A, under that part of the RBS spectrum from the thin film can also be used to specify the film thickness. In the surface approximation this area, A_0, is given by

$$A_0 = \sigma(E_0) \cdot \Omega \cdot Q \cdot \frac{Nt}{\cos \theta_1}. \tag{33}$$

In equation (33), absolute values for Ω and Q must be available.

3.6 Multilayered films

Consider a target that consists of a thin film of element A on top of a thin film of element B, both of these being on a light mass substrate and with A heavier than B. The signal from A can be analysed as above for the case of a single film on a light substrate. The signals from film B are influenced by the outer film A. The beam reaching the AB interface is reduced in energy by passing through the film A. Similarly, the scattered beam passes through the film A on the outward path and this also reduces the projectile energy. The spectrum from the film B can thus be considered *shifted* to lower energies by the presence of the outer film A as shown in Fig. 9.10. To a first approximation the energy shift in the leading edge of

Fig. 9.9. RBS spectrum from a thin film on a lower mass substrate.

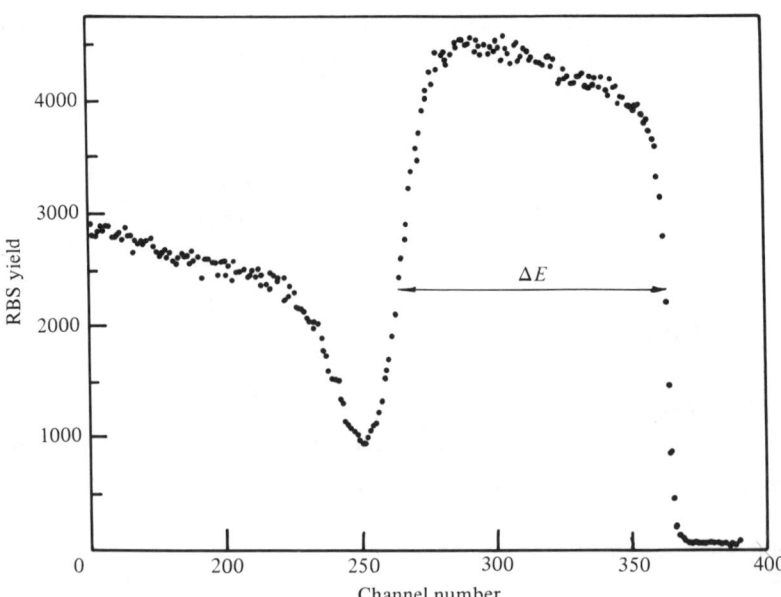

Rutherford back-scattering spectrometry 317

Fig. 9.10. Backscattering from a layer of heavy element, A, on top of a film of element, B. The presence of film A causes a shift ΔE_B^A.

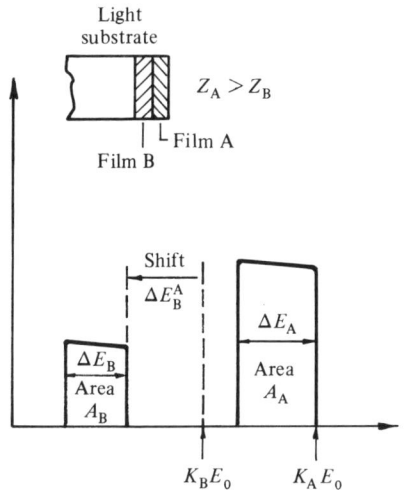

Fig. 9.11. Scattering from an SiO_2 film on a carbon substrate.

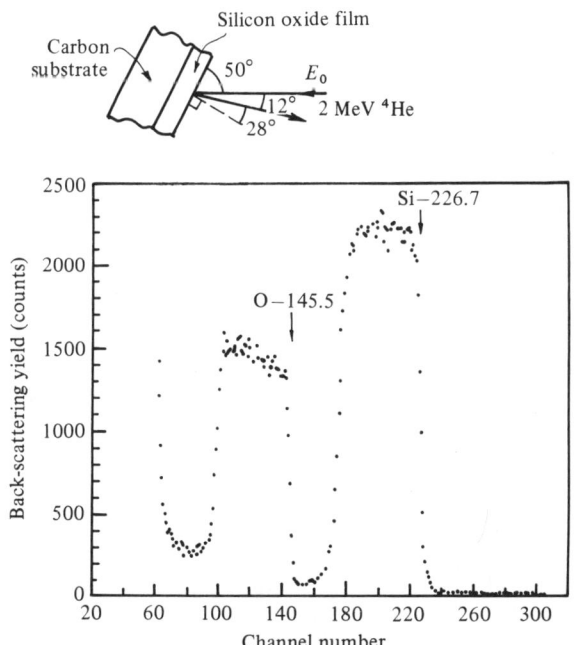

B is

$$\Delta E_B^A = N_A t_A [\varepsilon_0]. \tag{34}$$

As a final example consider the case of a thin film consisting of a homogeneous mixture of two elements A and B in the atomic ratio $A_m B_n$. If A_A and A_B are the total number of counts in the signals from the constituents A and B respectively then

$$N_A^{AB} \cdot t = (A_A/\sigma_A(E_0) \cdot \Omega \cdot Q) \cos \theta_1 \tag{35}$$

and

$$N_B^{AB} \cdot t = (A_A/\sigma_B(E_0) \cdot \Omega \cdot Q) \cos \theta_1. \tag{36}$$

Now

$$N_A^{AB} = m N^{AB} \text{ and } N_B^{AB} = n N^{AB}$$

and so

$$N_A^{AB}/N_B^{AB} = m/n = [A_A/\sigma_A(E_0)]/[A_B/\sigma_B(E_0)] \tag{37}$$

4 Back-scattering spectra

Various areas in which RBS analysis can be applied are illustrated in this section by a number of examples.

4.1 Compound thin film

Semiconductors are commonly protected during annealing by a surface film such as SiO_2. Fig. 9.11. shows a silicon oxide film deposited by RF sputtering onto a polished vitreous carbon substrate. The RBS analysis geometry is shown on the inset to Fig. 9.11. The incoming beam makes an angle of 50° to the normal (i.e. $\theta_1 = 50°$) in order to increase the path length (and hence the energy widths of the spectra) without causing the oxygen and silicon signals to overlap.

The kinematic factors, k_{Si} and k_o, calculated using equation (2) are $K_{Si} = 0.567$ and $k_o = 0.364$. The multichannel analyser was set at 5 keV/channel and so the leading edges of the silicon and oxygen signals should appear at 1.134 MeV (channel 226.7) and 0.727 MeV (channel 145.5) respectively. These positions are marked on Fig. 9.11. and agree well with the half-height positions of the leading edges of the silicon and oxygen signals.

Assuming that the atomic ratio of the constituents of the films is given by $Si_m O_n$, this ratio can be calculated from equation 37.

$$\frac{N_{Si}}{N_O} = \frac{m}{n} = \frac{A_{Si}}{A_O} \cdot \frac{\sigma_O(E_0)}{\sigma_{Si}(E_0)} \tag{37}$$

$$= \frac{1}{1.86}$$

which indicates that the film is not stoichiometric SiO_2 but is silicon rich. Knowing the atomic ratio m/n it is now possible to calculate the stopping cross-section, namely

$$\varepsilon^{AB} = m\varepsilon^A + n\varepsilon^B$$

and the number of molecular units per unit area is then given by

$$N^{AB} \cdot t = \Delta E_A / [\varepsilon]_A^{AB}$$
$$= \Delta E_B / [\varepsilon]_B^{AB}.$$

Alternatively, an approximate value for the thickness of the silicon oxide films can readily be obtained by assuming a stoichiometry of SiO_2 and assuming a bulk density 2.28×10^{22} SiO_2 molecules/cc. For the data of Fig. 9.11 this gives a film thickness of approximately 5000 Å.

Fig. 9.12. RBS spectrum from a silicon sample with an overlay film of ~ 2600 Å of SiO_2. The signal from the silicon substrate is shifted from channel 226.7 to channel 144.5.

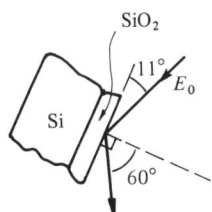

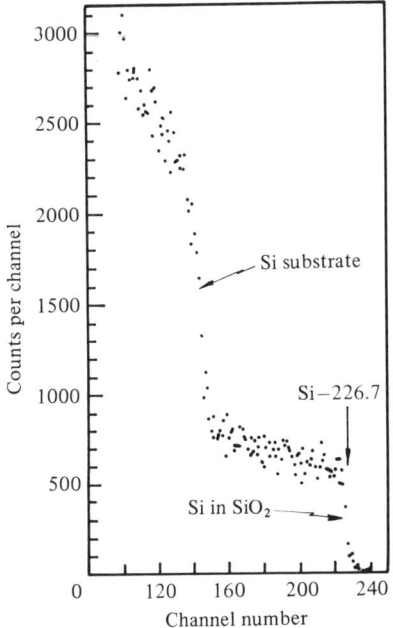

4.2 Diffusion of ion implanted impurity

Various dielectric materials are used as thin film protective coatings on semiconductors during thermal processing. Common encapsulants for GaAs are SiO_2 and Si_3N_4 and their ability to prevent outdiffusion of Ga is of some importance. Fig. 9.12 shows an RBS spectrum from SiO_2 on a silicon substrate. The incoming beam makes a small angle with the sample surface ($\theta_1 = 10°$) and leaves at an angle of 68° to the normal ($\theta_2 = 68°$). The signal from the silicon substrate is shifted due to the overlay of SiO_2 from channel 226.7 to channel 144.5. This shift of ~ 82 channels or 410 keV can be converted to an SiO_2 thickness of ~ 2600 Å.

Fig. 9.13 shows the RBS spectrum after implanting a nominal dose of 4×10^{16} ions cm^{-2} of 20 keV Ga. The Ga penetrates a very small distance into the SiO_2 and can be considered to be on the surface. Gallium on the surface should appear at channel 318.6 as indicated. Fig. 9.14 is a further RBS spectrum obtained after the addition of a further layer of SiO_2 of normal thickness 1000 Å. The Ga peak in Fig. 9.14 is shifted by ~ 40 channels and this shift, due to the second SiO_2 overlay, can be converted to a thickness of ~ 1100 Å. The dip in the spectrum near channel 185 is due to the Ga buried within the two SiO_2 layers. The high local concentration of Ga causes a local increase in stopping power and, since the yield is inversely proportional to stopping power, there is a consequent reduction in backscattering yield.

Fig. 9.13. RBS spectrum after implanting 20 keV Ga into an SiO_2 layer on top of Si.

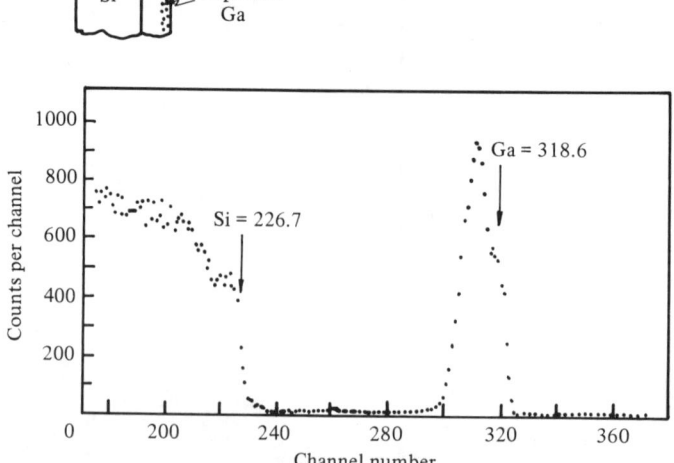

Fig. 9.14. RBS spectrum for sample consisting of Ga between a sandwich of two SiO$_2$ layers. The second SiO$_2$ layer causes the Ga signal to shift from channel 318.6 to channel 278 which is the position of the interface between the two SiO$_2$ layers.

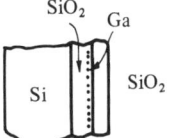

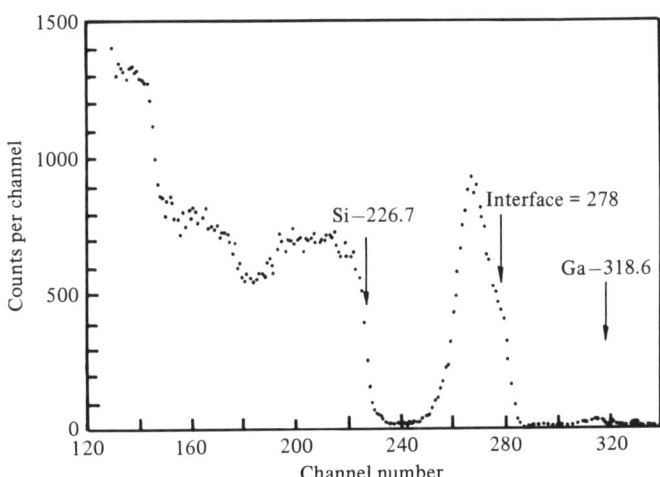

Fig. 9.15. RBS spectrum for the sample shown in Figure 9.14 after annealing for 30 minutes at 850 °C.

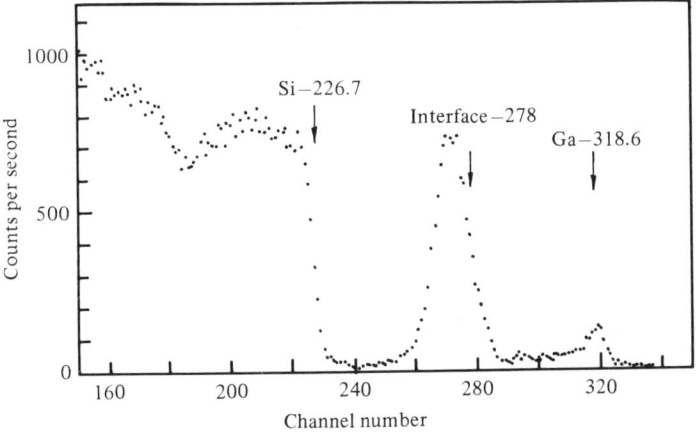

Fig. 9.15 shows the effect of annealing the sample of Fig. 9.14 for 30 min at 850 °C. The main changes in the RBS spectrum are (i) a shift in the position of Ga towards the interface (at channel 278) and (ii) the appearance of a second peak near channel 318.6. The second peak can be interpreted as Ga that has moved through the outer SiO_2 to the surface. Gallium is also distributed throughout the total thickness of the outer SiO_2 layer.

4.3 Redistribution of impurities during oxidation

It is well documented that many impurities redistribute during thermal oxidation of silicon. Impurities may, for example, be incorporated

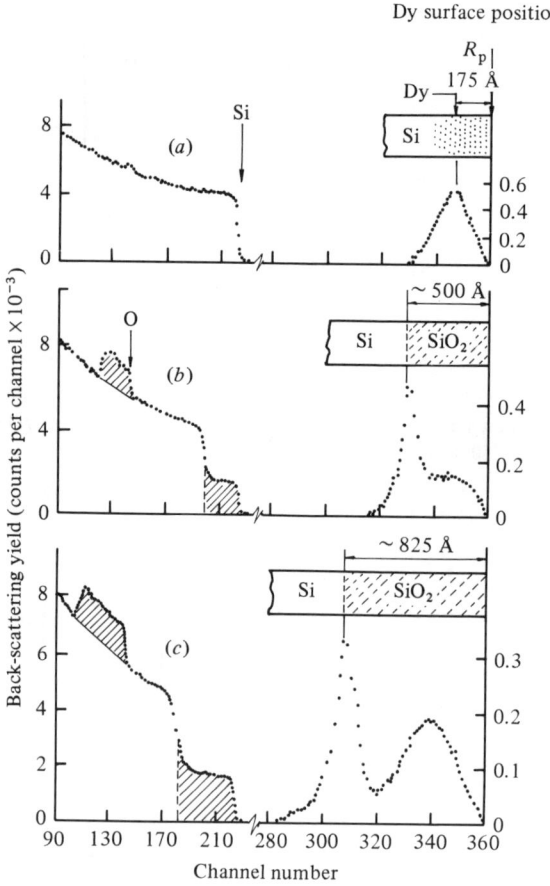

Fig. 9.16. RBS spectrum from Si implanted with Dy^+ ions: (a) immediately after implantation, (b) after growth of ~ 500Å of SiO_2 and (c) after growth of ~ 825 Å of SiO_2.

within the oxide or be pushed ahead by the advancing oxide–silicon interface. RBS provides a simple technique for examining the thermal oxidation of silicon implanted with a heavy impurity, as illustrated in Fig. 9.16. Immediately following implantation, Fig. 9.16(a), the RBS spectrum shows the Si substrate together with a well-resolved peak from implanted Dy^+ ions. The projected range, R_p, of the Dy can be calculated at 175 Å. A thermal oxidation step results in the spectra shown in Fig. 9.16(b). The shaded areas can be used to calculate the oxide thickness of ~ 500 Å. Some of the Dy impurity is pushed ahead by the advancing oxide and some is incorporated within the oxide. A second oxidation stage sufficient to take the oxide layer completely past the original impurity distribution is shown in Fig. 9.16(c). The Dy retreats further beyond the oxide–silicon interface, leaving a proportion as a clearly resolved peak located at the original, implanted position.

5 RBS and channelling
5.1 Physical concepts

Under certain circumstances the trajectory of an analysing beam is greatly influenced by the crystal structure of the target. A beam incident in a low index crystallographic direction can be channelled, i.e. the probe atoms are steered by a succession of gentle, large-impact-parameter collisions with the atoms that form the lattice rows and planes, as shown schematically in Fig. 9.17. The beam strikes the surface at normal incidence and separates into two fractions: a random fraction and a channelled

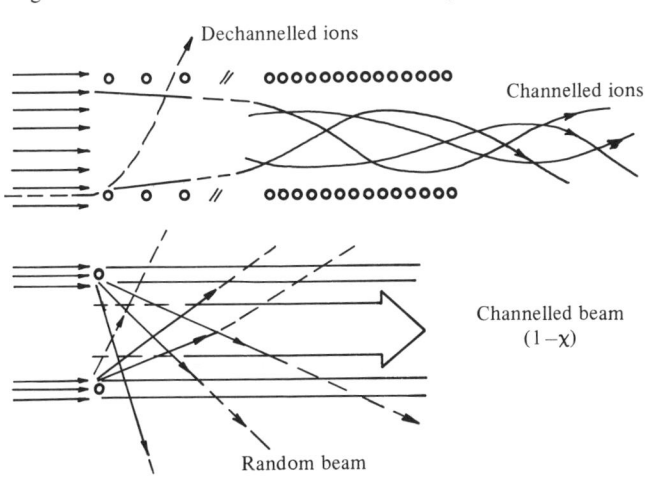

Fig. 9.17. Channelled and non-channelled ion trajectories in a single crystal.

fraction. That part of the beam that passes within a critical distance, r_{min}, of the surface atoms is scattered through angles greater than ψ_c, the critical angle beyond which channelling does not occur. Particles that pass outside r_{min} are channelled and spiral down the channel in a corkscrew fashion with a wavelength for MeV $^4He^+$ ions of several hundred Angstroms. Channelled ions do not make close encounter collisions with the target atoms and hence the RBS yield is reduced at all depths as illustrated in Fig. 9.18. The channelled fraction of the beam can easily be 0.95 for MeV $^4He^+$ ions (Carter & Grant, 1976).

If the probe beam enters a crystal far from a low index direction the RBS yield will be that for a random distribution of target atoms, i.e. the yield is the 'random yield'. When the beam enters in a low index direction, the surface atoms shadow the rest of the atoms in the row and the reduced RBS yield is called the 'aligned yield.' Because the outermost layer of target atoms is always visible to the incoming beam, the yield from this layer is identical for both random and aligned exposure. But due to the

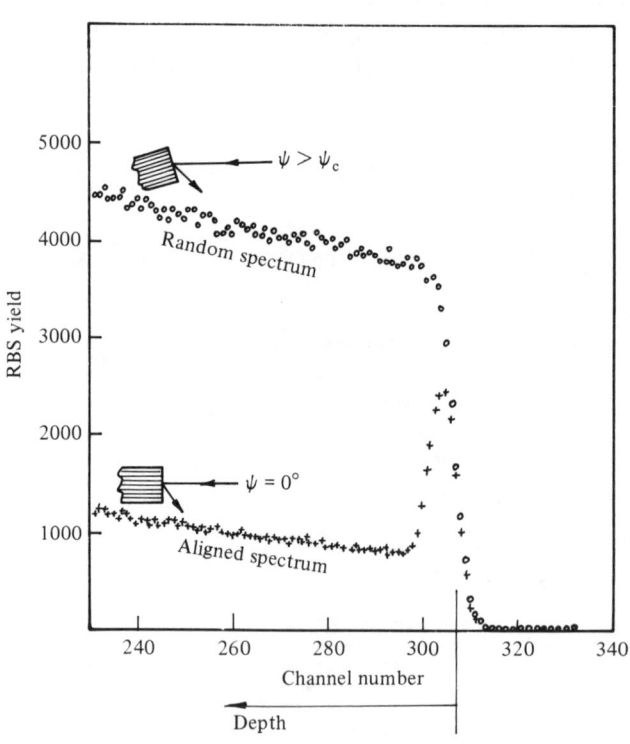

Fig. 9.18. Random and aligned RBS spectra from a single crystal.

limitations of depth resolution the aligned and random spectra do not coincide at the surface and instead the aligned yield shows a small surface peak, spread over the first few channels of the spectrum.

If the crystal is tilted so that the analysing beam enters the channel at an angle $\psi > 0°$, the yield will rise above the value for $\psi = 0°$. If the angle is large (say $\sim 5°$) no channelling occurs and the RBS spectrum has a 'random' value. The height of the spectrum at a given depth is a function of the angle, as illustrated in Fig. 9.19(a). The normalised yield from a thin scattering layer just below the surface is shown as a function of the tilt angle in Fig. 9.19(b). At $\psi = 0°$ the yield has a value $\chi_{min} = Y_A/Y_R$. This minimum yield χ_{min} can be calculated from (Mayer & Rimini, 1977):

$$\chi_{min} \sim Nd\pi(2u^2_1 + a^2), \tag{38}$$

where N is the atomic density, u_1 is the atomic vibrational amplitude and a is the Thomas–Fermi screening radius. For 2 MeV ^{4}He$^+$ ions incident on silicon at room temperature in the $\langle 110 \rangle$ direction $u_1 = 0.075$ Å, $a = 0.194$ Å, $d = 3.840$ Å and $N = 0.0499$ Å^{-3}, giving $\chi_{min} = 0.029$.

As the tilt angle is increased from $0°$, the yield rises from Y_A, the aligned value, to Y_R the random value. At a value midway between these two extremes, i.e. at $\frac{1}{2}(Y_A + Y_R)$, the tilt angle has a value of $\psi_{\frac{1}{2}}$ (the half-angle or critical angle). For MeV light ions the value of $\psi_{\frac{1}{2}}$ is

$$\psi_{\frac{1}{2}} = 0.8 F_{RS}(\xi)\psi, \tag{39}$$

Fig. 9.19. The relationship between (a) random and aligned spectra and (b) an angular yield curve.

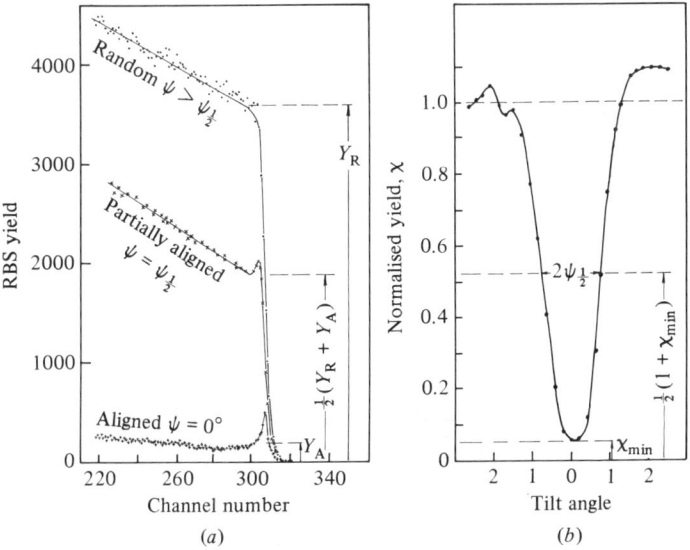

where

$$\psi = 0.307 \left\{ \frac{Z_1 Z_2}{Ed} \right\}^{1/2} \text{ (degrees)} \quad (40)$$

$\xi = 1.2 u_1/a$ and $F_{RS}(\xi)$ = square root of a dimensional string potential using Molière's screening function. For 2 MeV ^{4}He$^+$ ions incident at room temperature in the $\langle 110 \rangle$ direction on silicon $\psi = 0.586°$, $\xi = 0.4639$ and $F_{RS}(\xi) = 0.878$, giving $\psi_{\frac{1}{2}} \sim 0.4°$.

The combination of RBS and channelling can be used to measure the degree of crystal perfection in a given target and to locate the position of foreign impurities within a target lattice.

5.2 Damage measurements

RBS combined with channelling can be used to measure departures from crystallinity in single crystals. As an example consider the case

Fig. 9.20. RBS spectra from a single crystal with an amorphous surface layer.

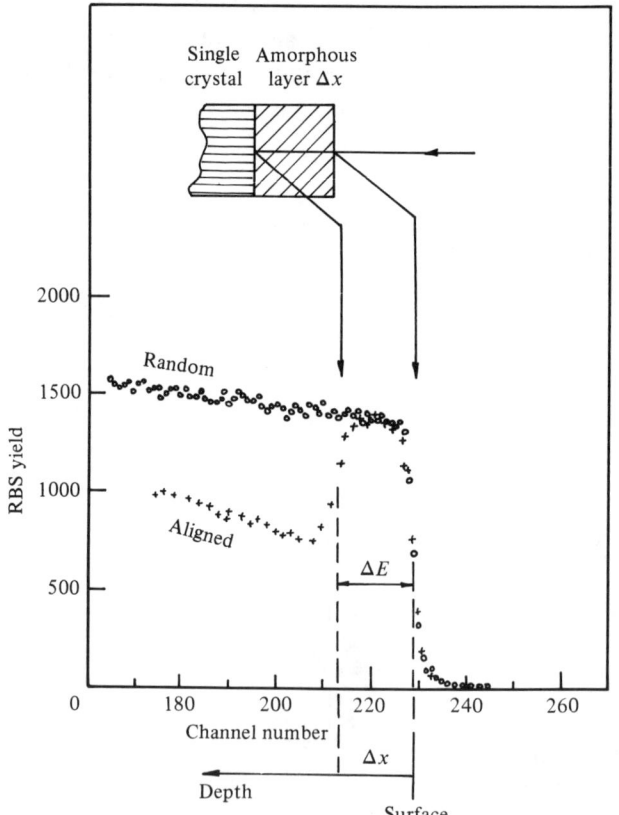

of a thin amorphous layer on the surface of a single crystal, a situation often found in ion-implanted silicon. The implanted ions cause radiation damage and the silicon lattice completely loses its regularity over the depth of implantation. The RBS spectrum from such a specimen is shown in Fig. 9.20. The aligned and random spectra are identical for the near-surface, damaged region. The aligned spectrum falls at the interface with the underlying crystal but not to the value for a perfect crystal. This is because a fraction of the beam that penetrates the amorphous layer emerges having been deflected through an angle greater than the critical angle, i.e. the random fraction increases. The thickness Δx of the amorphous layer can be calculated from the energy width ΔE, as sketched in Fig. 9.20.

The amorphous layer of silicon can be treated as a thin film on a substrate which is also silicon. The amorphous/crystalline interface is made visible by the reduction in yield at this interface because of channelling. This allows ΔE, and hence Δx, to be found.

A more general case of damage measurement in single crystals occurs when the damaged layer is not completely amorphous but retains a degree of crystallinity. This is often the case when a low dose of heavy ions is

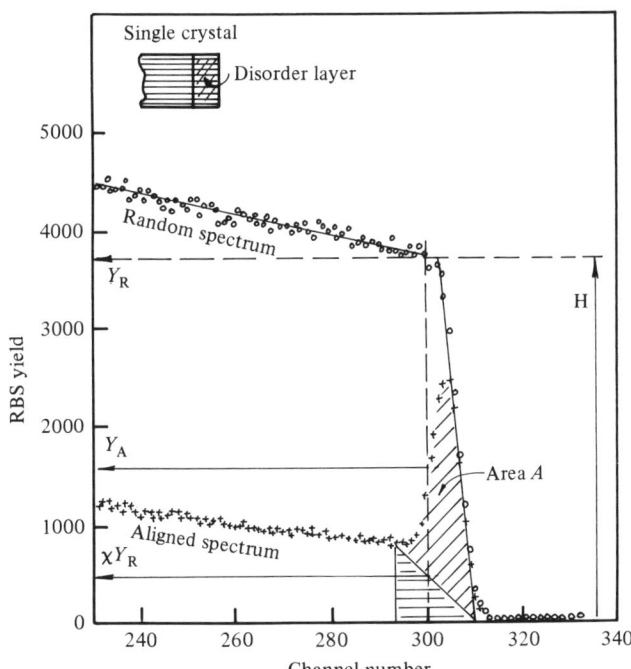

Fig. 9.21. RBS spectra from a single crystal with a disordered surface layer.

ion-implanted into silicon. Under these circumstances the aligned and random spectra are as shown in Fig. 9.21. The area under the surface peak can be converted to a measure of the number of displaced atoms. At any point within the damaged layer the aligned yield, Y_A, is made up of two components:

(a) scattering from displaced lattice atoms;
(b) scattering from atoms still on lattice sites.

The random part $\chi \cdot Y_R$ of the analysing beam strikes both displaced atoms and lattice atoms and so contributes to (a) and (b). The aligned part $(1 - \chi)Y_R$ of the beam can only strike displaced atoms and so contributes to (b). In order to separate the components (a) and (b) of the total yield it is necessary to know χ and its variation throughout the damaged layer. Various methods are available to calculate or compute χ. A simple, straight-line approximation to χ is shown in Fig. 9.21. And this allows extraction of the area A, i.e. the scattering due to displaced atoms. This area can be converted to a number of N^* of displaced silicon atoms by

$$N^* = \frac{A}{H} \cdot \frac{\delta E}{[\varepsilon]}. \tag{41}$$

The RBS yield, A, is converted to an area density of displaced lattice atoms in the same manner as described earlier for a heavy impurity located near the surface of a sample. In the present case the 'impurity' has the same mass as the target but is detectable because of the use of channelling.

5.3 Atom location

RBS combined with channelling can be used to locate the lattice positions of impurity atoms in a single crystal. Consider the case of a crystal containing a heavy impurity, as shown in Fig. 9.22. The location of the impurity can be investigated by methods similar to those described above for radiation damage. Exposure of the sample in both random and aligned directions will produce both types of spectra for the substrate and for the heavy impurity. In a random exposure the area A_R under the impurity peak is a measure of the total number of impurity atoms. In an aligned exposure any impurity located in substitutional sites (or within the crystal rows that form the channel wall) will be shadowed and the backscattering yield will fall. This reduction in yield is a measure of the fraction of impurity that is shadowed.

A schematic diagram of impurities in an f.c.c. crystal is given in Fig. 9.23. For ions entering in the $\langle 100 \rangle$ direction the substitutional foreign atoms and those in octahedral interstitial sites lie within the channel wall. If a

Fig. 9.22. RBS spectra from a crystal containing a heavy impurity.

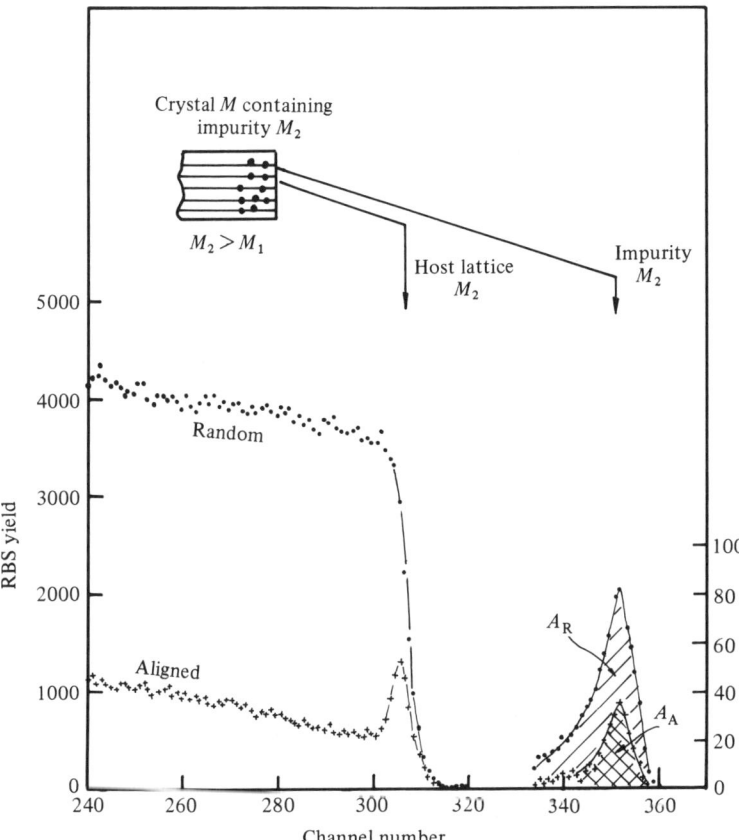

Fig. 9.23. Plane of an fcc single crystal showing the position of lattice impurities.

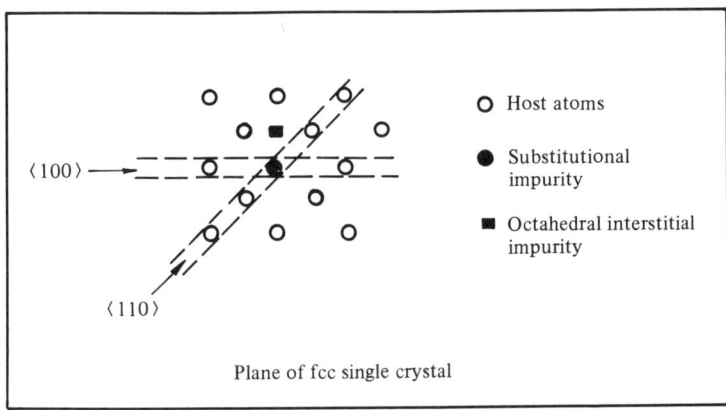

fraction, F, of impurity atoms sits within the channel wall then a fraction $(1 - F)$ sits within the channels. The analysing beam consists of a random component χ and a channelled component $(1 - \chi)$. The area, A_A resulting from an aligned exposure of the crystal to the analysing beam is composed of three components:

(i) the channelled fraction of the beam $(1 - \chi)$ interacting with atoms $(1 - F)A_R$ that lie within the channel;
(ii) the random fraction χ of the beam interacting with the same atoms $(1 - F)A_R$; and
(iii) the random fraction χ of the beam interacting with atoms FA_R that lie within the channel wall. The total yield A_A is

$$A_A = (1 - \chi)(1 - F)A_R + \chi(1 - F)A_R + \chi F A_R,$$

i.e.

$$F = \frac{1 - A_A/A_R}{1 - \chi}. \tag{42}$$

If we write χ_{host} for the minimum yield from the crystal lattice and $\chi_{\text{impurity}} = A_A/A_R$ for the minimum yield from the impurity then

$$F = \frac{1 - \chi_{\text{impurity}}}{1 - \chi_{\text{host}}}. \tag{43}$$

The fraction, F, determined from equation (43) is for impurities shadowed when the probe beam enters in a particular crystallographic direction. Further information on the exact position of impurities can be obtained if other crystallographic directions are used. A beam incident in the $\langle 110 \rangle$ direction in Fig. 9.23 would discriminate between dopants in substitutional sites (along the channel wall and therefore shadowed) and octahedral interstitial sites (which block the channel). This discrimination is not possible in the $\langle 100 \rangle$ direction.

Further information on the lattice site location of impurities can be gained from angular yield curves. Small displacements of impurities from substitutional sites result in a narrowing of the angular yield curve. The magnitude of the displacement can be estimated from the angular width at half-minimum. Interpretation of channelling spectra in terms of the exact position of impurities is, however, complicated for several reasons. One complicating factor is the phenomenon of flux peaking (Carter & Grant 1976). The channelled fraction of the analysing beam is not uniformly distributed as it passes down the channel as shown schematically in Fig. 9.24. The flux peaks towards the centre of the channel and there is also a variation in the magnitude of this flux peak with distance along the channel. The back-scattering signal from an interstitial impurity will therefore depend on its exact position within the channel.

Atoms located at the channel centre are struck by a great flux (2–3 times larger) in an aligned exposure compared to a random exposure. The angular scan under these circumstances is as shown in Fig. 9.25 where the yield shows a narrow peak above the random level. Although a complicating factor, flux peaking can be used to locate interstitial impurities with a high degree of precision.

Fig. 9.24. Flux peaking in a single crystal.

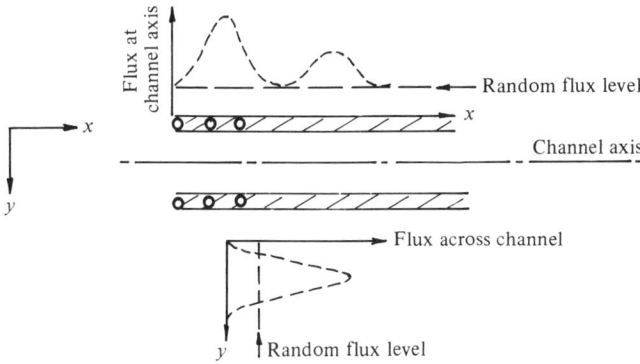

Fig. 9.25. The angular scan curve for an interstitial impurity located at the centre of a channel.

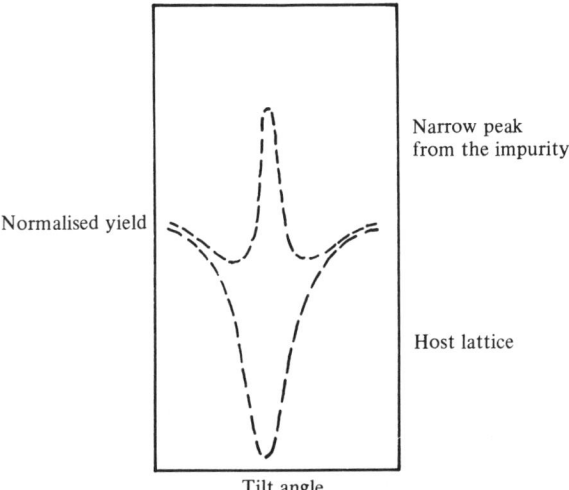

5.4 Dechannelling by defects

As mentioned in 5.2., dechannelling is increased by the presence of defects in a single crystal. The channelled fraction of the probe beam is slightly deflected as it passes a defect and may be dechannelled. Different types of defect lead to different forms of dechannelling, particularly its dependence on the energy of the analysing beam. The probability, dP/dx, of dechannelling per unit distance along the channel depends on the defect density N_D and a dechannelling factor σ_D (the value of which depends on the type of defect), i.e. $dP/dx = N_D \sigma_D$. The dechannelling factor, σ_D can be shown (Feldman, Mayer & Picraux, 1982) to depend on the energy, E, of the analysing beam as sketched in Fig. 9.26. For axial dechannelling by point defects $\sigma_D \propto Z_1 Z_2 d/E$, i.e. $\sigma_D \propto E^{-1}$. For axial dechannelling by

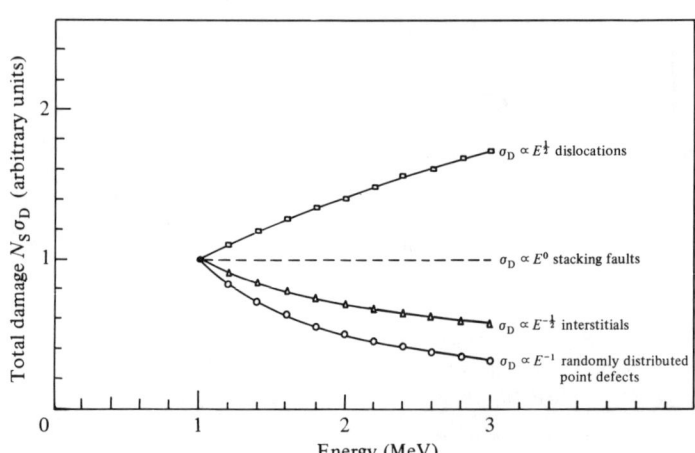

Fig. 9.26. The energy dependence of dechannelling for different types of defects.

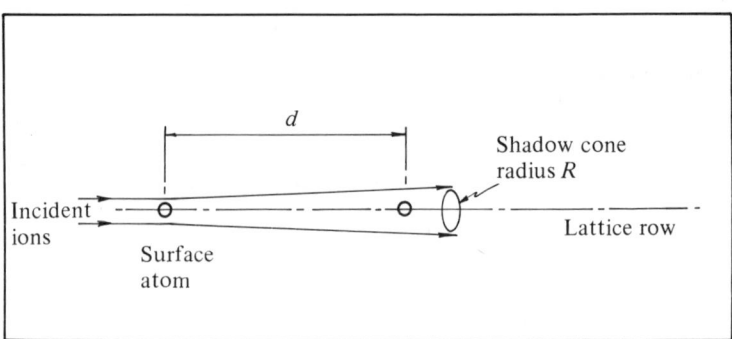

Fig. 9.27. The shadow cone radius: the first atom in the row shadows the second atom.

screw dislocations, $\sigma_D \propto E^{1/2}$, and for stacking faults, σ_D, is independent of the analysing beam energy. It is therefore possible to gain further insight into the type of defect present in a single crystal by examining the energy dependence of dechannelling.

5.5 Surface structure
(a) Surface shadow cone

In channelling experiments the surface atoms shadow the rest of the atoms in the row, as illustrated in Fig. 9.27. This leads to the concept of a shadow cone (Feldman, Mayer & Picraux, 1982) of radius R. For 1 MeV ^{4}He$^+$ ions incident in the $\langle 100 \rangle$ direction onto Si, the shadow cone $R = 0.089$ Å.

In a static silicon lattice the second atom in a row would be shadowed by the first atom. The second atom only sees probe ions at impact parameters greater than the shadow cone radius. At room temperature the silicon atoms vibrate with a two-dimensional root-mean-square amplitude $\rho = 0.106$ Å and this means that there is now the possibility of the second atoms in the row interacting strongly with the analysing beam. The degree of interaction for a given beam/target combination depends on the relative magnitudes of R and ρ. The cone radius, R, can be altered by varying the beam energy and is controlled by target temperature.

In channelling experiments on single crystals the surface atoms shadow the underlying atoms in the rows. Any direct backscattering is a measure of the number of atoms at the ends of these rows.

(b) Surface reconstruction

Surface atoms are known to behave differently to atoms located in the bulk of a crystal. In a number of crystals the surface atoms 'reconstruct'. i.e. change their position to form a different lattice as shown in Fig. 9.28(a). In the reconstructed surface the first atom is now aligned with the underlying row and so the shadowing effect is reduced with a consequent increase in the surface peak in an aligned RBS experiment. Measurements on reconstructed surfaces, such as the Si(001) surface, using RBS/channelling techniques have been reported (Stensgaard et al., 1981).

(c) Surface relaxation

The interatomic spacing between a surface atom and the underlying atom is also expected to differ from the spacing between atoms in the bulk as shown in Fig. 9.28(b). The outermost atom is shown at a larger,

Fig. 9.28. (a) Scattering from (A) a perfect surface and (B) a reconstructed surface. In (B) shadowing of the second atom in a row by the first atom is not complete and there is a subsequent increase in the back-scattered yield. (b) Scattering from (A) a perfect surface and (B) a relaxed surface. In (B) the shadowing of the second atom in a row is incomplete for non-normal incidence and the back-scattering yield increases. (c) Scattering from (A) a perfect surface and (B) a surface covered with an epitaxial heavy impurity. In (B) the outermost substrate atoms in a row are shadowed by the heavy impurity and this reduces the yield in the surface peak.

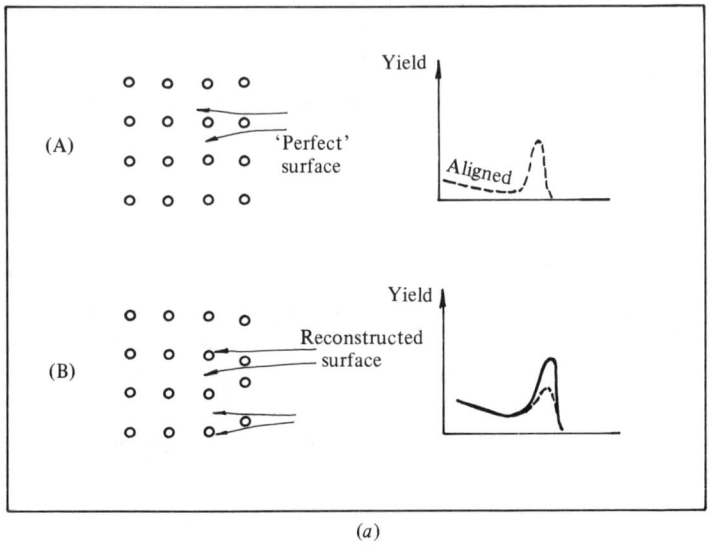

(a)

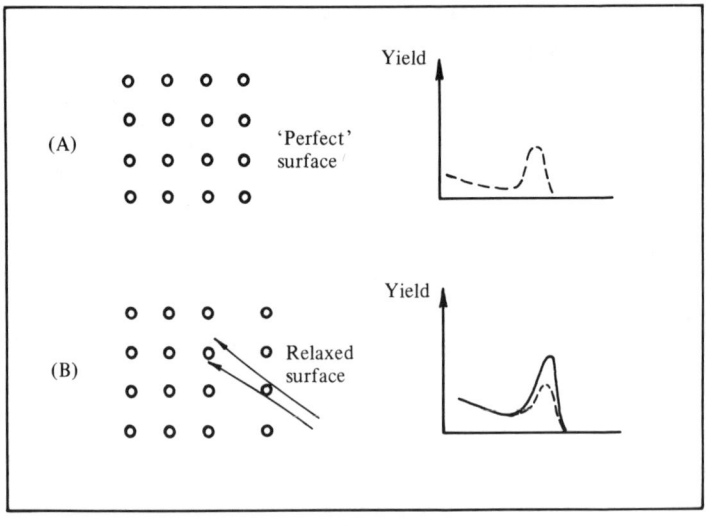

(b)

Rutherford back-scattering spectrometry

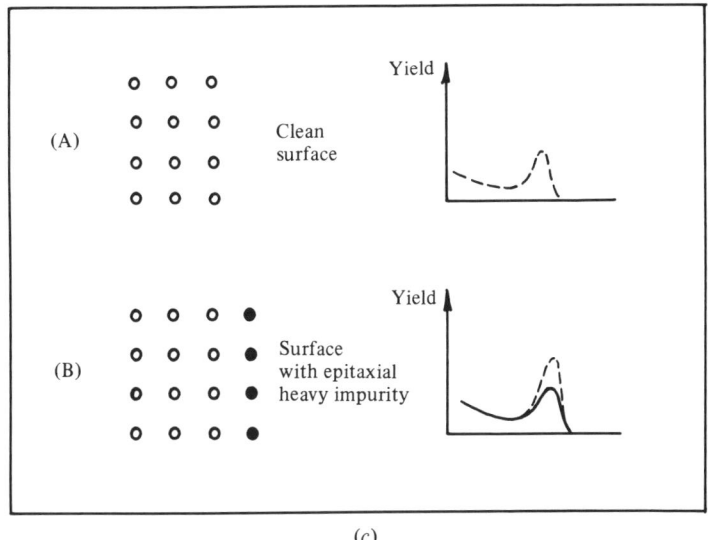

(c)

i.e. relaxed, position. In a normal incidence channelling experiment this outermost atom would still shadow the next atom in the row but for a non-normal beam the second row is outside the shadow cone radius and so contributes to the back-scattering yield. Measurements of surface relaxation have been obtained in a number of systems (Davies et al., 1981).

(d) Epitaxy

The shadowing caused by a surface impurity can be used to investigate the initial stages of epitaxy as shown in Fig. 9.28(c). If the impurity is aligned with the lattice rows it will shadow these rows and so the yield from the substrate will decrease as the epitaxial process proceeds. The epitaxial growth of Au on Ag has been monitored by this RBS/channelling technique (Culbertson et al., 1981).

6 Conclusions

The above sections have given a short introduction to RBS and its use as a surface analysis technique. It is convenient to finish the chapter by summarising the more important aspects of the technique. Table 9.1 gives a summary of the major parameters for RBS.

Under most conditions RBS can be considered as a non-destructive technique although in some circumstances a target may suffer radiation damage. The semiconductor GaAs, for example, can be rendered semi-insulating by light particle irradiation. Similarly it has been shown that the lattice site position of certain dopants in semiconductors can be

Table 9.1. *Rutherford back-scattering spectrometry*

Probe beam	$^1H^+$, $^4He^+$, other light ions
Probe energy	1–3 MeV
Beam diameter	~ 0.5–1.0 mm ($\sim 2\,\mu m$ with microbeam)
Beam current	~ 2–20 nA
Analysis time	~ 5–30 min
Integrated charge	~ 1–40 μC (6×10^{12} – 2.5×10^{14} ions)
Scattering angle	170°
Energy analyser	Surface barrier detector 15–25 keV energy resolution
Probing depth	~ 1–2 μm
Depth resolution	20–30 nm (3–4 nm with tilted targets)
Mass resolution	Isotope resolution up to ~ 40 amu
Sensitivity	10^{-2}–10^{-4} monolayers for heavy surface impurities
	10^{-1}–10^{-2} monolayers for light surface impurities
Accuracy	3–5% (typical)

influenced by RBS analysis. The method of analysis is relatively quick, with a typical analysis time of less than 30 min. Perhaps the most important aspect of RBS is the *quantitative* nature of the data. Thus RBS can provide, for example, quantitative measurement of film thickness and impurities within thin films without the need for calibrated standards. The second most attractive aspect of RBS is that it provides *simultaneously* both mass and depth analysis. A single RBS spectrum shows, for example, both the amount of an impurity in a thin film and its distribution throughout that film. The sensitivity with which an element can be detected depends principally on the atomic number, Z_1, of that element but is also influenced by the competing elastic scattering from the substrate Z_2. RBS is most sensitive to heavy elements since the back-scattering yield is proportional to Z^2. The minimum concentration, N_{min}, of Z_1 located as a surface impurity in a substrate Z_2 is given approximately by $N_{min} \sim (Z_2/Z_1)^2 \times 10^{14}$ atoms cm^{-2}, and this gives minimum detection limits of 10^{-2}–10^{-4} monolayers for heavy impurities and 10^{-1}–10^{-2} monolayers for light impurities. The detection of light impurities can be improved by combining RBS with channelling since the signal from the substrate is considerably reduced in channelled spectra and the impurity signal is consequently easier to pick out. An alternative approach to the detection of light elements is to use nuclear reaction analysis NRA. The elements oxygen, nitrogen and carbon, for example, can be detected by the resonance reactions $^{18}O(p,x)$ $^{14}N(d,x)$ and $^{12}C(d,p)$. In each of these examples the reaction product is emitted at an energy much higher than the incident beam energy and so can be easily resolved from elastically scattered particles.

The depth that can be probed in RBS analysis is typically 1–2 μm. The depth resolution, using the surface energy approximation outlined earlier, is typically between 100 Å and 300 Å for near-surface analysis using MeV ^{4}He and for a system with a 15 keV energy resolution and a standard beam/target geometry. The depth resolution for very small depths (< 1000 Å) can be considerably improved by tilting the sample such that the analysing beam enters and leaves at small angles (< 10°) to the surface plane. A depth resolution of $\sim$ 30 Å for silicon substrates can be obtained by this technique but the improvement is rapidly lost for depths greater than $\sim$ 1000 Å because of energy straggling.

For crystalline substrates the combination of RBS and channelling is particularly informative. In addition to composition analysis, information can be obtained on the structural order of the crystal lattice and the lattice site position of impurities. Lattice reordering and the movement of impurities can be monitored by in situ annealing measurements. Dechannelling measurements provide information on the type of defect present in the lattice. If RBS/channelling is carried out under uhv conditions it becomes possible to examine effects such as surface reconstruction and surface relaxation and to monitor processes such as the epitaxial growth of films.

References

Bird, J.R. & Clark, G.J. (1981). *Proceedings of the 5th International Conf. on Ion Beam Analysis, Nuclear Instr. and Methods*, **191**, 1–597.

Carter, G. & Grant, W.A. (1976). *Ion Implantation of Semiconductors*, Edward Arnold.

Chu, W.K., Mayer, J.W. & Nicolet, M-A (1978). *Backscattering Spectrometry*, Academic Press.

Culbertson, R.J., Feldman, L.C., Silverman, P.J. & Boehm, H. (1981). *Phys. Rev. Lett.*, **47**, 657.

Davies, J.A., Jackman, T.E., Jackson, D.P. & Norton, P.R. (1981). *Surf. Sci.*, **109**, 20.

Feldman, L.C., Mayer, J.W. & Picraux, S.T. (1982). *Materials Analysis by Ion Channelling*, Academic Press.

Goldstein, H. (1959). *Classical Mechanics*, Addison-Wesley.

Harris, J.M. & Nicolet, M-A (1975). *Phys. Rev.*, **B11**, 1013.

Mayer, J.W. & Rimini, E. (1977). *Ion Beam Handbook for Materials Analysis*, Academic Press.

Stensgaard, I., Feldman, L.C. & Silverman, P.J. (1981). *Surf. Sci.*, **102**, 1.

Ziegler, J.F. & Chu, W.K. (1974). *At. Nucl. Data Tables*, **13**, 463–89.

Index

Numbers in **bold type** denote pages that are wholly concerned with the subjects to which they refer. Page numbers in *italics* denote tables.

adhesion 1, 117
adsorption 6, 15, 145, 173, 188–97, 284–6, 288, 289, 291–3, 296–7
alkali ion beams 279, 282, 284
alloys 7, 51, 193, 200–1, 257, 288, 289
angle-resolved methods 6
appearance potential spectroscopy (APS) 7
atom probe (field-ion microscope) 7, 11–14
atomic mixing (cascade) 9, 21, 40, 48, 53, 55, 218, 241, 248, 267
atomic resolution 11
atomic step 12
Auger effect 87–9
Auger electron(s) 58, 87–8, 107, 109
   crystallographic effect 109–11
   scattering 89–93
   X-ray excited 140
Auger electron spectroscopy (AES) 2, 6, 13, *14*, 15, 18, 26, **58–73**, 74, **87–126**, 188, 222, 284, 288, 291
   depth profiling 4, 98, 101, 108, 112, 117–18, 120, 156
   element mapping 14, 57, 95, 106, 108, 112, 122–5
   escape depth 94
   line scan 112, 122
   point analysis 113–17, 125
   primary electron beam 89, 95, 100, 102
   primary electron beam energy 95–8
   spatial resolution 5, 13, 57, 87, 89, 93–5, 97, 100, 101, 103, 108, 112
Auger spectrum 3, 63, 88, 102
   chemical effects 111–12
   crystallographic effects 109–11
   differentiation 103, 107
   peak shift 89, 111
   peak intensities 107–9

ball cratering 118–20
binding energy 88, 128, *129–31*, 135, 149–50, 267, 270
   measurement and calibration 149
   see also surface binding energy
biological materials 6, 18, 127, **252–5**
bonding 1, 113, 117, 201, 289
Bremsstrahlung radiation 133, 140

caesium ion source 22, 28, 178, 219, 228, 242
calibration standards 6, 14, 18, 55, 225, 336
carbide 20, 112
catalysis 1, 20, 146, 188–9, 191, 200–4, 289
channel plate 11, 163, 287
channelling *see* ion channelling
charge coupled device 72, 288
charge neutralisation (electron) 16, 170, 206, 212, 228, 232, 288
charging (sample) 14, 97, 100, 170, 206–7, 212
chemical microscopy 170, 214
cluster ions 172, 185, 188, 194, 202, 204, 214
cold cathode discharge ion source 22, 23–4, 156, 228
collision cascade 30–1, 33, 36, 172, 176, 217–18, 241, 267
composition-depth profiling *see* depth profiling
computer datasystems 51, 103–6, 120, 138, 143, 148, 151–2, 158, 166, 184–5
concentric spherical sector analyser *see* hemispherical energy analyser
cones 40, 43, 48
constant analyser energy (CAE) mode 74, 135

Index

constant retard ratio (CRR) mode 74, 105, 135
corrosion 1, 5, 128
cryopumps 4
cryoshield 304
cylindrical mirror analyser (CMA)
 AES **59–66**, 68, 71, 73, 98, **99**, 100, 103, 105
 double pass CMA 75
 étendue 75, 77, 80, 163
 ISS 80, 283, 286, 289
 operation 62–6
 positional sensitivity 62, 143
 SIMS 181
 transmission 61–2, 75
 XPS **74–6**, 143

datasystems *see* computer datasystems
depth of analysis (surface sensitivity) 6, 9–10, 13, 17
depth profiling 4–5, 23
 non-destructive 140, 152–6
 depth resolution 9, 14, 17, 46, 48–9, 113, 118, 216, **240–2**, 252, 263–4, 304, 311–13, 325, 337
 see also sputter depth profiling
deuterium 17, 195, 197
diffusion pump(s) 4
discrete anode(s) 72
dopant profile(s) 5, 6, 20, 54
duoplasmatron ion source 22, 27, 156, 228, 231, 285
dynamic range 237–40

Einstein relation 128, 140, 152
electrical contact failure 20
electrical stress 12
electron backscattering coefficient 93
electron band structure 7
electron beam charge neutralisation *see* charge neutralisation
electron beam damage 14
electron beam decomposition 112, 170
electron beam post-ionisation 7
electron bombardment heating 284
electron channelling 109
electron energy analyser(s) 57–84, 98–9, 143
 see also CMA; HSA
electron gun 63, 75, 98, 99
electron impact ion source 21, 25–7, 156, 176–7, 284, 286
electron multiplier 60, 63, 73, 103, 104, 143, 283, 287
electron probe microanalysis (EPMA) 7, 88, 106, 122, 288
electron scattering 14, 89–93
electron spectroscopy for chemical analysis (ESCA) *see* X-ray photoelectron spectroscopy (XPS)
electron stimulated desorption (ESD) 206
electrostatic ion source 23, 286
energy dispersive X-ray analysis 7, 125
epitaxy 248, 250, 289, 335, 337
étendue 75, 77, 80, 163

fast atom beam 16, 23, 172, 177–8, 208
fast atom bombardment mass spectroscopy (FABMS) 170, 200, 201–4, 207–10, 213–14
Fermi level 149, 174
field emission source 100
field evaporation 12, 21, 29
field-ion microscope (atom probe) 7, 11–14
field ionisation 21
Firsov screening length 36, 269
fixed analyser transmission (FAT) mode 74
fixed retard ratio (FRR) mode 74
focussed collision sequence 33
fracture surface(s) 20, 87, 110, 114–17, 123, 145
fragment valency 172
fragmentation pattern 169, 191, 195, 210
functional group(s) 6

gallium arsenide 10, 14, 219–22, 241, 320–2, 335
gallium ion beams 5, 29, 178
 see also liquid metal ion sources
geology 252
germanium 51, 81–2, 150, 152
glass surfaces 212, 213, 222, 224–5, 227
goniometer, three axis 304

hemispherical energy analyser (HSA)
 AES **66–73**, 98, **99–100**, 103, 105
 energy resolution 100, 143
 étendue 77, 80
 ISS 286, 287
 input lens 66, 67, 68–71, 77–9, 99, 143, 160, 162
 multichannel detection 71–3, 80, 99, 143, 163
 operation 66–8, 76–7
 positional sensitivity 71
 transmission 68
 XPS **76–80**, 138, 143, 147
hot filament ion source 22

impurity atom location 328–31
ion beam etching *see* sputtering
ion beam machining 29
ion beam sputtering *see* sputtering
ion channelling 17, 18, 33, 38, 301, 303–4, 323–6, 336–7
ion de-channelling 33, 332–3

340   *Index*

ion erosion *see* sputtering
ion implantation   6, 21, 35, 37, 48, 220, 243–8, 320–2, 327–8
ion-induced effects
  amorphisation   327
  association   21
  chemical decomposition   21, 55, 156, 206
  structural effects   33–5
  topography   21, 35, 40–8, 51, 55, 156, 241
ion mapping   185
ion microprobe *see* secondary ion mass spectrometry
ion optics   21, 179
ion pumps   4
ion reflection   44, 46, 268
ion scattering spectroscopy (ISS)   2, 4, 6, 13, 15, *17*, 20, 27, 57, 80–1, **263–297**
  detection limit   264, 267, 283
  element mapping   283, 287
  energy resolution   282, 284
  escape depth (depth resolution)   263–4
  ion yield   267, 274–6, 280, 282, 283, 287, 288, 293, 297
  mass resolution   265, 267, 273, 282, 287, 289
  multiple scattering   273, 276–80
  scattering cross-sections   265, 267, 269, 274, 275, 276, 283, 294
  sensitivity   267, 273, 283, 284, 287
ionisation loss spectroscopy (ILS)   7
ionisation potential   21, 28, 198–9, 225
isotopes   16, 185, 195, 216, 226, 233, 235, 248
isotopes, radioactive   255
isotope ratio measurements   252

Kaufmann ion source   22, 24–5
kinematic factor *see* Rutherford backscattering
knock-on   30, 31, 33, 49

laser microprobe mass spectrometry (LAMMS)   7, 9–11
laser post-ionisation   7
laser, pulsed   9, 12
lattice valency   172
Lindhard velocity stopping parameter   36
linear cascade   30, 31, 33, 48
liquid metal ion source   16, 22, **28–30**, 170, **178–9**, 186, 212, 231
local thermal equilibrium model (LTE)   225
lock-in amplifier (phase sensitive detector)   64, 103
low dimensional structures (multilayer structures)   2, 4, 20, 289, 316–18
low energy electron diffraction (LEED)   7, 35, 58–9, 91, 98, 284, 291–3, 295, 297
low energy electron loss spectroscopy (LEELS)   91
low energy ion scattering spectroscopy (LEISS) *see* ion scattering spectroscopy

magnetic sector mass spectrometer   179, 183, **229–30**
  energy filter   230, 233, 243
  mass resolution   230, 235–7
mapping
  AES   14, 57, 95, 106, 108, 112, 122–5
  ISS   283, 287
  SIMS   5, 16, 29, 212
mass filter
  magnetic   228, 301
  quadrupole   177, 179–80
  Wien   27, 177
mass interferences   233
matrix effects
  AES   107
  ISS   274, 286
  SIMS   173, 218, 220–2, 224
  SNMS   7
memory effect   240
metallo-organic chemical vapour deposition (MOCVD)   248
modulation voltage   104
molecular beam epitaxy (MBE)   4, 250, 289
molecular fragments   6, 16
multichannel analyser   303, 311, 318
multichannel scaling mode (MCS) *see* Rutherford back-scattering
multilayer structures *see* low dimensional structures

nitride   20, 112
nuclear reaction spectroscopy   17, 288, 336

optical coatings   5
organic mass spectrometry   170
organic surface analysis   6, 11, 15, 127, 156, 170, 185, 188, 204–12
optical gating   239
oscillating electron ion source (twin anode electrostatic ion source)   22, 23, 286
oxidation   1, 20, 32, 125, **197–200**, 212–14, 318
  effect on ion yield   175, 218
  experiments   145
  impurity redistribution   322–3
  needle   29
oxidation state   5
oxygen flood   29, 232
oxygen ion beam   219

phase sensitive detector   64, 103

Index

phosphor screen 11, 12, 72, 232
photoelectric effect 128–32, 137
photoelectron spectromicroscope (PESM) 163
photoionisation 128, 134–5, 136, 140
photoionisation cross-sections 132–3, 157, 159
plasma discharge 7, 21
plasma post-ionisation 7, 9
plasmon loss peaks 91
polymers 6, 15, 18, 55, 127, 146, 156, 204–12
position sensitive detectors (PSD) 72, 166, 287
powders 15, 127, 142, 200
preferential sputtering 21, 48, 51–3, **222–4**, 289
preparation vessel 145, 176, 185
pulse height analysis (RBS) *see* Rutherford back-scattering
pulse pile-up 303

quadrupole mass spectrometer 18, 58, **179–83**, 207, **229**, 239–40, 243
  energy filter 180, 183, 235
  mass resolution 182–3, 229
  operating conditions (SIMS) 182–3
  pole bias 181, 183
  target potential 183
quantitative analysis 6, 11, 14, 55, 64
  AES **106–7**, 112
  ISS 266, 274–5, 277–9, **288**, 293, 295
  RBS 6, 17, 301, 304, 336
  SIMS 18, 187–8, 191, 216, **224–7**
  XPS 135, **156–7**

radiation-enhanced diffusion 21, 54–5
radio-frequency ion source 286, 302
random shot noise 65
raster gating, electronic 239, 285
reactive ion (beam) etching 289
recoil analysis 17
recoil implantation 48–9, 51, 53, 289
redeposition 46
reflection high energy electron diffraction (RHEED) 7, 91
resistive anode network 72
resistive filament 72
retarding field analyser (RFA) 58–9, 98
Rutherford back-scattering 2, 4, 6, 13, *17–18*, 20, 21, 57, 81–4, 263, 269, 271, 291, **299–337**, *336*
  damage measurements 326–8
  depth profiling 301, 304, 310–13, 314–15
  energy straggling 304, 309–10
  kinematic factor 304–6, 314–15, 318
  mass resolution 304, 306
  multichannel scaling mode (MCS) 303
  pulse height analysis 303

scattering cross-sections 304, 306–8, 311, 313, 315
stopping cross-sections 304, 308–9, 314, 319

scanning electron microscope (SEM) 92, 95, 100, 101, 117, 231
scanning transmission electron microscope (STEM) 7
scanning tunnelling microscope (STM) 7
secondary electron imaging
  electron-induced 92, 100, 112, 115, 122
  ion-induced 29, 186, 212
secondary ion mass spectrometry (SIMS) 2, 4, 6, 7–9, 13, 15, *16*, 20, 21, 27, 32, 53–4, 58, **169–262**, 267, 284
  depth profiling 48, 52, 185, 201, 240–2, 285
  detection limits 240
  escape depth 205, 217
  line-scan 259
  primary ion beam **176–7**, 219, 228, 240, 241, 242
  secondary ion current **218**, 225, 226, 227, **242–3**
  secondary ion emission 9, 169, 172, 225
  secondary ion energy distribution 172, 180, 207, 224, 230
  secondary ion imaging 185, 212, 230–2, 255
  secondary ion yields (intensities) 6, 29, 171, 177, 190, 192, 193, 197, 200, 201, **218–20**, 225, 228, 232, 242
  sensitivity 6, 18, 187–8, 228, 242, 243, 255
  spatial resolution 5, 13, 16, 18, 29, 170, 176, 185, 186, 212, 231
  stigmatic imaging 232, 243, 255
semiconductor materials
  AES 87
  atom probe microanalysis 14
  dopant profiles 5, 16, 20, 54
  energy filtering 233
  ISS 289
  implantation 29, 320–2
  ion beam damage 35
  ion induced topography 43
  low dimensions 1–2
  RBS 81–2, 299, 318, 320–2
  SIMS 169, 212, 222, 225, 227, 228, 243–52
  XPS 146
  see also germanium; silicon; gallium arsenide
semiconductor metallisation 5
signal-to-background ratio 5
signal-to-noise ratio 57, 59, 64, 65, 71, 96, 142, 148, 205, 287, 303
silicon 14
  AES 112

342 Index

amorphisation 35, 327
depth profiles 48, 117, 120
dopant profiles 54
energy filtering 233–5
ISS 284, 291
ion implantation 37, 237, 326–8
ion induced topography 43–4
oxidation 175, 318–23
preferential sputtering 223–4
RBS 81–3, 309, 311, 333
SIMS 205, 219
solid state detector 57, 81, 84, 301, 303, 306
specimen stage (manipulator) 101
ISS 286
XPS 138, 143, 166–7
spin–orbit coupling 134
spike regime 30, 31, 48, 55
spike removal 151
sputter-depth profiling 5, 9, 11, 16, 20, 33, 38, 40, 48, 55, 113, 118, 145, 156
  see also depth profiling
sputtered neutral mass spectrometry (SNMS) 7–9, 18
sputtering (ion beam etching) 3, 4, 20, 24, 170–6, 216–18
  chemical 32
  electronic excitation 32, 35
  mechanisms 21, 30–2
  rates 26, 101
  structural effects see ion-induced effects
  two ion beams 46
sputtering yield 31–2, 36, 40, 43, 51, 171, 242
  angle-of-ion incidence 38, 46–8
  angular dependence 33, 37–40
  energy dependence 35, 37, 40, 49
surface binding energy 31, 37
surface cleaning 23, 284
surface coatings 4, 14, 120, 204, 320
surface energy approximation 309, 311, 314
surface ionisation ion source 22, 28
  see also caesium ion source
surface reactions 195–7
surface reconstruction 297, 301, 333, 337
surface relaxation 301, 333–5, 337
surface segregation 201, 289
surface sensitivity (depth of analysis) 6, 9–10, 13, 17
surface shadow cone 333
surface structure 7, 15, 17, 333
surface vitrification 214

taper sectioning 5, 113, 118–22
thermionic electron source 100, 284
thermionic emission 1, 25
thin films 4, 14, 20, 25, 193–5, 315–19, 327

time of flight mass spectrometer 9, 12, 183–4, 212, 287, 297
trace element distributions 252
turbomolecular pumps 4
twin-anode electrostatic ion source (oscillating electron ion source) 22, 23, 286

ultra high vacuum (UHV) **2–4**, 9, 20, 21, 28, 29
  AES 87, 92–3, 101
  ISS 284, 292
  RBS 301, 303, 337
  SIMS 176, 179, 232, 240
  XPS 128, 138–9
ultra-violet photoelectron spectroscopy (UPS) 7, 127, 163
useful ion yield 243

Van der Graaf generator 302
valence electrons 111, 128, 135
vibrational spectroscopy 191

wear 1
work function 28, 132, 173, 190, 193, 195, 199, 201, 288

X-rays 15, 74, 128, 138, 139–42, 146–7
  monochromated 74, 76, 79–80, 133–4, 140–2, 160–3
  satellites 139, 140, 151
X-ray fluorescence spectroscopy 88
X-ray photoelectron spectroscopy (XPS) 2, 5–6, 7, 10, 13, *15*, 18, 26, 55, 57, 58, 66, **74–80, 127–67**, 188, 200, 201, 204, 206, 222
  angle resolved XPS 153
  atomic sensitivity factors 133
  energy resolution 133, 140, 142
  escape depth 136, 152, 205
  operation 145–9
  photoelectron imaging 128, 138, 162
  sample preparation 142–3, 145
  sensitivity 87, 154–5, 163–6
  small area 79, 128, 138, 159–63
  spatial resolution 15, 159–63
  spectrum acquisition 147–9
XPS spectrum 53, 55, 57
  baseline (background) removal 151
  chemical shifts 111, 135, 201
  deconvolution 151
  depth profiling 25, 145, 152–6, 166
  differentiation 151
  fine structure 133
  lineshapes 133, 135–6, 147, 158
  peak assignment 149–51
  peak integration 151
  peak intensity measurements 157
  satellite removal 151
  smoothing 151